Communications in Computer and Information Science   **718**

Commenced Publication in 2007
Founding and Former Series Editors:
Alfredo Cuzzocrea, Dominik Ślęzak, and Xiaokang Yang

More information about this series at http://www.springer.com/series/7899

Piotr Gaj · Andrzej Kwiecień
Michał Sawicki (Eds.)

Computer Networks

24th International Conference, CN 2017
Lądek Zdrój, Poland, June 20–23, 2017
Proceedings

 Springer

Editors
Piotr Gaj (iD)
Silesian University of Technology
Gliwice
Poland

Andrzej Kwiecień
Silesian University of Technology
Gliwice
Poland

Michał Sawicki
Silesian University of Technology
Gliwice
Poland

ISSN 1865-0929   ISSN 1865-0937 (electronic)
Communications in Computer and Information Science
ISBN 978-3-319-59766-9   ISBN 978-3-319-59767-6 (eBook)
DOI 10.1007/978-3-319-59767-6

Library of Congress Control Number: 2017943004

Printed on acid-free paper

This Springer imprint is published by Springer Nature
The registered company is Springer International Publishing AG
The registered company address is: Gewerbestrasse 11, 6330 Cham, Switzerland

Preface

Computer networks are one of the most important elements of our technical life, i.e., the technical means we use every day. A great number of devices around us communicate via computer networks and, moreover, all online services we use need to be connected to a network to operate properly. This applies to professional activities as well as private ones. Computer networks are part of the field of computer science and this is one of the most intensively developed branches with a very important impact on world economy. Research in computer networks has an influence on other branches of technical science and contributes to the development of completely new areas as well. Therefore, the domain of computer networks has become one of the most important fields of research.

The area of computer networks and the entire field of computer science are the subject of constant change. It is caused by the general development of IT technologies, by overall technical progress, and by the strong need for innovations in the sphere of how we communicate with each other, how we work, and how we perform our daily activities. This results in a very creative and interdisciplinary interaction between computer science technologies and other technical activities, and leads to perfect solutions. New methods, together with tools for designing and modeling computer networks, are regularly extended. Above all, the essential issue is that the scope of computer network applications is increased thanks to the results of new research and to new applications. Such solutions were not taken into consideration in the past few decades. Whereas the requirements of contemporary markets and the creative applications of existing network facilities stimulate the progress of scientific research, the extensive use of new solutions leads to numerous problems, both practical and theoretical, which need to verified, solved, and improved.

24th International Science Conference *Computer Networks*

This book collates the research work of scientists from numerous notable research centers. The chapters refer to the wide spectrum of important issues regarding the computer networks and communication domain. It is a collection of topics presented at the 24th edition of the International Conference on Computer Networks, which was held in Stonemout Castle, located near Lądek Zdrój, the famous health resort in southern Poland, during June 20–23, 2017. The conference, organized annually since 1994 by the Institute of Informatics of Silesian University of Technology together with the Institute of Theoretical and Applied Informatics in Gliwice, is the oldest event of its kind in Poland. The current edition was the 24th such event, and the international status of the conference was attained nine years ago, with the tenth international edition taking place in 2017. Just like previous events in the series, the conference took place under the auspices of the Polish section of IEEE (technical co-sponsor), and the conference partner was iNEER (International Network for Engineering Education and Research).

In 2017 the total number of submissions was 80. The presented papers were accepted after careful reviews made by at least three independent reviewers in a double-blind way. The acceptance level was below 45%, and thus the proceedings contains only 35 full papers. The chapters are organized thematically into several areas in the following tracks:

- Computer Networks
  This group of papers is the largest one. General issues of networks architecture, analyzing, modeling, and programming are covered in 16 papers. Topics on wireless systems and wireless sensor networks, fault-tolerant algorithms, security concerns, indoor localization issues, Internet technologies, and redundancy in industrial networks, among others, are included.
- Teleinformatics and Communications
  This section refers the general communications theory and related issues. It contains five papers related to interesting topics on overflow study in multi-tier cellular networks, the WebRTC technology, efficient calculation of radiation in wideband transmission systems, transmission range estimation for vehicular ad hoc networking, and usage of convolution algorithms for modeling network systems.
- New Technologies
  The chapter of new technologies used in the networking contains four papers which are connected with brand new areas of computer networks research, usage, and applications. There are topics on quantum direct communication, construction of firewall for SDNs and Qutrit Switch for quantum networks, and SLA life cycle management for cloud computing.
- Queueing Theory
  The domain of queueing theory is usually one of the most strongly represented areas at the Computer Network conference. This year, five papers are included, e.g., a paper on a performance model for studying distributed Web systems with usage of queueing Petri nets, a paper on the performance of fractional order PID controller as an AQM mechanism and the impact of traffic self-similarity on network utilization, a paper on applying a fluid limit approach methodology to find a sufficient and necessary stability condition for the Basic Collaboration system with feedback

allowed, a paper on the investigation of the Erlang service system with limited memory space under control of an AQM mechanism, and a paper on the investigation of queueing systems with demands of random space requirements and limited buffer space, in which queueing or sojourn time is limited by some constant value.

– Innovative Applications
The five papers in this section refer to research in the area of innovative applications of computer networks theory and facilities. There are contributions on innovative usage of in-vehicle communication, indoor positioning systems based on magnetic fields, reactive auto scaling models in order to improve sensitivity on load changes in cloud infrastructure, management of dynamic network models and the optimization criterion in the example distributed system.

Each chapter includes highly stimulating studies that may interest a wide readership.

In conclusion, on behalf of the Program and Organizing Committee of the Computer Network Conference, we would like to express our gratitude to all authors for sharing their research results as well for their assistance in developing this volume, which we believe is a reliable reference in the computer networks domain.

We also want to thank the members of the Technical Program Committee for their participation in the reviewing process.

If you would like to help us make the conference more attractive and interesting, please send us your opinions and proposals at cn@polsl.pl.

April 2017   Piotr Gaj
                                                                 Andrzej Kwiecień

Organization

CN 2017 was organized by the Institute of Informatics from the Faculty of Automatic Control, Electronics and Computer Science, Silesian University of Technology (SUT) and supported by the Committee of Informatics of the Polish Academy of Sciences (PAN), Section of Computer Network and Distributed Systems, in technical co-operation with the IEEE and consulting support of the iNEER organization.

Executive Committee

All members of the Executive Committee are from the Silesian University of Technology, Poland.

Honorary Member	Halina Węgrzyn
Organizing Chair	Piotr Gaj
Technical Volume Editor	Michał Sawicki
Technical Support	Aleksander Cisek
Technical Support	Jacek Stój
Office	Małgorzata Gładysz
Web Support	Piotr Kuźniacki

Co-ordinators

PAN Co-ordinator	Tadeusz Czachórski
IEEE PS Co-ordinator	Jacek Izydorczyk
iNEER Co-ordinator	Win Aung

Program Committee

Program Chair

Andrzej Kwiecień	Silesian University of Technology, Poland

Honorary Members

Win Aung	iNEER, USA
Adam Czornik	Silesian University of Technology, Poland
Bogdan M. Wilamowski	Auburn University, USA

Technical Program Committee

Omer H. Abdelrahman	Imperial College London, UK
Anoosh Abdy	Realm Information Technologies, USA
Olumide Akinwande	Imperial College London, UK
Iosif Androulidakis	University of Ioannina, Greece

Francesco Malandrino	Politecnico di Torino, Italy
Aleksander Malinowski	Bradley University, USA
Marcin Markowski	Wroclaw University of Science and Technology, Poland
Przemysław Mazurek	West-Pomeranian University of Technology, Poland
Kevin M. McNeil	BAE Systems, USA
Agathe Merceron	Beuth University of Applied Sciences, Germany
Jarosław Miszczak	IITiS Polish Academy of Sciences, Poland
Vladimir Mityushev	Pedagogical University of Cracow, Poland
Evsey Morozov	Petrozavodsk State University, Russia
Włodzimierz Mosorow	Lodz University of Technology, Poland
Sasa Mrdovic	University of Sarajevo, Bosnia and Herzegovina
Diep N. Nguyen	Macquarie University, Australia
Sema F. Oktug	Istanbul Technical University, Turkey
Michele Pagano	University of Pisa, Italy
Nihal Pekergin	Université de Paris, France
Maciej Piechowiak	University of Kazimierz Wielki in Bydgoszcz, Poland
Piotr Pikiewicz	College of Business in Dabrowa Górnicza, Poland
Jacek Piskorowski	West Pomeranian University of Technology, Poland
Bolesław Pochopień	Silesian University of Technology, Poland
Oksana Pomorova	Khmelnitsky National University, Ukraine
Sławomir Przyłucki	Lublin University of Technology, Poland
Tomasz Rak	Rzeszow University of Technology, Poland
Stefan Rass	Alpen-Adria-Universität Klagenfurt, Austria
Silvana Rodrigues	Integrated Device Technology, Canada
Przemysław Ryba	Wroclaw University of Science and Technology, Poland
Vladimir Rykov	Russian State Oil and Gas University, Russia
Wojciech Rząsa	Rzeszow University of Technology, Poland
Dariusz Rzońca	Rzeszow University of Technology, Poland
Alexander Schill	Technische Universität Dresden, Germany
Artur Sierszeń	Lodz University of Technology, Poland
Akash Singh	IBM Corp., USA
Mirosław Skrzewski	Silesian University of Technology, Poland
Tomas Sochor	University of Ostrava, Czech Republic
Maciej Stasiak	Poznań University of Technology, Poland
Janusz Stokłosa	Poznań University of Technology, Poland
Zbigniew Suski	Military University of Technology, Poland
Bin Tang	California State University, USA
Kerry-Lynn Thomson	Nelson Mandela Metropolitan University, South Africa
Oleg Tikhonenko	Częstochowa University of Technology, Poland
Mauro Tropea	University of Calabria, Italy
Homero Toral Cruz	University of Quintana Roo, Mexico
Leszek Trybus	Rzeszów University of Technology, Poland
Adriano Valenzano	National Research Council of Italy, Italy
Bane Vasic	University of Arizona, USA

Peter van de Ven   Eindhoven University of Technology, The Netherlands
Miroslaw Voznak   VSB-Technical University of Ostrava, Czech Republic
Krzysztof Walkowiak   Wrocław University of Technology, Poland
Sylwester Warecki   Intel, USA
Jan Werewka   AGH University of Science and Technology, Poland
Tadeusz Wieczorek   Silesian University of Technology, Poland
Lukasz Wisniewski   Hochschule Ostwestfalen-Lippe, Germany
Józef Woźniak   Gdańsk University of Technology, Poland
Hao Yu   Auburn University, USA
Grzegorz Zaręba   University of Arizona, USA
Zbigniew Zieliński   Military University of Technology, Poland
Liudong Zuo   California State University, USA
Piotr Zwierzykowski   Poznań University of Technology, Poland

Reviewers

Olumide Akinwande	Zbigniew Huzar	Przemysław Ryba
Iosif Androulidakis	Jacek Izydorczyk	Vladimir Rykov
Tülin Atmaca	Sergej Jakovlev	Wojciech Rząsa
Zbigniew Banaszak	Jerzy Klamka	Dariusz Rzońca
Robert Bestak	Wojciech Kmiecik	Alexander Schill
Grzegorz Bocewicz	Zbigniew Kotulski	Artur Sierszeń
Leoš Bohac	Stanisław Kozielski	Mirosław Skrzewski
Amlan Chatterjee	Henryk Krawczyk	Tomas Sochor
Ray-Guang Cheng	Andrzej Kwiecień	Janusz Stokłosa
Erik Chromý	Piotr Lech	Zbigniew Suski
Andrzej Chydziński	Aleksander Malinowski	Bin Tang
Tadeusz Czachórski	Marcin Markowski	Kerry-Lynn Thomson
Dariusz Czerwiński	Przemysław Mazurek	Oleg Tikhonenko
Waltenegus Dargie	Agathe Merceron	Mauro Tropea
Andrzej Duda	Jarosław Miszczak	Adriano Valenzano
Alexander N. Dudin	Vladimir Mityushev	Peter van de Ven
Peppino Fazio	Włodzimierz Mosorow	Miroslaw Voznak
Max Felser	Sasa Mrdovic	Krzysztof Walkowiak
Holger Flatt	Michele Pagano	Sylwester Warecki
Jean-Michel Fourneau	Nihal Pekergin	Jan Werewka
Janusz Furtak	Maciej Piechowiak	Tadeusz Wieczorek
Rosario G. Garroppo	Piotr Pikiewicz	Lukasz Wisniewski
Natalia Gaviria	Jacek Piskorowski	Józef Woźniak
Erol Gelenbe	Oksana Pomorova	Hao Yu
Roman Gielerak	Sławomir Przyłucki	Zbigniew Zieliński
Mariusz Głąbowski	Tomasz Rak	Liudong Zuo
Edward Hrynkiewicz	Stefan Rass	Piotr Zwierzykowski

Sponsoring Institutions

Organizer: Institute of Informatics, Faculty of Automatic Control, Electronics and Computer Science, Silesian University of Technology
Co-organizer: Committee of Informatics of the Polish Academy of Sciences, Section of Computer Networks and Distributed Systems
Technical co-sponsor: IEEE Poland Section

Technical Partner

Conference partner: iNEER

Contents

Teleinformatics and Telecommunications

New Technologies

Queueing Theory

Innovative Applications

Computer Networks

Traffic Flows Ateb-Prediction Method with Fluctuation Modeling Using Dirac Functions

Ivanna Dronyuk$^{(\boxtimes)}$ and Olga Fedevych

Lviv Polytechnic National University, 12 Bandera Street, Lviv, Ukraine
ivanna.droniuk@gmail.com, olhafedevych@gmail.com
http://www.lp.edu.ua/ikni

Abstract. In this paper a mathematical model for predicting traffic in computer networks based on heterogeneous differential equations describing the oscillatory motion was proposed. It is suggested that the solution to the equation should be modified on the basis of the sum of Dirac functions which have different arguments at different points of time to simulate bursts of network traffic. Based on this model, the software implementing network monitoring and forecasting of traffic on the network on the grounds of mathematical calculations and collected data was developed. Testing of models based on real network data from LPNU ACS Department was presented. Also, the simulation results have been compared with real traffic data. The results are illustrated in the form of both tables and plots.

Keywords: Dirac functions · Ateb-functions · Traffic modeling · Traffic monitoring and analysis

1 Introduction

Networks are an integral part of business, education, government and home communications. A lot of home, business and mobile IP networking are developing mainly due to the combination of traffic, social networking traffic and advanced collaboration applications, which is called a visual network. The forecasting load intensity of the nodal network equipment allows to provide reliability of its work condition and rational use of network resources to ensure effective adaptive management of its equipment. Therefore, it is urgent to develop the methods of modeling the traffic intensity [1] and decision-making for adaptive traffic management of computer networks.

Development of a method to predict the intensity of traffic in computer networks aims at improving the operation of the components, in particular switches, routers, and label switches of different types. The program for collecting, processing and forecasting of the global network traffic – Cisco Visual Networking Index – was taken as the basis for the justification of the research. According to Cisco forecasts [2], the volume of data traffic by 2020 will increase significantly.

© Springer International Publishing AG 2017
P. Gaj et al. (Eds.): CN 2017, CCIS 718, pp. 3–13, 2017.
DOI: 10.1007/978-3-319-59767-6_1

2   Improved Mathematical Model of Ateb-Prediction of Traffic Flows

Let us consider the changing traffic in the network over time, as a nonlinear oscillating system with one degree of freedom and with small perturbation. In known work [3] traffic model is one-dimensional, so it is advisable to use a differential equation of oscillating movement with one degree of freedom for simulation. Modeling behavior of the network traffic $x(t)$ is generated by an ordinary differential equation with a small parameter ε in the form

$$\ddot{x} + \alpha^2 x^n = \varepsilon f(t, x, \dot{x}), \tag{1}$$

where $x(t)$ – is the number of packets in the network at time t; α – a constant that determines size of the period of traffic oscillation; $f(t, x, \dot{x})$ – any analytical function that is used to simulate small deviations from the main component of the traffic fluctuations, n – a number that determines the degree of nonlinearity of the equation that affects the period of the main component of fluctuations. In the performance of such conditions on α and n : $\alpha \neq 0, n = \frac{2k_1+1}{2k_2+1}, k_1, k_2 = 0, 1, 2 \ldots$ it is proved [4], that the analytical solution of equation (1) is represented by Ateb-functions. It is known [5], that Ateb-functions are defined as inversion of the incomplete Beta-function. P. Senyk [5] discovered, that Ateb-functions for value $\alpha > 0$ are periodic and for $\alpha > 0$ – hyperbolic (aperiodic).

In order to construct Ateb-functions let us consider a dependency between ν and ω that is a function of n and m

$$\omega = \frac{n+1}{2} \int\limits_{0}^{-1 \leq \nu \leq 1} (1 - \bar{\nu})^{-\frac{m}{m+1}} d\bar{\nu} \tag{2}$$

where n, m are defined by equations

$$n = \frac{2\theta_1' + 1}{2\theta_1'' + 1}, m = \frac{2\theta_2' + 1}{2\theta_2'' + 1}, (\theta_1', \theta_1'', \theta_2', \theta_2'' = 0, 1, 2, \ldots). \tag{3}$$

This function is called Ateb-sine function and is denoted as: $\nu = \mathrm{Sa}(n, m, \omega)$. Ateb-cosine is given similarly.

To solve the Eq. (1) asymptotic method is applied [6]. The first approximation of the Eq. (1) solution is obtained as [8]:

$$\begin{cases} x(t) = A\mathrm{Ca}(n, 1, t) + \varepsilon f(t); \\ \dot{x}(t) = A^{\frac{1+n}{2}} h\mathrm{Sa}(1, n, t); \end{cases} \tag{4}$$

where A and h are certain constants that depend on α and n, $\mathrm{Ca}(n, 1, t), \mathrm{Sa}(1, n, t)$ – Ateb-cosine and Ateb-sine accordingly.

For problems of forecasting behavior of the traffic in computer network or mobile network it is important to choose the type of function f, as these terms take into account the specific features of particular network. Previously, the authors examined a small perturbation in the form of periodic functions [7]. This approach is appropriate for modeling a network with smooth traffic changing. In the present work the perturbation as a sum of delta functions was proposed. For numerical traffic simulation let us consider the functions $f(t, x, \dot{x})$ in more simpler form. We suppose that traffic and changing traffic value do not affect the small perturbation. This description corresponds to a network with abrupt changes of traffic. In our previous work analytical method for solving of the differential Eq. (1) based on Ateb-functions [8] was described in details. Thus, we construct the perturbation function in following form. Thus, the function $f(t, x, \dot{x})$ was considered and perturbation is shown in the following form:

$$f(t) = \sum_{i=1}^{N} a_i \delta(t_i), \tag{5}$$

where δ - Dirac function, N is the number of disturbances over the period $[0, T]$, a_i – amplitude of perturbation, $-A \leq a_i \leq A$, A – the maximum amplitude of perturbation (randomly generated in simulation), f_i – perturbation function, t_i – time in which comes off the i-th perturbation, which is generated randomly. Here T stands for the simulation time. Perturbation function f_i – in our case does not depend on the number of packets x and the speed of packets changing $\dot{x}$.

3   Developed Software for Traffic Flows Analysis

Main features of the developed traffic flow analyzer are: capturing samples of network traffic of a computer network with the ability to record the results of capturing in a file of a special format *.pcap; the analysis of pre-recorded network traffic. Processed traffic data are stored in a relational database with a resolution of 1 s. The software also provides a possibility of calculating optimal network data routes based on the graph model using the modified Dijkstra's algorithm [9,10]. Figure 1 shows the visualization of observations made in the main window of the application according to the parameter of the total traffic on the server. In addition, the analyzer of the traffic flow is designed to provide an automated collection of information from network devices and provide control of the communication channels. It envisages automation of the collection and analysis of network characteristics and their display in a convenient format to the user. The use of the developed computer network software enables:

- to automatically and continuously collect traffic data from devices in a computer network;
- real-time monitoring of the network performance;
- real-time forecasting of network load based on the collected data and mathematical modeling. For this purpose the database management system PostgreSQL [11] is used. There exists an ability to connect to any compatible database and change connection during the program work. Network monitoring in computer network is an important practical task. To this end, we collect sample data to build a predictive traffic flow to ensure efficient use of resources of a computer network.

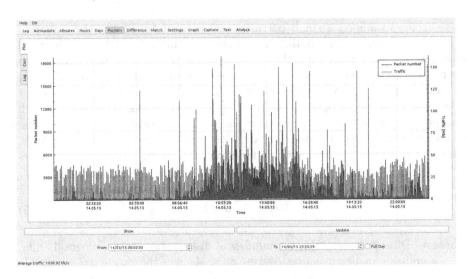

Fig. 1. Sample traffic flow of a computer network of the Department of ACS, obtained with the developed software.

Developed network traffic flow analyzer consists of a server side (C, C++, Haskell) and a client side (C++/Qt). The server side deals with the capturing and storage of data. In turn, the client side conducts data gathering and processing and displays the results. Computations on client side are conducted in multithreaded environment in order to prevent user interface blocking. This software product has been written on programming languages C++ and Haskell and is licensed under the GNU GPL.

Traffic capturing is implemented in C programming language as a standalone shared library and uses libpcap/WinPcap library in UNIX/Windows operating systems. The shared library is then linked to the main program and can be linked to other applications via the standard C foreign function interface. Traffic capturing is available as a distinct window (Fig. 2) and enables user to set the network interface to capture from, capturing time (value −1 indicates infinite time), number of packets to capture (value −1 indicates infinite number of packets), possible dump file name to store capturing results and a filter string for packet filtering.

Traffic forecasting is implemented as a maximization of the sample correlation coefficient between real traffic and selected Ateb-function. Calculation of Ateb-function values are time consuming. Since Ateb-functions are periodical, their values are precomputed on a range of the function period with a step equal to 0.01 and stored as a constant vectors inside the software executable binary file. The software supports a set of Ateb-funtions with different n, m parameters which represent typical shapes of Ateb-functions. Linearly interpolated values are used then in actual computations. The forecast is generated as a Ateb-function with

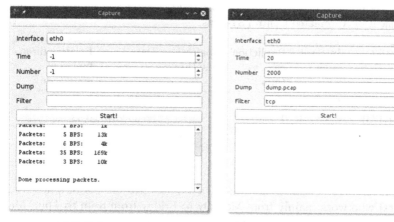

(a) Infinite capturing

(b) Capture either for 20 minutes or for 2000 packets with filtering only tcp packets, dump captured data to file "dump.pcap"

Fig. 2. Traffic capturing window of developed software traffic analyzer.

parameters specified during the maximization stage. This forecasting method is an improvement of method described in [13].

4   Experiments

Based on the theory of Ateb-functions and collected samples of computer network traffic flow, traffic flow values in the network have been predicted. The advantage of the proposed method is the use of a single analytical formula to calculate the values of the traffic.

The effectiveness of the proposed method was confirmed by experimental studies using a computer network monitoring software. Software testing was conducted using samples of the traffic flow collected from computer networks at the Department of Automated Control Systems of the Lviv Polytechnic National University (ACS LPNU) and at the Institute of Theoretical and Applied Informatics of the Polish Academy of Sciences (Gliwice).

To test the theoretical calculations, it was necessary to conduct experimental studies of traffic in computer networks. To collect experimental network data load we used a network of the ACS Department of LPNU (May 2015). Monthly data traffic (May 2012) of a computer network of the Institute of Theoretical and Applied Informatics of the Polish Academy of Sciences (ITAI PAS), in cooperation with the LPNU was also processed. Data was captured during May 2012 on the online gateways of the Institute of Theoretical and Applied Computer Science in Gliwice, Poland [12].

Table 1. Description of parameters to forecast traffic flow in a computer network of the Department of Automated Control Systems

No	Function	Interval (min.)	Prediction step (min.)	Monitoring time (hrs.)	Date	Relative dirac function amplitude
1	Sa(0.01, 0.1)	60	1	6	07.05.2015	0.10
2	Sa (1,1/3)	60	1	6	06.05.2015	0.33
3	Sa (1/5,1)	60	1	6	08.05.2015	0.23

Traffic ITAI PAS contains the data of about several dozens of office users (researchers) who work mainly from Monday to Friday from 8 am to 4 pm. May 1–3 is a national holiday in Poland, so the traffic might be less. A break in the data traffic was recorded on May 22 14:53:32 2014 CET. IP packet size was limited to 64 bytes, in most cases packet contains all the headers plus a few bytes of the payload of the transport protocol. Local DNS traffic is made invisible using pre-implemented specific network settings. IP addresses are not anonymous.

The ACS network contains about 20 working computers of employees which are loaded on average from 8:30 am to 5:30 pm, about 4 computers which are loaded before 9:00 pm, and 3 computer labs with 32 workstations, which are loaded with on average from 8:30 am to 4:00 pm.

Sampling of source data corresponds to the parameter values from Table 1. The amplitude of Dirac function is considered as a point great number, which was selected randomly. This table shows three examples of seven available in the developed software Ateb-sine functions used for forecast generation.

To predict the traffic, Eq. (1) is used, in particular, a first approximation of solution in the form (4).

On Figs. 3, 4, 5, 6, 7 and 8, the red line corresponds to the real values of the traffic flow, while the black graphs represent the calculated values of the predicted traffic flow. These plots represent solutions of Eq. 1. The observations were carried out continuously from 1 to 30 May 2015 in the computer network at the Department of Automated Control Systems, but the observation is presented for the period of 3 days.

Traffic analysis has been carried out to predict the trend of traffic for the next 3–5 min to optimize the load of network equipment.

The described software has been tested not only on the sample traffic flow of a computer network of the ACS Department, but also on the traffic samples in ITAI PAS [12], which were obtained using a network protocol analyzer like Wireshark, and the developed traffic flow analyzer to ensure the reliability of the results.

To calculate the reliability of the results of the experiment, the value of the maximum correlation and the coefficient showing the ratio of standard deviation to maximum were chosen.

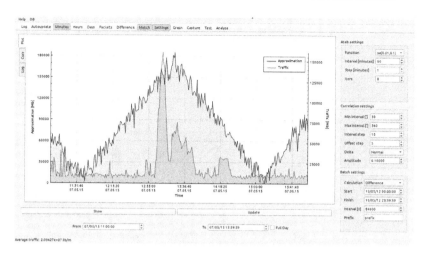

Fig. 3. Comparison of real traffic sample and its prediction for 07.05.15 with Dirac function (calculated using (5) with apmlitude $A = 0.1$), where $n = 1, m = 0.01$ - Ateb-sine parameters. (Color figure online)

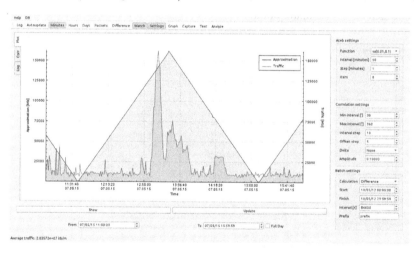

Fig. 4. Comparison of real traffic sample and its prediction for 07.05.15 without Dirac function (calculated using (5) with apmlitude $A = 0.1$), where $n = 1, m = 0.01$ - Ateb-sine parameters. (Color figure online)

Coefficient K of the ratio of the maximum to the standard deviation:

$$K = \frac{s_{max}}{s}, \tag{6}$$

where $s = \sqrt{\frac{N}{N-1}\sigma^2}$ and $s_{max} = \max_s s$.

Comparison of simulation experiment results was carried out according to the criterion of maximum correlation ρ, k – factor as the ratio of standard deviation

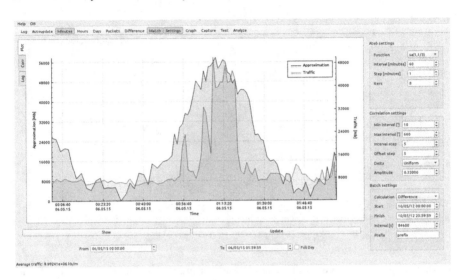

Fig. 5. Comparison of real traffic sample and its prediction for 06.05.15 with Dirac function (calculated using (5) with apmlitude $A = 0.33$), where $n = 1, m = 1/3$ - Atebsine parameters. (Color figure online)

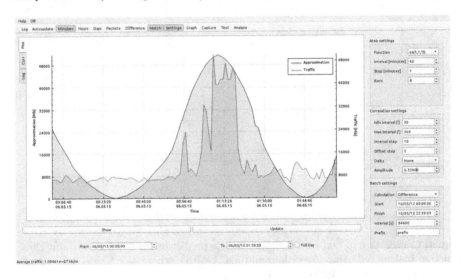

Fig. 6. Comparison of real traffic sample and its prediction for 06.05.15 without Dirac function (calculated using (5) with apmlitude $A = 0.33$), where $n = 1, m = 1/3$ - Atebsine parameters. (Color figure online)

to maximum. The computed results of the comparison are presented in Tables 2 and 3. Forecast improvement obtained by using of Dirac function is presented in Table 4.

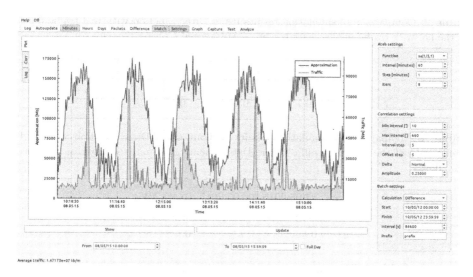

Fig. 7. Comparison of real traffic sample and its prediction for 08.05.15 with Dirac function (calculated using (5) with apmlitude $A = 0.23$), where $n = 1, m = 1/5$ - Atebsine parameters. (Color figure online)

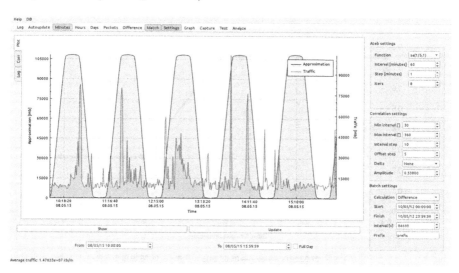

Fig. 8. Comparison of real traffic sample and its prediction for 08.05.15 without Dirac function (calculated using (5) with apmlitude $A = 0.23$), where $n = 1, m = 1/5$ - Atebsine parameters. (Color figure online)

There was calculated a dependency between the relative amplitude of the Dirac function and the correlation value. The dependency is presented in Fig. 9 and it is seen that the uniform distribution of a function (solid line) increases fluctuation amplitude more rapidly than the normal distribution which causes the correlation value to decrease.

Table 2. Comparison of simulation results using Dirac function with real traffic data of the network traffic of the Department of Automated Control Systems for 1 day.

Date	07.05.2015	06.05.2015	08.05.2015
K	97.0%	74.9%	99.1%
ρ	0.7	0.75	0.67

Table 3. Comparison of simulation results without using Dirac function with real traffic data of the network traffic of the Department of Automated Control Systems for 1 day.

Date	07.05.2015	06.05.2015	08.05.2015
K	90.0%	65%	67%
ρ	0.69	0.73	0.64

Table 4. The improvement of prediction using Dirac function.

Date	07.05.2015	06.05.2015	08.05.2015
ΔK	8%	15%	50%
$\Delta \rho$	1.5%	3%	5%

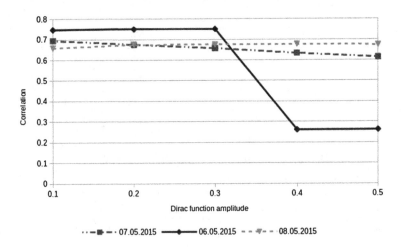

Fig. 9. The dependency between the relative amplitude of the Dirac function and the correlation value (Color figure online)

5   Conclusions

The method of analysis and forecasting of traffic flow in a computer network based on the differential equations of vibrational motion with the right parts of the equations as Dirac functions is quite effective. Based on the proposed model, a software was developed for the analysis and forecasting of traffic flow in a computer network, which is tested on the real data of the network. The criterion of adequacy selected for the model was maximum correlation. The experiments were conducted where the deviation of predicted and actual traffic, on average, was around 25–30%. In general, the software shows the positive results of the prediction method in the node of computer network. The next stage of the study is to redistribute the traffic on the basis of predicted overload. This will effectively reduce the intensity of the load in hub equipment and improve the operations of computer networks in general.

References

1. Czachórski, T., Pekergin, F.: A diffusion approximation model of an electronic-optical node. In: Bravetti, M., Kloul, L., Zavattaro, G. (eds.) EPEW/WS-FM-2005. LNCS, vol. 3670, pp. 187–199. Springer, Heidelberg (2005). doi:10.1007/11549970_14
2. Cisco VNI Forecast Widget. http://newsroom.cisco.com
3. Czachórski, T.: Queueing models for performance evaluation of computer networks - transient state analysis. In: Mityushev, V.V., Ruzhansky, M. (eds.) Analytic Methods in Interdisciplinary Applications. Springer Proceedings in Mathematics and Statistics, vol. 116, pp. 55–80. Springer, Heidelberg (2015). doi:10.1007/978-3-319-12148-2_4
4. Cveticanin, L., Pogany, T.: Oscillator with a Sum of noninteger-order nonlinearities. J. Appl. Math. Article No. 649050, 20 pages (2012). doi:10.1155/2012/649050 (in Ukrainian)
5. Senyk, P.M.: About Ateb-functions: AofS USSR Reports, Series A - 1, pp. 23–27 (1968). (in Ukrainian)
6. Bogolyubov, N.N., Mitropolsky, Y.A.: Asymptotic methods in the theory of nonlinear oscillations. M. Izd. Fiz-Mat. Lit. 407 (1963). (in Russian)
7. Medykovsky, M., Droniuk, I., Nazarkevich, M., Fedevych, O.: Modelling the pertubation of traffic based on ateb-functions. In: Kwiecień, A., Gaj, P., Stera, P. (eds.) CN 2013. CCIS, vol. 370, pp. 38–44. Springer, Heidelberg (2013). doi:10.1007/978-3-642-38865-1_5
8. Dronjuk, I., Nazarkevych, M., Fedevych, O.: Asymptotic method of traffic simulations. In: Vishnevsky, V., Kozyrev, D., Larionov, A. (eds.) DCCN 2013. CCIS, vol. 279, pp. 136–144. Springer, Cham (2014). doi:10.1007/978-3-319-05209-0_12
9. Dijkstra, E.W.: A note on two problems in connexion with graphs. Numer. Math. **1**, 269–271 (1959). doi:10.1007/BF01386390
10. Nurminen, J.N.: Using software complexity measures to analyze algorithms - an experiment with the shortest-paths algorithms. Comput. Oper. Res. **30**, 1121–1134 (2003). Elsevier Science
11. PostgreSQL. https://www.postgresql.org/
12. Foremski, P.: Mutrics: multilevel traffic classification. http://mutrics.iitis.pl/
13. Fedevych, O., Droniuk, I., Nazarkevych, M.: Monitoring and analysis of measured and modeled traffic of TCP/IP networks. In: Gaj, P., Kwiecień, A., Stera, P. (eds.) CN 2016. CCIS, vol. 608, pp. 32–41. Springer, Cham (2016). doi:10.1007/978-3-319-39207-3_3

Improving Accuracy of a Network Model Basing on the Case Study of a Distributed System with a Mobile Application and an API

Wojciech Rząsa[1], Marcin Jamro[2], and Dariusz Rzonca[1(✉)]

[1] Department of Computer and Control Engineering,
Rzeszow University of Technology,
al. Powstancow Warszawy 12, 35-959 Rzeszow, Poland
{wrzasa,drzonca}@kia.prz.edu.pl
[2] TITUTO Sp. z o.o. [Ltd.], Zelwerowicza 52G Street, 35-601 Rzeszow, Poland
marcin@tituto.com
http://kia.prz.edu.pl
http://tituto.com

Abstract. Nowadays, many IT products are created as distributed solutions that consist of many parts, such as mobile applications, web-based back-ends, as well as APIs that connect various parts of the system. It is a crucial task to apply a suitable architecture to provide users of mobile applications with satisfactory operation, especially when the Internet connection is necessary to get or send some data. The simulation of network architecture and configuration using a high-level model of the system described with dedicated Domain-Specific Language (DSL), enabled by the Timed Colored Petri Nets (TCPNs) formalism is a beneficial approach that could be applied in real case studies. The already proven research method has been applied to one of the scenarios regarding the system offered by TITUTO Sp. z o.o. [Ltd.] company (Rzeszow, Poland). The first obtained results were not sufficiently precise for detailed analysis of the system. Thus, the case study was used to improve the simulation method in order to more accurately model data transmissions over the network. After modifications were implemented in the simulation tool, significantly better results have been received, as discussed in the paper.

Keywords: Simulation · Petri Nets · Performance · Distributed system · API · Mobile application

1 Introduction

While preparing a distributed solution, it is an important task to choose a suitable architecture that provides sufficient performance and correctly handles scenarios with diverse number of users who get or send some data via the Internet. For this reason, introducing research methods into the process of choosing an architecture is a beneficial approach. There are several approaches that could be used for this purpose. One of possible solutions is modeling network to check

© Springer International Publishing AG 2017
P. Gaj et al. (Eds.): CN 2017, CCIS 718, pp. 14–27, 2017.
DOI: 10.1007/978-3-319-59767-6_2

whether the given system architecture and configuration could properly handle the operation by a specified number of users.

As mentioned in the title of this paper, it is based on the real case study. It is related to one of the products developed by TITUTO Sp. z o.o. [Ltd.][1] (Rzeszow, Poland). This company offers a set of own IT products dedicated to hotels and tour operators. One of solutions, which is analyzed in this paper, is the mobile trip assistant TOURISER[2]. It uses smartphones and tablets of trip participants to present a set of information useful during the organized tour. Of course, the product consists of many parts forming together the distributed system:

- the mobile applications running on smartphones and tablets with Android, iOS, and Windows operating systems
- web applications, available via web browser, to manage data presented in the mobile applications, as well as to perform other tasks
- Application Programming Interface (API) to connect various system parts.

The mobile applications are equipped with a set of features, such as presentation of a trip program. Of course, they should operate also in the off-line mode, so it is crucial to provide users with a possibility to download all necessary data, including textual content, as well as a set of photos and audio recordings. Such a process is further referred as update. It can be performed by many people simultaneously, especially when the update process is forced by the tour guide who informs the group of trip participants to launch the application to get the latest data for the current trip. What is more, the update process could be performed from various devices with diverse performance, as well as using the broad range of speed of the Internet connection, even with a limited availability.

To provide users with satisfactory update times, ensure that the initial architecture of the system is sufficient, as well as prepare the product in an expandable way, the company has decided to cooperate with researchers from Department of Computer and Control Engineering from Rzeszow University of Technology. At the beginning, the authors of this paper used the already available tools and research methods. Unfortunately, while performing experiments, it has been necessary to introduce modifications in order to improve accuracy of a network model to present results more proper for the real-world scenario.

In this paper, the detailed analysis of results, together with a set of graphs, is presented. In the second chapter, the initial approach is described. Then, the doubts regarding received results are discussed. The fourth chapter shows the problem of parallel data transfers, together with improving the previously mentioned model, and estimating the improvement. In the following chapter, the corrected simulator is applied to the case study with a detailed comparison and explanation of the received results.

---

[1] http://tituto.com.

[2] http://touriser.com.

2 Initial Approach to Analysis

Different approaches are described in the literature as suitable for analyzing distributed applications and network impact on performance. It is possible to simulate computer networks behavior in so-called "network simulator" (e.g. widely used family of ns-2 and ns-3 [1] open-source simulation tools). Another approach involves creating models of network components in one of the formalisms dedicated to analyze and solve concurrency problems. Among these formalisms various classes of Petri nets seem to be the most popular, due to their ability to model activities of concurrent and distributed systems. For example Fuzzy Petri Nets may be used as a formal representation of a concurrent control algorithm [2]. Queueing Petri Nets can be used as a performance prediction tool during the software engineering process of a distributed component-based system [3] or for response time analysis of distributed web systems [4]. Timed Coloured Petri Nets are also used for simulation and performance analysis of distributed internet systems [5]. Another Petri net class, RTCP-nets are used for modeling and analysis of embedded and real-time systems [6]. Analysis of such Petri net models is possible in strictly formal way [7–9], but also the model can be simulated to observe behavior of the modeled system [10], and perform statistical performance analysis. Both ways provide reliable results and can be used to verify some properties and validate the requirements [11].

In the case considered in this paper, the analysis of application was based on the method that allows to obtain precise results, as described in [12–14]. The method assumes that a high-level model of application should be created. The model is then a basis for simulation. It is described with a Domain Specific Language (DSL) designed especially for this purpose. Reliability of simulation is ensured by the formalism of Timed Colored Petri Nets (TCPN) [15]. The concept of connecting a high-level model with TCPN is described in [16] and also applied in [17].

2.1 DSL Concepts

The DSL allows presenting model of application using concepts of programs that describe actions performed by processes running on nodes. Each pair of nodes can be connected through one or more net segments that allow remote communication. In the model used in this research, there were two programs, namely describing (1) an API server and (2) mobile clients. There was one instance of the API server program running in each simulation. The number of processes that run client programs depended on configuration of specific simulation. Each process in the simulation run on a separate node.

Network communication between clients and the server was possible through network segments that modeled server and clients Internet links. Network route between a client and the server always led through two network segments that represented (1) the client's Internet link and (2) the servers Internet link. Thus, it was possible to model limited bandwidth (1) between the server and the Internet (shared between all clients) and (2) that of each client Internet link. The network links enabled full-duplex communication.

2.2   Processes

The server process served two types of requests, as follows:

1. :config that returned content of the update configuration file. The configuration described all files that should be downloaded during the update.
2. :update with specific file name as its parameter. It returned a given file to the client.

Each file included in the update configuration had its specific size and, therefore, required a given network transmission time.

A client process, after it was started, sent the :config request to the server in order to obtain information about files that should be updated. Thereafter, it downloaded subsequent files from the received list by sending the :update requests for each of them. Downloading of each file was accompanied with time-out. If a timeout occurred, it was considered that the whole update process for this client failed. There was a startup delay between the moment the client program started and the moment it begun its update. It was set to a random value from a specified range. Consequently, all clients did not start their updates at the same moment in time. It is consistent with the real world behavior, even if application update is triggered by a person suggesting this update to a group of people.

2.3   Gathering Statistics

The model was used to gather performance-related statistics about the application. The most important factor was the update time for clients. This was measured for each client as time difference between the moment the client started update and the moment the last file from the update was received. Excessive number of simultaneous updates could result in the number of clients not being able to complete the process and timeout due to the overloaded server network. Therefore, the number of timeouts during the update processes was the second parameter measured during simulations.

The simulations were carried out for three types of clients, characterized by the following bandwidth of Internet link:

:slow – 512 kbps
:normal – 2 Mbps
:fast – 20 Mbps

In accordance with parameters of an exemplary scenario of the analyzed application, each client performed an update process consisting of 308 files of diverse sizes. Total size of each update was almost 70 MB. Network bandwidth for the

Table 1. Average update time for a client in different configuration of simulation.

Number of clients of each type	Average update time [s]
20 × :slow	1145.91 s
20 × :normal	293.56 s
40 × :fast	195.18 s
40 × :fast, 20 × :normal, 20 × :slow	546.33 s
80 × :fast, 20 × :normal, 20 × :slow	675.23 s
20 × :fast, 20 × :normal, 20 × :slow	551.17 s

API server was set to 100 Mbps. Simulations were performed for different numbers of clients of each type, as shown in Table 1, which presents also an average update time for a client.

For these configurations, no timeouts were observed and all update processes finished successfully. The results show that for the considered configuration of resources the application should behave correctly. The update times for :slow clients were significant, but such a situation was caused by their own slow Internet links and could not be changed. The update times for other groups of clients were on the limit of acceptance. This was a result of insufficient bandwidth of the Internet link of the API server. However, the value used in the experiments was deliberately low to verify behavior of the application in unfavorable conditions. The real application was supposed to work on a server with 1 Gbps Internet link. It seemed that the application was not threatened by transmission timeouts, since the first indicators of too slow Internet link of the API server were significant client update times. The update times would become unacceptably long before insufficient bandwidth of server link could cause transmission timeouts.

3   Assessing Simulator Accuracy

3.1   Doubts Concerning Results

The analysis of application provided satisfactory results. However, unlike in case of previous research [12–14], additional consideration was required for this specific case. The modeled and simulated application performance proved to significantly depend on the network transmission efficiency. However, the simulator used a basic network data transmission model. Data sent between two processes in a single message were treated as a whole and transmission time over a network was modeled as a delay computed for the whole data package. If a route between two nodes consisted of more than a single network segment, a data package was transmitted over another segment only after it had been completely transmitted over the previous one.

This approach allowed the very efficient simulation with accuracy sufficient for applications where data processing had important impact on performance.

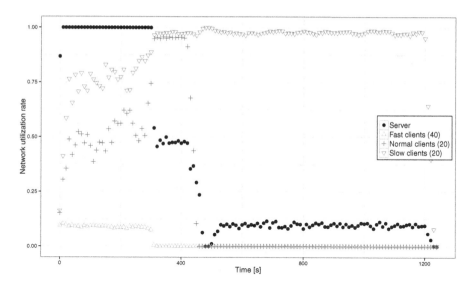

Fig. 1. Network utilization rate for multiple clients of each type.

However, applying this simplified model to the application presented in this paper required consideration of possible inaccuracies. Therefore, more detailed results of simulation concerning network usage were analyzed.

Figure 1 presents mean rate of network utilization for individual types of Internet links for clients and for the API server. At the beginning, the server link was fully loaded since all the clients performed their update processes. After the :fast clients finished their update processes, the server's network utilization decreased, since the :normal and :slow clients' Internet links did not allow them to fill the servers network bandwidth. This was an expected behavior. However, the rate of network utilization of :normal and :slow clients while the :fast ones still performed updates raised justified doubts. The results of additional simulations performed for single clients of each type, presented in the Fig. 2, had the same characteristics: slower clients did not fill bandwidth of their network links despite the fact that their share of the server's network bandwidth should be sufficient for this.

3.2   Comparison with a Real-World Experiment

The results of the basic simulations with individual clients of each type could be verified by comparison with the real-world experiments. This would be hardly feasible for the case in which a larger number of clients is considered.

The experiments were performed on three virtual machines running Linux. Three clients (fast, normal, and slow) were running on the virtual machines and downloading data from the host. The operating system's traffic control capabilities configured with the Linux tc program were used to limit bandwidth available for the transfers. The server's bandwidth was limited with Token Bucket Filter

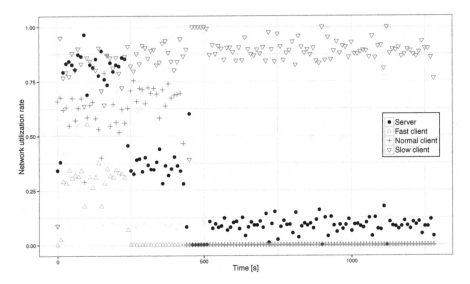

Fig. 2. Network utilization rate for single clients of each type.

(TBF) attached to outgoing network interface. Incoming traffic for each client was shaped with a TBF attached to Intermediate Functional Block devices – virtual network interfaces enabling control of ingress network traffic in Linux. Consequently, each stream of downloaded data was passing two network segments with different bandwidth: one on server's network interface (shared by all clients) and the other on the clients network interface. The bandwidth of the server network was set to 5 Mbps and bandwidths for the clients were configured as follows:

- slow – 0.5 Mbps
- normal – 2 Mbps
- fast – 8 Mbps

Consequently, slower clients were not able to fill the bandwidth of server's network, and the fast client had a faster network link than the server. This is consistent with the configuration of the original simulations with multiple clients if the total bandwidth available for each type of clients is considered.

As observed during the experiment, average bandwidths used by the clients in two-minute download processes were as follows:

- slow – 0.47 Mbps
- normal – 1.9 Mbps
- fast – 2.2 Mbps

The results clearly show that in the analyzed configuration the slower clients almost fully utilized bandwidths of their own network links. This was a significant difference in comparison with the simulation results.

4 Improving Precision of the Network Model

4.1 The Problem of Parallel Data Transfers

A more precise estimation of network transmission time could be achieved by combining existing simulator with a precise model of Transmission Control Protocol (TCP). It is even feasible to include crucial parts of the real TCP implementation into the simulator, as shown in [18]. However, such a solution requires significant effort not only to modify existing simulator, but also – what is even more important – to obtain reliable information about details of TCP versions and configurations of particular servers and clients. The second problem seems to be especially difficult for real-world business applications. For this reason, another – less precise, but more practical – solution was chosen.

As already mentioned, the network model in the simulator assumed that a data package (e.g., a file) was transmitted as a whole, generating a proper delay for delivery of the data package and occupying a network segment for the whole transmission time. Consequently, the next data transfer was delayed until the previous one was finished. Thus, in the simulations described above, data packages designated for slower clients were frequently delayed on the server link by other transfers, especially the ones sent to the faster clients. Thus, the slower client's network links were idle for some periods of time, waiting for data from the faster link. Therefore, although the server link was significantly faster than the links of normal and slow clients, the crude network model prevented the transfers from filling bandwidths of the slower networks.

This situation does not occur in the real-world scenarios, because transmitted data are divided into segments. Thus, even large data transfers do not delay the others until they are completed. First, the data transfers over one network overlap. Second, segments already transmitted over one network are passed to another one and a single data package (e.g. file) can be concurrently transmitted over more than one of subsequent networks on a route. Thus, faster network links, even if significantly loaded, usually can transfer sufficient number of data segments to fill bandwidth of the slower ones ensuring as efficient data transmission as possible.

4.2 Improving the Simplified Model

In order to mitigate the problem without including the complete TCP model in the simulator, an intermediate solution was devised. In order to ensure a lower grained and more precise simulation of parallel transmissions, the transmitted data packages (e.g., files) were divided into a number of fragments. Each fragment was transmitted separately and the simulator reported data reception only after all fragments were received. This solution also ensured that a single data package could be transmitted through more than one network on a route. Thus, with a larger number of smaller data fragments to transfer, the faster network links should have been able to fill bandwidth of the slower ones even if the faster links were busy.

The solution with data fragmentation was implemented in the simulator as an additional option. The maximum number of fragments for a data package can be configured for each simulation. The actual number of fragments for a specific package is selected according to (1), where M is the maximum number of fragments configured for simulation and s is size of the whole data package in bytes. The value of 1500 is derived from the Maximum Transmission Unit (MTU) for the most popular Ethernet protocol. Consequently, the fragments are not smaller than 1500 B and with value of M simulator may be configured to use more or less fine grained model.

$$min(M, \lfloor s/1500 \rfloor) \tag{1}$$

4.3 Estimation of Achieved Improvement

The results provided by the new version of the simulator were again compared with the results of the experiment described in Sect. 3.2. The maximum number of fragments was set to different values in order to find a setting that ensured the sufficient precision. As presented in Fig. 3, setting the maximum number of fragments to 5, significantly improved accuracy of results. Figure 4 shows additional improvement for the maximum number of fragments set to 10, but in this case the difference is not significant. Similarly, increasing the maximum number of fragments to 30 did not cause meaningful improvement (Fig. 5).

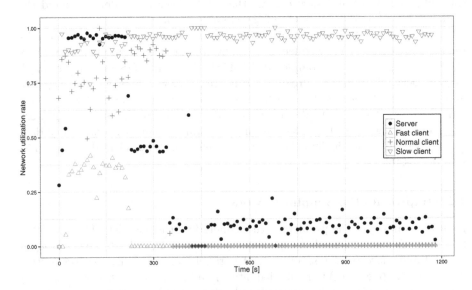

Fig. 3. Results of simulations for the maximum number of data fragments set to 5.

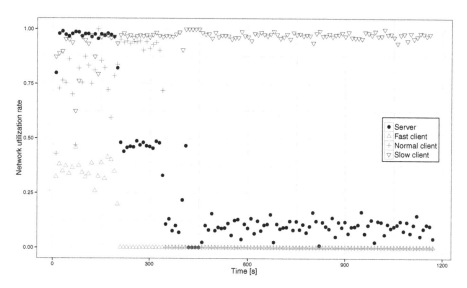

Fig. 4. Results of simulations for the maximum number of data fragments set to 10.

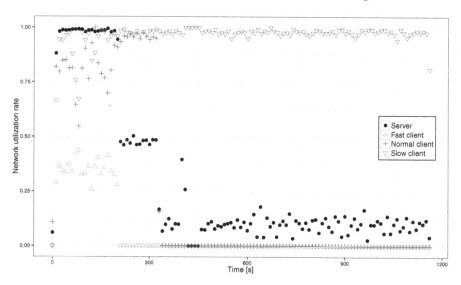

Fig. 5. Results of simulations for the maximum number of data fragments set to 30.

5   Applying Improved Simulator to the Case Study

The improved simulator was used again to analyze the real case-study application in order to check how the more precise simulation affected results. The simulations were carried out for two values of the maximum number of data fragments set to 5 and 10. Average update times are summarized in Table 2. Graphs of the network usage are presented in Figs. 6 and 7.

Table 2. Average update time for a client for different values of the maximum number of data fragments.

Number of clients of each type	Average update time [s]		
	No fragmentation	5 fragments	10 fragments
40 × :fast, 20 × :normal, 20 × :slow	546.33 s	542.26	541.05
80 × :fast, 20 × :normal, 20 × :slow	675.23 s	683.10	679.02
20 × :fast, 20 × :normal, 20 × :slow	551.17 s	541.28	540.74

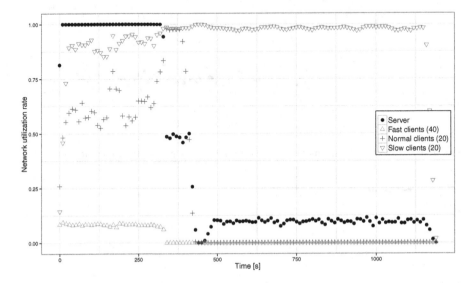

Fig. 6. Results of simulations for the maximum number of data fragments set to 5.

Comparison of results from Table 2 does not show significant differences in update times for different configurations of simulator. This is due to the fact, that in average, shorter update time for :slow and :normal clients is compensated by longer update times for :fast clients. Therefore, analysis of the less aggregated results of simulation presented in Figs. 6 and 7 more clearly shows differences in obtained results.

To illustrate these differences between the initial result of simulation (without fragmentation, Fig. 1) and the final one (maximum 10 fragments, Fig. 7), such scenarios are directly compared in Fig. 8. For clarity of the graph, only results for slow clients are shown.

It is clearly seen that the network utilization rate is higher when fragmentation is considered, especially for the first part, when fast clients were still downloading the update. Thus, the obtained simulation results more closely correspond to the experiment results described in Sect. 3.2.

For the considered application, the aggregate metrics, i.e., the average client update time and the number of client timeouts were the required results. These

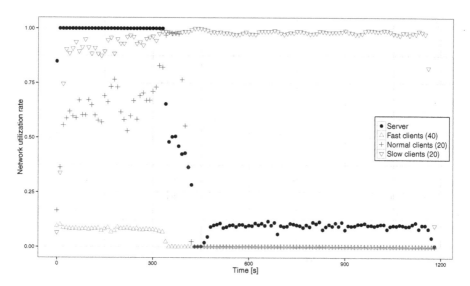

Fig. 7. Results of simulations for the maximum number of data fragments set to 10.

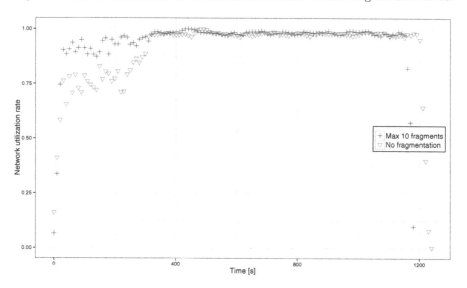

Fig. 8. Comparison of simulation results for slow clients depending on fragmentation.

values were not significantly affected by the basic model of network data transmission, despite of the fact that the application is network-bounded. However, simulations are frequently used also to observe behavior of a system in a more detailed way. In such the case, where not only aggregated results are important and when performance of an application depends mainly on network transmission time, the more precise network model may be an important improvement for the simulator.

6 Summary

Based on the experiments described in the paper, using simulation enabled by the formalism of Timed Colored Petri Nets (TCPNs) to analyze a high-level model described with a Domain-Specific Language (DSL) allows to obtain reliable results regarding network-related performance of distributed systems. The aim of the case study presented in this paper was to examine performance of one of the systems developed and offered by TITUTO Sp. z o.o. [Ltd.] company (Rzeszow, Poland).

The initial version of the simulation method did not provide satisfactory results while more detailed analysis of the real-world scenario was performed. For this reason, the research approach has been adjusted to properly handle transmission of files over the network, by adding the possibility of dividing them into fragments. In the paper, a few experiments are described, together with a detailed analysis of results. At the end, the already checked and described tool has been enhanced with additional features.

References

1. Riley, G.F., Henderson, T.R.: The ns-3 network simulator. In: Wehrle, K., Güneş, M., Gross, J. (eds.) Modeling and Tools for Network Simulation, pp. 15–34. Springer, Heidelberg (2010)
2. Gniewek, L.: Sequential control algorithm in the form of fuzzy interpreted Petri net. IEEE Trans. Syst. Man Cybern. Syst. **43**(2), 451–459 (2013)
3. Kounev, S.: Performance modeling and evaluation of distributed component-based systems using queueing Petri nets. IEEE Trans. Softw. Eng. **32**(7), 486–502 (2006)
4. Rak, T.: Response time analysis of distributed web systems using QPNs. Math. Probl. Eng. **2015**, 1–10 (2015). doi:10.1155/2015/490835. Article ID 490835
5. Rak, T., Samolej, S.: Simulation and performance analysis of distributed internet systems using TCPNs. Informatica Int. J. Comput. Inform. **33**(4), 405–415 (2009)
6. Szpyrka, M., Szmuc, T.: Integrated approach to modelling and analysis using RTCP-nets. In: Sacha, K. (ed.) Software Engineering Techniques: Design for Quality. IIFIP, vol. 227, pp. 115–120. Springer, Boston (2006). doi:10.1007/978-0-387-39388-9_11
7. Gniewek, L.: Coverability graph of fuzzy interpreted Petri net. IEEE Trans. Syst. Man Cybern. Syst. **44**(9), 1272–1277 (2014)
8. Murata, T.: Petri nets: properties, analysis and applications. Proc. IEEE **77**(4), 541–580 (1989)
9. Szpyrka, M.: Analysis of RTCP-nets with reachability graphs. Fundamenta Informaticae **74**(2), 375–390 (2006)
10. Wells, L., Christensen, S., Kristensen, L.M., Mortensen, K.H.: Simulation based performance analysis of web servers. In: Proceedings 9th International Workshop on Petri Nets and Performance Models, pp. 59–68 (2001)
11. Girault, C., Valk, R.: Petri Nets for Systems Engineering. A Guide to Modelling Verification, and Applications. Springer, Heidelberg (2003)
12. Rząsa, W.: Simulation-based analysis of a platform as a service infrastructure performance from a user perspective. In: Gaj, P., Kwiecień, A., Stera, P. (eds.) CN 2015. CCIS, vol. 522, pp. 182–192. Springer, Cham (2015). doi:10.1007/978-3-319-19419-6_17

13. Rząsa, W., Rzonca, D.: Event-driven approach to modeling and performance estimation of a distributed control system. In: Gaj, P., Kwiecień, A., Stera, P. (eds.) CN 2016. CCIS, vol. 608, pp. 168–179. Springer, Cham (2016). doi:10.1007/978-3-319-39207-3_15
14. Rząsa, W.: Predicting performance in a PaaS environment: a case study for a web application. Comput. Sci. [S.l.] **18**(1), 21–39 (2017). http://dx.doi.org/10.7494/csci.2017.18.1.21
15. Jensen, K., Kristensen, L.: Coloured Petri Nets. Modeling and Validation of Concurrent Systems. Springer, Heidelberg (2009)
16. Rząsa, W.: Timed colored Petri net based estimation of efficiency of the grid applications. Ph.D. thesis, AGH University of Science and Technology, Kraków, Poland (2011)
17. Jamro, M., Rzonca, D., Rząsa, W.: Testing communication tasks in distributed control systems with SysML and Timed Colored Petri Nets model. Comput. Ind. **71**, 77–87 (2015). http://dx.doi.org/10.1016/j.compind.2015.03.007
18. Rząsa, W.: Combining timed colored Petri nets and real TCP implementation to reliably simulate distributed applications. In: Kwiecień, A., Gaj, P., Stera, P. (eds.) CN 2009. CCIS, vol. 39, pp. 79–86. Springer, Heidelberg (2009). doi:10.1007/978-3-642-02671-3_9

Method for Determining Effective Diagnostic Structures Within the Military IoT Networks

Jan Chudzikiewicz[✉], Tomasz Malinowski[✉], Zbigniew Zieliński[✉],
and Janusz Furtak[✉]

Faculty of Cybernetics, Military University of Technology,
ul. S. Kaliskiego 2, 00-908 Warszawa, Poland
{jan.chudzikiewicz,tomasz.malinowski,zbigniew.zielinski,
janusz.furtak}@wat.edu.pl
http://www.wat.edu.pl/

Abstract. In the paper, the method for determining the diagnostic structures for effective fault detection within military IoT networks is proposed. The method is based on partitioning IoT network among disjoint clusters and applying comparison approach for diagnosis of misbehaving nodes within network clusters. Common communication and diagnostic structures are determined on the basis of the nodes attainability characteristics and subsequent generation of subset of dendrites. The procedure for choosing the most effective diagnostic structures is proposed in the paper. In the procedure for selecting the diagnostic structure the following measures are used: the global dendrite attainability, the number of elementary comparison tests and the number of stages for performing all required tests. Also, influence of diagnostic tests on the values of performance metrics for some real-time services is investigated. The usefulness of the method has been verified through simulation experiments prepared and performed in Riverbed Modeler environment.

Keywords: Fault-tolerant system · Fault-diagnosis · WSN · Internet of Things

1 Introduction

The adaptation of the Internet of Things concept in the military domain (Military Internet of Things MIoT) emphasizes on connectivity of military things (e.g. military vehicles, weapons, materials, sensors, soldier devices, etc.) without any human participation [1–3]. The MIoT should be used in safety critical real-time applications and have to deal with faults arising out of unreliable hardware, limited energy, connectivity interruption, and harsh or even hostile physical environment. These reasons force to consider fault tolerance techniques while designing the MIoT application, in order to achieve of MIoT resilience against failures. Wireless Sensor Networks (WSNs) are the most significant part of MIoT and so, WSN's resilience against failures have to be also considered [4,5].

The MIoT can be seen as a collection of collaborating objects. Each of them is equipped with a number of sensor nodes (SN) and transmitter-receiver devices,

© Springer International Publishing AG 2017
P. Gaj et al. (Eds.): CN 2017, CCIS 718, pp. 28–43, 2017.
DOI: 10.1007/978-3-319-59767-6_3

which will be referred to as collecting sensor nodes (CSN). For efficiency and security reasons it is assumed that WSN will be partitioned among disjoint groups (clusters) with a limited number of objects on the basis of communication capabilities. The overall structure of WSN is illustrated in the Fig. 1.

We also assume that interested area of the MIoT comprises of a large number of self-organized sensor devices. Sensor devices usually consist of a number of physical sensors gathering environmental data, a microcontroller processing the data, and a radio interface to communicate with other nodes.

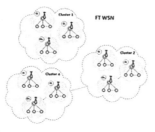

Fig. 1. The overall structure of fault-tolerant WSN for military application [5]

We will focus on a one cluster, which includes a number of objects that are able to perform similar usable functions (Fig. 2). An example of the object could be a wire or wireless body area network installed on a soldier body.

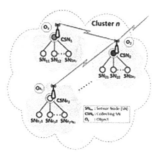

Fig. 2. Structure of the sensor nodes cluster [4]

In order to guarantee the quality of service and performance of the IoT applications, it is essential to detect faults, and to perform something akin to healing and recovering from events that might cause faults or nodes misbehavior in the network. Some of military applications are used in real-time mode (or near real-time) and require both very high reliability and high efficiency of data processing throughout all the network life cycle. Furthermore, assuming harsh or even hostile physical environment for the system, a number of factors including self-diagnosing, fault tolerance and reconfiguration should be seriously considered in the designing phase. One of the possible ways to achieve fault tolerance is to reconfigure the network to a smaller system after the fault has been diagnosed. The new (i.e. degraded) network continues its work after resources reassigning and under the condition that it meets a special requirements.

We consider the MIoT network partitioned among different clusters within each of them some security techniques are implemented according to work [5]. Each cluster of the WSN is an autonomous space of CSN nodes with its own communication structure. Each cluster creates one security domain. In the domain there is one node, which plays a role of Master (the M node) and stories the authentication data of all nodes belonging to the domain (Fig. 3). All other CSN_k nodes, which also belong to the security domain, play the role of replicas of Master (the rM nodes). In case of detection of the malfunction of the M node by diagnostic procedures, the role of the M node can be taken over by another rM node. In the domain the central node and the recipient of the sensory data is the Gateway node (the G node) (Fig. 3). Transfer of the sensory data is cryptographically protected using a common symmetric key, which is securely distributed using mechanisms implemented insecurity domain.

To establish security during the diagnostic procedure and during the transfer of sensory data, in the solution was adopted a communication structure, which enforces authentication of CSN_k nodes by the Master node, before these nodes will be able to start building routing table. It allows to avoid several types of routing attacks, as for instance routing table overflow attack.

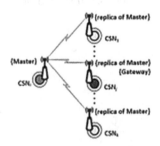

Fig. 3. CSN_i node placement within a cluster

Fault tolerance in WSNs has been the subject of active research in the past few decades. Recently, fault tolerance and management in WSNs have drawn researches' attention [5–8]. Some approaches, energy efficient detection schemes and algorithms for failure detection, faulty sensor identification have been proposed [7,8]. The mutual testing method [9] at the processor level was adopted among them, in which each processing element (sensor node) is capable of testing its neighbors [9].

In the paper [4] we proposed some techniques of fault diagnosing. The purpose of these techniques (FTT) is to use available means to detect, identify, and isolate possible sensor faults, actuator faults, processing elements faults, communication links and system faults. In the design of a fault-tolerant system the fault model should be defined. The great majority of works in the field of system level diagnosis is devoted to localization of faulty units in processor networks. Applying known algorithms of identification of faulty nodes which were proposed for distributed systems might be difficult within IoT networks due to limited resources of IoT devices and could consume too much energy in the necessity of performing periodical tests. Our approach is based on detectability notion and applying t_d-detectable structures for the IoT network which is divided into relatively small clusters. This approach allow us periodically confirm reliability of all tested nodes within a cluster assuming that the number of faulty nodes not exceed a given value of t_d. If after applying all tests from the detectable structure the reliability state of cluster is determined as unfit we might decide to apply localization procedures or simply to reject part of data from a faulty cluster.

At the sense layer of the MIoT, the components of each node can be divided into two groups. The first group consists of the nodes with a processor/microcontroller, storage subsystem and power supply infrastructure. The second group can include sensors and actuators. However simultaneous occurrence of microprocessor's and sensor faults couldn't be excluded. Faulty nodes and its sensors should be diagnosed and isolated from the system. At the network layer of the MIoT we assume communication links faults, communication nodes (CSN, gateway faults, etc.).

According to approach called MM method [9], the diagnosis is performed by sending the same input to pairs of processors and comparing their responses. Based on the collective results of comparisons, one may be able to claim the faulty/fault-free status of the nodes in the cluster.

We assume, that the diagnosis of a cluster CSN nodes is performed with a use of the MM* method and consists of conjunction of all possible results of comparison tests for the logical structure G of the cluster. Three nodes take part in each comparison test: a comparator (e_k) and a comparison pair $\{e', e''\}$. Comparator orders neighboring node e' and e'' the same task and checks whether the results of the task are identical. We will denote the comparison test by $\psi = (e_k; e', e'')$.

Let us denote by E^1 and E^0, accordingly, the set of faulty as well as fault-free network nodes, and $d\left(\psi, E^1\right)$ - result of the comparison test ψ for the set of faulty nodes E^1. We assume that $d\left(\psi, E^1\right) = 0$ means that the difference between obtained results from both compared sensor nodes not exceed given threshold ϑ_k and $d\left(\psi, E^1\right) = 1$ - in the contrary.

The efficiency of diagnostic procedures will depend both on computational complexity of determining diagnostic structures and their capability for detecting faulty nodes. The diagnostic tests performed periodically for each MIoT cluster might adversely affecting on the functioning of MIoT applications. This applies in particular to real-time applications or operating in a near real-time mode. A generalized cost of a network traffic with a specified tests scheduling and workload of a network is usually tested using an experimental methods or simulation [11].

The main goal of the paper is to propose the method for determining the diagnostic structures for effective fault detection within military IoT networks. The procedure for determining diagnostic structures should be simplified for its possible implementations in devices with limited computational capability. We limited our considerations to a given partitioning of MIoT among disjoint clusters assuming that the clusters have been already determined. The next aim of the paper is to investigate the impact of diagnostic tests on the values of performance metrics of some real-time services taking Voice over IP as an example.

We see our contribution as follows: (1) We extend our approach presented in the work [4,5] for application t_d-detectable structures within IoT; (2) An analytical way for determining the effective diagnostic structure under the MM-model (in terms of minimization of the number of elementary comparison tests and the number of stages for completing all required test) was elaborated; (3) We proved by simulation experiments usefulness of the approach for military IoT applications, and particularly its small influence on some real-time services was confirmed.

The rest of the paper consists of three sections and a summary. The second section provides a basic definitions of notions and properties of diagnostic structures used in the paper. In the third section the method for determining diagnostic structures for a logical structure of given network cluster is presented. The basis of this method constitute analytical construction of possible dendrites and calculation of some measures proposed for selecting the best diagnostic structure. The results of simulation studies (implemented in Riverbed Modeler environment) are described in the Sect. 4. Finally some concluding remarks are given in the last section.

2   The Basic Definitions and Properties

Definition 1. *The structure of a network is described by coherent ordinary graph* $G = \langle E, U \rangle$ *(E - set of network nodes, U - set of bidirectional data transmission lines).*

Let $d(e, e' \mid G)$ be the distance between nodes e and e' in a coherent graph G, that is the length of the shortest chain (in the graph G) connecting node e with the node e'.

Definition 2. *We say ([10,11]) that the graph* $\langle G; \dot{E} \rangle$ *for* $\left| \dot{E} \right| \geq 1$ *is* $(m, d \mid G)$ - *perfect placement for* $m \in \{1, \ldots, \mu(G)\}$, $d \in \{1, \ldots, D(G)\}$, $D(G)$ - *diameter of the graph* G *if there exists the set* $\dot{E}$ *of minimum cardinality such that*

$$\left[\forall_{e \in \{E(G) \setminus \dot{E}\}} : \left| \left\{ e' \in \dot{E} : d(e, e' \mid G) \leq d \right\} \right| \geq m \right] \wedge \left[\forall_{\{e^*, e^{**}\} \subset \dot{E}} : d(e^*, e^{**} \mid G) > d \right]$$

$$\wedge \left[(\mu(e'' \mid G) = 1) \Rightarrow (e'' \notin \dot{E}) \right].$$

Let $r(e,G) = \max\limits_{e' \in E(G)} d((e,e') \mid G)$ be the radius of a node, and $D(G) = max\{d((e',e'') \mid G) : \{e',e''\} \subset E(G)\}$ be the diameter of a graph G.

Denote by $E^{(d)}(e \mid G) = \{e' \in E(G) : d(e,e' \mid G) = d\}$ for $d \in \{1,\ldots,D(G)\}$ and by

$$\varsigma(e,G) = (\varsigma_1(e,G),\ldots,\varsigma_{r(e,G)}(e,G)) \text{ for } \varsigma_d(e,G) = \left|E^{(d)}(e \mid G)\right|. \quad (1)$$

Definition 3. Let $\varphi(e,G) = \sum\limits_{e' \in E(G)} d(e,e' \mid G)$ for $e \in E(G)$ be the attainability of the node e in the network G and $\Phi(G) = \sum\limits_{e \in E(G)} \varphi(e,G)$ be the attainability of the network G.

Using (1) we have

$$\varphi(e,G) = \sum_{d=1}^{r(e,G)} d\varsigma_d(e,G). \quad (2)$$

Example 1. Consider the exemplary network structure presented in the Fig. 4a. In the Table 1 the radius (A), and attainability using (1) (B) for all nodes of G are presented.

In Table 1 the nodes e_2 and e_7 have the highest value of the attainability. Based on this node's, the diagnostic structure for G will be determined.

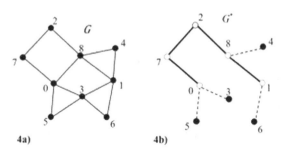

4a)   4b)

Fig. 4. Exemplary cluster of network (4a) and diagnostic structure of the structure G (4b)

Definition 4. Let $T\langle E,U^* \rangle$ be the dendrite i.e. such coherent acyclic partial graph of $G = \langle E,U \rangle$ that:

$$\exists \langle e',e'' \rangle \in U \implies \langle e',e'' \rangle \in U^* \iff [(d(e_i,e')) \neq d(e_i,e'') \wedge d(e',e'') \neq 1].$$

$$\text{for } r(e_i) = \min\limits_{e \in E(G)} r(e).$$

Table 1. The $r(e|G)$ (A), $\varphi(e,G)$ (B), and $\varPhi(G)$ for the nodes of the structure G presented in the Fig. 4a

A		B				
$e \in E(G)$	$r(e|G)$	$e \in E(G)$	$d(e,e'|G)$			$\varphi(e,G)$
			1	2	3	
e_0	2	e_0	4	4	0	12
e_1	3	e_1	4	3	1	13
e_2	3	e_2	2	3	3	17
e_3	3	e_3	4	3	1	13
e_4	3	e_4	2	4	2	16
e_5	3	e_5	2	4	2	16
e_6	3	e_6	2	4	2	16
e_7	3	e_7	2	3	3	17
e_8	2	e_8	4	4	0	12
$\varPhi(G)$						132

Denote by $T(e)$ for $e \in E(G)$ the set of all possible dendrites determined for node e.

Let $\varPhi_{max}(T(e)) = \underset{t \in T(e)}{max} \sum_{e \in E(t)} \varphi(e|t)$ be the value of maximal attainability for $T(e)$.

The dendrite t is a basis to determine the diagnostic and communication structures of G. The algorithm for determined the dendrite t is presented in [11]. The method for determined the optimal diagnostic structure based on the dendrites determination is presented in Sect. 3.

Property 1. $r(e,G) = D(G) = \frac{|E(G)|}{2}$ if the structure G is a Hamiltonian cycle.

Property 2. If $G' \subset G''$ that $\{E(G'') \setminus E(G')\} = e^*$, then $\forall_{e \in \{E(G'') \setminus \{e^*\}\}}$: $\varphi(e,G'') = \varphi(e,G') + d(e,e^*|G'')$.

Definition 5. *A cluster is t – diagnosable if each sensor node can be correctly identified as fault-free or faulty based on a valid collection of comparison results, assuming that the number of faulty nodes does not exceed a given bound t.*

Definition 6. *A cluster is t_d – detectable if each faulty subset of CSNs nodes can be correctly indicated, assuming that the number of faulty nodes does not exceed a given bound t_d.*

According to the work [4] the cluster with a logical structure G is t_d – detectable if for any fault situation there exist at least one comparison test $\varPsi$ such that $d(\psi, E^1) = 1$ provided that the number of faulty nodes does not exceed t_d i.e. $|E^1| \leq t_d$.

Let ψ be the comparative test, and $\Psi(G)$ be the set on possibility comparative tests in G, and $E(\psi)$ be the set of nodes involved in comparative test ψ.

Denote by $K(\psi)$ and by $P(\psi)$ respectively comparator node and comparative pair in ψ.

$$\psi = (e^*; e', e'') \text{ for } K(\psi) = e^* \text{ and } P(\psi) = \{e', e''\}.$$

Definition 7. *Let $\Psi' \subseteq \Psi(G)$ be the coverage of the set of nodes if and only if G if*

$$\bigcup_{\psi \in \Psi'} P(\psi) = E(G).$$

Definition 8. *The set $\Psi' \subset \Psi(G)$ is the coverage of the set of nodes if and only if*

$$\underset{e \in E}{\forall} \underset{\psi \in \Psi}{\exists} e \in P(\psi).$$

Definition 9. *The set of comparisons tests $\Psi' \subset \Psi(G)$ is called a diagnostic structure of G if it is the coverage of $E(G)$ providing that the network is $t_d -$ detectable $(t_d > 0)$ or $t - diagosable$ $(t > 0)$.*

Further we consider only detectable diagnostic structures i.e. structures which are $t_d - detectable$. For representation a diagnostic structure we will use introduced in [4] simplified comparison graph (SCG) as a graph $G(G; E^*)$, where the set of comparators E^* will be distinguished with bold nodes, a pair of anti-symmetric arcs will be represented by the edge, and arcs leading to nodes which aren't comparators will be represented by dotted lines. The arc $\langle e', e'' \rangle$, where $e', e'' \in E^*$, denotes that the node e'' (a comparator) is an element of a comparable pair of comparator e'. An example of diagnostic structure representation in the form of SCG is presented in the Fig. 4b.

3   Determining the Diagnostic Structure

The method of determining the diagnostic structures for a given network cluster G consists of three stages. In the first stage all possible dendrites $T(e)$ for $e \in E(G) : \varphi(e) = \underset{e \in E(G)}{max} \varphi(e)$ are determined. In the second stage, for every dendrites determined in the first stage, the value of attainability $\Phi(T(e))$ is calculated. In the third stage, the dendrite t as an effective diagnostic structure (simplified comparisons graph) is selected on the basis of three measures. The first measure, for selected diagnostic structure, is the maximal attainability of t_i. The selected structure t_i fulfills the condition $\Phi(t_i) = \underset{t_i \in T(e)}{max} \Phi_{max}(t_i)$ for $i = \{1 \cdots |T(e)|\}$. The second measure is a number of comparative tests $|\Psi'(t_i)|$, and the third is a number of stages in which comparative tests can simultaneously run. The structure presented in the Fig. 4a have two nodes $\{e_2, e_7\}$ (marked respectively in the Figs. 5 and 6 by bold circles) with the maximal value of attainability. In the Figs. 5 and 6 respectively the possible diagnostic structures $T(e_2)$ and $T(e_7)$ are presented. The nodes of $K(\psi)$ are marked as circle.

In the Table 3 one of possibility sets Ψ' of comparative tests of structures, which are shown in Fig. 5, are presented. Sets Ψ' of comparative tests, which can be simultaneously run, for structures $\{t_1, t_2\}$ are divided into three stages S_i, and for structure $\{t_3\}$ are divided into four stages. The stages in the Table 3 are marked by the colors, which correspond to color of arcs in Fig. 5 (Table 2).

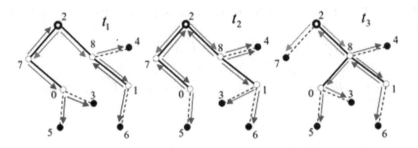

Fig. 5. The possible diagnostic structures $T(e_2)$ determined for the node e_2 (Color figure online)

Table 2. Results of determining the attainability for structures presented in the Fig. 5

$T(e_2)$	t_1	t_2	t_3
$\Phi(t)$	208	204	172

Table 3. One of possible sets of comparative tests for the structures shown in Fig. 5

S_i/Ψ'	$\Psi'(t_1)$	$\Psi'(t_2)$	$\Psi'(t_3)$
	(0; 3, 5)	(1; 3, 6)	(0; 3, 5)
S_1	(1; 6, 8)	(8; 2, 4)	(1; 6, 8)
		(0; 5, 7)	
S_2	(8; 1, 4)	(8; 1, 4)	(8; 1, 4)
	(7; 0, 2)	(7; 0, 2)	
S_3	(2; 7, 8)	(2; 7, 8)	(8; 0, 2)
S_4			(2; 7, 8)

In the Table 5 one of possibility sets Ψ' of comparative tests of structures $T(e_7)$, which are shown in the Fig. 6, are presented. Sets Ψ' for structures $\{t_1, t_2\}$ are divided into four stages, and for structures $\{t_3, t_4\}$ are divided into three stages (Table 4).

3.1 Remarks

Acceleration of selecting the diagnostic structure is the goal of the proposed method. The method assumes a reduction of the structure's selection process for

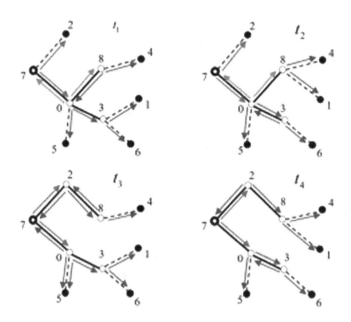

Fig. 6. The possible diagnostic structures determined for the node $T(e_7)$

Table 4. Results of determining the attainability for structures presented in the Fig. 6

$T(e_2)$	t_1	t_2	t_3	t_4
$\Phi(t)$	172	172	204	208

Table 5. Results of determining the attainability for structures presented in Fig. 6

S_i/Ψ'	$\Psi'(t_1)$	$\Psi'(t_2)$	$\Psi'(t_3)$	$\Psi'(t_4)$
S_1	(3; 1, 6)	(8; 1, 4)	(3; 1, 6)	(8; 1, 4)
	(8; 0, 4)	(3; 0, 6)	(0; 7, 5)	(3; 0, 6)
			(8; 2, 4)	
S_2	(0; 3, 5)	(0; 3, 5)	(0; 3, 5)	(0; 3, 5)
			(2; 7, 8)	(2; 7, 8)
S_3	(7; 0, 2)	(7; 0, 2)	(7; 0, 2)	(7; 0, 2)
S_4	(0; 7, 8)	(0; 7, 8)		

consideration the structures created for the node $e \in E(G)$: $\varphi(e) =_{e\in E(G)}^{max}$ $\varphi(e)$. Then $T(e)$: $r(e,t) = r(G)$ for $t \in T(e)$ are determined. Next $t : \Phi(t) =_{t\in T(e)}^{max} \Phi_{max}(t)$ is selected as the diagnostic structure, while the structure $t : \Phi(t) =_{t\in T(e)}^{min} \Phi_{min}(t)$ is selected as the communication structure. The problem is when the number of nodes, or the number of structures, with maximum (or minimal for choose the optimal communication structure) attainability is greater than one, like in the cases presented in Sect. 3. In that cases an additional selection criteria (for choose the optimal diagnostic structure), that must be met by generated structure, must be considered. Minimal number of the comparative tests is the first of selection criteria, and the minimal number of the stages of comparative test performing is the second one. It is assumed that at a given moment the node may participate in only one comparative test.

4   The Results of Simulation Studies

Procedure for determining the best set of diagnostic tests in the wireless network, specified in the Sect. 3, has been verified through simulation tests. The aim of the simulation studies was to confirm the correctness of the theoretical considerations, arguments, and evaluation of influence of diagnostic tests on functioning of the motionless wireless sensor network (WSN). We assumed that the tests affect the performance of the selected network service, and simulation study will indicate the test of the "smallest" impact on that service. The simulation experiment confirmed the accuracy of selecting the optimal diagnostic test, which was indicated by the analytical calculations.

During the experiment it was assumed that wireless nodes communicate using the VoIP (Voice over IP service). During the exchange of voice data, the procedure of node's testing was performed, according to the selected schedule tests S_i/Ψ', as defined in the Tables 3 and 5.

4.1   WSN Model and Simulation Scenarios

Communications structure of wireless sensor network, corresponding to graph G (Fig. 4) is shown in the Fig. 7.

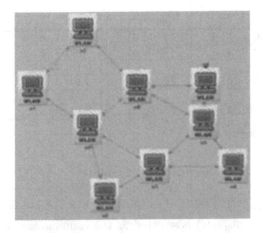

Fig. 7. WSN communication structure

Flows between nodes illustrate, which nodes are in direct Wi-Fi range of its neighbors. For example, node e_2 is in the range of node e_7 and node e_8, and because of the excessive distance from other nodes remains beyond their immediate range (which has been proven experimentally).

Simulation studies have been prepared and implemented in Riverbed Modeler environment. In simulation scenarios, models of wireless workstations available in Riverbed Modeler, working according to the 802.11 g standard, have been used. All workstations use OLSR (Optimized Link State Routing Protocol).

A single scenario, showing the structure for which the test $\Psi'(t_1)$ is carried out, is shown in the Fig. 8.

Each simulation scenario corresponds to the one of diagnostic structures shown in the Fig. 5 (diagnostic structures for the node e_2) and in the Fig. 6 (diagnostic structures for the node e_7).

Flows (shown in the Fig. 8) illustrate partial tests of $\Psi'(t_1)$, carried out according to schedule from Table 3. For example, partial test $(0; 3, 5)$ corresponds to two unidirectional flows (the force and the response sequence) between nodes e_0 and e_3 and between e_0 and e_5 (node e_0 is the comparator). Unidirectional flow model is very simple, as shown in Fig. 9. Two seconds after the start of the sending the force sequence, tested node starts sending identical sequence, which is treated as a response to the force.

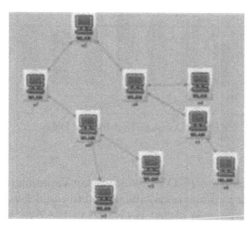

Fig. 8. Single simulation scenario corresponding to $\Psi'(t_1)$

4.2   The Results of Simulation Experiment

Some interesting results, confirming the correctness of the procedure for determining the optimal diagnostic structure, are shown in the Figs. 10, 11, 12 and 13. The dotted lines corresponds to the results obtained for the $t_1(e_2)$, which is, according to the procedure, the most effective test procedure.

For each simulation scenario, to assess the impact of test procedure on the functioning of the VoIP network, two characteristics were determined: a Voice End-to-End Delay (EE_Del), and a Voice Delay Variation (VDV).

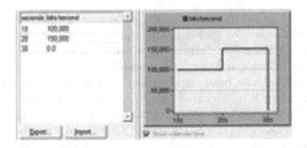

Fig. 9. Model of unidirectional flow

A. Voice End-to-End Delay (EE_Del). End-to-end delay is average delay in seconds for all network nodes communicating each other under VoIP. The lowest value of EE_Del is desired.

Results obtained during the simulation are presented in the Figs. 10 and 11. The Fig. 10 shows the average values of EE_Del.

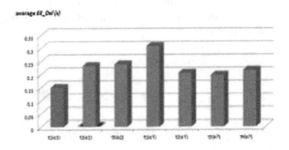

Fig. 10. Average End-to-End Delay

B. Voice Delay Variation (VDV). Voice Delay Variation is average variance in seconds (for all voice workstations) among end to end delays for voice packets. It is measured from the time packet is created to the time it is received. The lowest value of VDV is the best. The results are presented in the Figs. 12 and 13.

The results of evaluation of *EE_Del* and *VDV* are presented in the Table 6. The results confirm that t_1 (e_2) has the least impact on the functioning of the specific network with a VoIP service.

Table 6. Summary of simulation results

Test	t_1 (e_2)	t_2 (e_2)	t_3 (e_2)	t_1 (e_7)	t_2 (e_7)	t_3 (e_7)	t_4 (e_7)
Average *EE_Del(s)*	0,148353	0,232728	0,238334	0,308361	0,205328	0,19681	0,215851
Average *VDV(s)*	0,015298	0,083317	0,073897	0,105536	0,053993	0,035568	0,059377

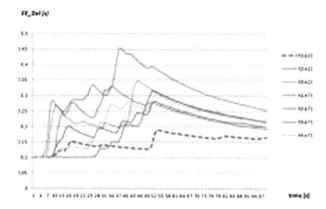

Fig. 11. End-to-End Delay for VoIP transmission

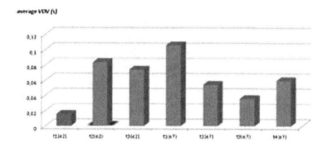

Fig. 12. Average Voice Delay Variation

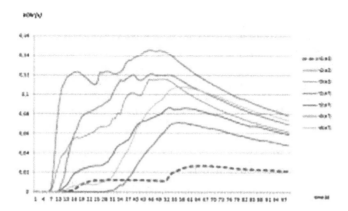

Fig. 13. Voice Delay Variation for VoIP transmission

5 Summary

In this paper the method for determining the diagnostic structures for effective fault detection within Military IoT networks is proposed. The proposed method is useful in military applications, where the replacement of the failed component is not possible. The structure of the network cannot be renewed (is in the process of a softly degradation). Simulation studies confirmed that the solutions for network diagnostics, based on the adoption of system-level comparison methods proposed in the paper, have a good efficiency with a minimal influence on some MIoT services such as VoIP. The proposed approach has low computational complexity and could be applied in relatively small cluster's devices with limited resources. The goal of the work to find simple analytical method for a MIoT cluster diagnostic structure determining based on the properties of its logical structure was achieved.

References

1. Zheng, D.E., Carter, W.A.: Leveraging the Internet of Things for a more efficient and effective military. A Report of the CSIS Strategic Technologies Program (2015)
2. Suri, N., et al.: Analyzing the applicability of Internet of Things to the battlefield environment. In: Proceedings of ICMCIS Conference, Brussel (2016)
3. Yishi, L., Fei, J., Hui, Y.: Study on application modes of military Internet of Things (MIOT). In: Proceedings of 2012 IEEE International Conference on Computer Science and Automation Engineering (CSAE), vol. 3, pp. 25–27 (2012)
4. Chudzikiewicz, J., Furtak, J., Zielinski, Z.: Fault-tolerant techniques for the Internet of Military Things. In: Proceedings of 2nd IEEE World Forum on Internet of Things WF-IoT15 (2015)
5. Chudzikiewicz, J., Furtak, J., Zielinski, Z.: Integrating some security and fault tolerant techniques for military applications of Internet of Things. In: Proceedings of 3rd IEEE World Forum on Internet of Things WF-IoT16 (2016)
6. Koushanfar, F., Potkonjak, M., Sangiovanni-Vincentelli, M.A.: Fault-tolerance in sensor networks. In: Mahgoub, I., Ilyas, M. (eds.) Handbook of Sensor Networks. CRC Press, Boca Raton (2004). Section VIII, no. 36
7. Jiang, P.: A new method for node fault detection in wireless sensor networks. Sensors **9**, 1282–1294 (2012)
8. Chen, J., Kher, S., Somani, A.: Distributed fault detection of wireless sensor networks. In: DIWANS 2006 Proceedings of the 2006 Workshop on Dependability Issues in Wireless Ad Hoc Networks and Sensor Networks, pp. 65–72 (2006)
9. Maeng, J., Malek, M.: A comparison connection assignment for self-diagnosis of multiprocessor systems. In: Digest International Symposium Fault Tolerant Computing, pp. 173–175 (1981)
10. Chudzikiewicz, J., Zieliński, Z.: On some resources placement schemes in the 4-dimensional soft degradable hypercube processors network. In: Zamojski, W., Mazurkiewicz, J., Sugier, J., Walkowiak, T., Kacprzyk, J. (eds.) Proceedings of the Ninth International Conference on Dependability and Complex Systems DepCoS-RELCOMEX. June 30 – July 4, 2014, Brunów, Poland. AISC, vol. 286, pp. 133–143. Springer, Cham (2014). doi:10.1007/978-3-319-07013-1_13

11. Chudzikiewicz, J., Malinowski, T., Zieliński, Z.: The method for optimal server placement in the hypercube networks. In: Proceedings of the 2015 Federated Conference on Computer Science and Information Systems. ACSIS, vol. 2, pp. 947–954 (2015). doi:10.15439/2014F159
12. Chudzikiewicz, J.: Sieci komputerowe o strukturze logicznej typu hiperszecianu. Instytut Automatyki i Robotyki, Wojskowa Akademia Techniczna, Warsaw (2002). (in Polish)
13. Chudzikiewicz, J., Zieliński, Z.: Reconfiguration of a processor cube-type network. Przegld Elektrotechniczny (Electr. Rev.) **86**(9), 149–153 (2010)
14. Kulesza, R., Zieliński, Z.: The life period of the hypercube processors' network diagnosed with the use of the comparison method. In: Monographs of System Dependability - Technical Approach to Dependability, pp. 65–78. Oficyna Wydawnicza Politechniki Wrocawskiej, Wrocaw (2010)
15. Sethi, A.S., Hnatyshin, V.Y.: The Practical OPNET User Guide for Computer Network Simulation. Chapman and Hall/CRC, Boca Raton (2012)

QoS-Based Power Control and Resource Allocation in Cognitive LTE-Femtocell Networks

Jerzy Martyna[✉]

Faculty of Mathematics and Computer Science, Institute of Computer Science,
Jagiellonian University, ul. Prof. S. Lojasiewicza 6, 30-348 Cracow, Poland
martyna@ii.uj.edu.pl

Abstract. This paper proposes a new joint power control and resource allocation algorithm in cognitive LTE femtocell networks. The presented algorithm minimizes the transmit power of each femtocell, while satisfying a maximum number of users with QoS requirements. The optimization problem is multi-objective NP-hard. Hence, in the paper a joint channel assignment and power control allocation scheme is proposed, solving the problem so formulated. Especially, a Gale-Shapley method is involved in this scheme to address the co-tier femtocell interference issue. In order to find a pareto optimal solution, a weighted sum is used. The performance of this method is evaluated by simulation.

1 Introduction

Cognitive radio (CR) systems are proposed to allow opportunistic use of unutilized licensed resources by sensing spectrum [12]. The sensed information is used by unlicensed users (secondary users, SUs) to take so-called spectrum holes without producing significant interference to licensed users (primary users, PUs).

Recent studies indicate that more than 60% of mobile traffic is generated indoors [11]. In order to improve coverage, cellular networks have integrated by use Femtocell Access Points (FAPs) [1]. They act as ordinary base stations (BSs) in wireless cellular systems. FAPs are small-coverage, inexpensive, low powered base stations deployed for indoors. They are connected to the wireline infrastructure. Thus, FAPs receive data from mobile stations (MSs) in their building and transfer to the network through the wireline connections. When MSs are outdoors, the wireless connections can be made with the macro base station (MBS).

Merging cognitive radio network in two-tier cellular networks integrated by FAPs can permit the successful and cost-effective deployment of cognitive femtocells. These cognitive radio femtocell networks have been studied in many papers. Among others, the dynamic spectrum sharing problem in cognitive radio femtocell networks with the application of decomposition theories was presented by J. Xiang et al. [16,17]. Power control and subcarrier allocation for OFDMA based underlay cognitive radio network was studied by N. K. Gupta et al. [6]. The interference study for cognitive LTE-femtocell has been carried out by

© Springer International Publishing AG 2017
P. Gaj et al. (Eds.): CN 2017, CCIS 718, pp. 44–54, 2017.
DOI: 10.1007/978-3-319-59767-6_4

Z. Zhao et al. [18]. The interference study for cognitive LTE-femtocell has been carried out by Z. Zhao et al. [18]. In turn a cost-effective approach in the cognitive femtocell networks based on the 4 G architecture has proposed by Y.-Y. Li et al. [9].

The main objective of the study is to introduce a new model of resource allocation in cognitive femtocell systems that use the LTE technology. The innovation consists in the use of the "effective capacity" calculation method known in the literature on the subject and applied in the study to propose an optimization problem for this type of systems that include two types of users: fixed and mobile ones. The original approach to identifying a suboptimal solution of the problem consists in the use of the Gale-Shapley algorithm to achieve a stable matching of power control.

The rest of paper is organized as follows. In Sect. 2 the system model is presented. Section 3 provides the formulation of the optimization problem. In Sect. 4 the use of the Gale-Shapley algorithm to solve the problem of joint power control and resource allocation in cognitive LTE-femtocell networks is presented. Section 5 gives the simulation results to show that the proposed method achieves comparable efficiency to classical methods. Section 6 concludes the paper.

2 System Model

It is assumed that a cognitive femtocell system consists of a macrocell and multiple femtocells, as shown in Fig. 1. The system also includes a Primary Base Station (PBS), which is here TV transmitter, and a Secondary Base Station (SBS) (here Macrocell Base Station, MBS). MBS is communicating with the femtocell access points (FAPs). Femtocell users treated as SUs share the same

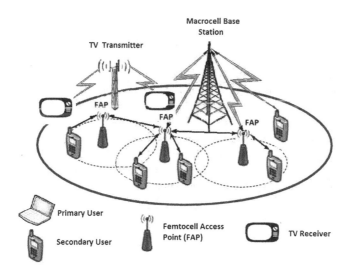

Fig. 1. Basic architecture of cognitive femtocell network.

spectrum with primary users (PUs), which are here TV system users. There is no interactions between femtocells. To avoid severe co-channel interference or occupation of femtocells by the same channels, it is here assumed they are separated at least for safety distance. Despite this the aggregate interference from multiple femtocells to a certain PU must sometimes be taken into account due to the large number of femtocells using the same channel.

The following model can describe the channel power gain of the subcarrier k of the i-th femtocell, namely

$$H_{k,i} = X_{k,j} \cdot 10^{-PL/10} \tag{1}$$

where $X_{k,j}$ is the effect of the fading and can be assumed to be Rayleigh distributed random variables with mean equal to one. PL is the path loss component that can be calculated by use the model given by P. Kysti [8]:

$$PL_{LOS}(dB) = 18.7 \log(d) + 46.8 + 20 \log(f_c/5) \tag{2}$$

$$PL_{NLOS}(dB) = 20 \log(d) + 46.4 + 20 \log(f_c/5) + L_W \tag{3}$$

where d denotes the distance in meters between the femtocell access point, f_c is the carrier frequency (GHz), L_W gives the wall penetration loss (dB) with $L_W = 5n_W$ for the light walls and $L_W = 12n_w$ for heavy walls where n_w is the number of walls between BS and MS.

The interference in the LTE-femtocell network introduced by the k-th subcarrier of the i-th femtocell to the PU, I_{ki}, is the integration of the power spectrum density of the k-th subcarrier of the i-th femtocell accross the PU band, B. Therefore, it is given by

$$I_{ki} = \int_{d_{ki}-\frac{B}{2}}^{d_{ki}+\frac{B}{2}} G_{ki} \Phi_{ki}(f) df = P_{ki} \Omega_{ki} \tag{4}$$

where G_{ki} is the channel power gain between the k-th subcarrier of the i-th femtocell and the PU receiver. d_{ki} is the spectral distance between the k-th subcarrier of the femtocell and the PU band. Φ_{ki} is the power spectrum density of the k-th subcarrier of the i-th femtocell.

Now to try to describe the dependence introduced by the interference in the cognitive LTE-femtocell network. It is assumed that the available channels are assigned to the femtocells based on the distance. Then the power budgets of different femtocell access points is decreased by the underlying interference. Such dependencies can be described as follows.

If femtocells are very close to each other, the spatial correlation of the neighbouring femtocell clusters, γ, can be defined as follows

$$\gamma = \frac{r}{D} \tag{5}$$

where r defines the radius of femtocell cluster which depends on the safety distance, and D denotes the distance between two clusters. To ensure the interference among the femtocells in the macrocell, the constraint concerning the spatial correlation of femtocells should be satisfied, namely

$$\gamma \le \gamma_0 \tag{6}$$

where $\gamma_0 = \frac{R}{2R+d_0}$ and R denotes the radius of the macrocell, d_0 is the safety distance. If R is defined as half of the safety distance d_0, then $\gamma_0 = 1/4$.

2.1 Effective Capacity of the Femtocell Fulfilling the QoS Requirements

The impact of time-varying fading channel is evident in the inaccessibility of deterministic QoS guarantees. A practical solution is to provide statistical QoS guarantees. Therefore, the effective bandwidth concept [2,3] can be used to define the probability that the packet delay violates the delay requirement, namely

$$Pr\{Delay > d_{\max}\} \approx e^{-\theta\delta d_{\max}} \tag{7}$$

where $d_{\max}$ is the delay requirements, δ is the constant jointly determined by arrival and service processes, θ is a positive constant referred to QoS requirement.

A duality of effective bandwidth is effective capacity which defines the maximum constant arrival rate that can be supported by the system subject to a given θ [14]

$$E_{\mathrm{cap}}(\theta) \stackrel{\triangle}{=} -\frac{\Lambda_c(-\theta)}{\theta} = -\lim_{t\to\infty}\frac{1}{\theta t}\log(E[e^{-\theta S[t]}]) \tag{8}$$

where $S[t] = \sum_{j=1}^{t} R[j]$ is the partial sum of the discrete-time stationary and ergodic service process given by $R[j], j = 1, 2, \dots$ and

$$\Lambda_c(\theta) = \lim_{t\to\infty}\frac{1}{t}\log(E[e^{\theta S[t]}]) \tag{9}$$

By applying the results given by Sh.-Y. Lien et al. [10] and D. Wu et al. [14], the effective capacity of the i-th femtocell SU is given by

$$E_{\mathrm{cap}}^{i,l} = -\frac{1}{\theta}\log(e^{-n\theta}) \tag{10}$$

where n is the identical number of bits carrying l resource blocks.

3   Optimization Problem of QoS-Based Power Control and Resource Allocation in Cognitive LTE Femtocell Networks

In the following, the above problem can be formulated mathematically. The downlink power allocation problem in the cognitive LTE-femtocell relies on interference minimization, which still guarantees the communication. So, the objective is to minimize the aggregate cross-tier interference from multiple femtocells to the total requirements and transmission constraints with the QoS requirements. Thus, the optimization problem for the cognitive LTE-femtocell consisting of N femtocells, with the M users in each femtocell and L mobile users that do not belong to femtocells, while K is the total number of subcarriers in femtocells, can be formulated as follows:

$$\min_{P_{kji}} \sum_{i=1}^{N} \sum_{j=1}^{M} \sum_{k=1}^{K} \rho_{kji} P_{kji} \Omega_{kji} \tag{11}$$

$$\min_{P_{kl}} \sum_{l=1}^{L} \sum_{k=1}^{K} \mu_{kl} P_{kl} \Omega_{kl} \tag{12}$$

subject to

$$\rho_{kji} \in \{0,1\} \quad \text{for} \quad \forall k,j,i$$

$$\sum_{j=1}^{M} \rho_{kji} \leq 1 \quad \text{for} \quad \forall k, \ \forall i \in \{1,2,\dots,N\}$$

$$\mu_{kl} \in \{0,1\}, \quad \text{for} \quad \forall k,l$$

$$\sum_{l=1}^{L} \mu_{kl} \leq 1, \quad \text{for} \quad \forall k$$

$$\sum_{j=1}^{M} \sum_{k=1}^{K} \rho_{kji} C_{kji} \geq E_{\text{cap}_i} \quad \forall i \in \{1,2,\dots,N\}$$

$$\sum_{l=1}^{L} \sum_{k=1}^{K} \mu_{kl} C_{kl} \geq E_{\text{cap}_l} \quad \forall l \in \{1,2,\dots,L\}$$

$$\sum_{j=1}^{M} \sum_{k=1}^{K} \rho_{kji} P_{kji} + \sum_{l=1}^{L} \sum_{k=1}^{K} \mu_{kl} P_{kl} \leq P_{\text{tot}_i} \quad \text{for} \quad \forall i \in \{1,2,\dots,N\}$$

$$P_{kji} \geq 0 \quad \text{for} \quad \forall k \in \{1,2,\dots,K\}, \quad \forall j, \ l$$

$$P_{kl} \geq 0 \quad \text{for} \quad \forall k \in \{1,2,\dots,K\}, \ \forall l$$

where ρ_{kji} and μ_{kl} are the subcarrier indexes. If the subcarrier k-th is allocated to the j-th SU belonging to the i-th femtocell, $\rho_{kji} = 1$, otherwise $\rho_{kji} = 0$. Similarly, if the k-th subcarrier is allocated to the l-th mobile SU which does not belong to femtocells, $\mu_{kl} = 1$, otherwise $\mu_{kl} = 0$. P_{kji} denotes the transmission power in the k-th subcarrier from the i-th FAP to the j-th SU. Ω_{kji} indicates the the interference factor of the j-th SU in the i-th femtocell in the k-th subcarrier and Ω_{kl} denotes the interference factor of the l-th mobile SU which does not belong to the femtocells in the k-th subcarrier. C_{kji} denotes the capacity of the j-th SU of the i-th femtocell in the k-th subcarrier and C_{kl} indicates the capacity of the l-th mobile user in the k-th subcarrier. E_{cap_i} denotes the effective capacity requirement of the i-th femtocell and E_{cap_l} gives the effective capacity requirement of the l-th mobile SU which does not belong to femtocells, respectively. P_{tot_i} denotes the total power budget of the i-th femtocell.

This is a multi-objective integer non-linear programming problem. However, it is assumed that each FAP knows only its own constraints and not those of other FAPs. Therefore, here a solution of this problem is provided using the Gale-Shapley algorithm assuming that each FAP can satisfy the QoS requirements of its own users.

4 An Algorithm to Solve the Problem of QoS-Based Power Control and Resource Allocation in Cognitive LTE-Femtocell Networks

This section introduces an algorithm to solve the problem of QoS-based power control and resource allocation in a cognitive LTE-femtocell network. The solution given is sub-optimal and uses the mapping from the elements of one set to the elements of the other called stable matching. As a solution, the Gale-Shapley algorithm introduced by D. Gale and L. S. Shapley [4] is used here.

The assumption here is in accordance with the specified dependencies given by Jang et al. [7] that the subcarriers can obtain the maximum transmission rate in downlink if the subcarriers are assigned to the SU with the best channel gain on the subcarrier. To take of this, first is estimated the number of subcarriers required for the j-th SU, namely

$$\mathcal{N}_j = \frac{r^{SU}}{\log_2(1 + \frac{f_k \overline{T}_k}{N_0})}, \quad j \in \{1, 2, \ldots, M + L\} \tag{13}$$

where r^{SU} is the transmission bit rate of the SU, $\overline{T}_k$ is the average transmission power of PU allocated to subcarrier k, e.g. $\overline{T}_k = \frac{T_k}{K}$, and K is the number of subcarriers. If the number of subcarriers allocated to the j-th SU satisfied the subcarrier requirement $\mathcal{N}_j$, the subcarriers assignment for it will stop. If the subcarrier requirement is met for every SU, the rest of subcarriers are allocated according to the necessity.

The channel gain on the k-th subcarrier, h_k is determined as follows:

$$h_k = \sum_{j=1}^{M+L} a_{kj} h_{kj}, \quad k \in \{1, 2, \ldots, K\} \tag{14}$$

where $a_{kj} \in \{0, 1\}$, $\forall k, j$, and denotes the subcarrier allocation indicator, i.e. $a_{kj} = 1$ only if the k-th subcarrier is allocated to the j-th SU.

According to the Shannon capacity formula the transmission rate of the k-th subcarrier is given by

$$R_k = \log_2(1 + \frac{p_k \mid h_k \mid^2}{N_0^2}) \tag{15}$$

Table 1. Simulation parameters

Parameter description	Value
System bandwidth	10 MHz
Subcarrier bandwidth	15 KHz
Channel bandwidth	150 KHz
No. of subcarriers per Resource Block	12
No. of RBs	100
Modulation	64 QAM
Target BER	10^{-6}
Penetration loss outer wall	20 dB
Penetration loss of inner wall	5 dB

where N_0 is the total noise power. Thus, the admissible transmission power P_i for the i-th FAP can be expressed by

$$P_i = \frac{(2^{R_k} - 1)N_0^2}{|h_k|^2}, \quad i \in \{1, 2, \ldots, N\} \tag{16}$$

where N is total number of FAPs.

Based on the analysis presented above, the proposed algorithm combining both stable matching method and heuristic to solve the QoS-based power control and resource allocation in cognitive LTE-femtocell network is as follows:

Step 1. Using Eq. (13) determine the number of subcarriers $\mathcal{N}_j$ to be initially assigned to each SU;

Step 2. Assign the subcarriers to each SU in a way that guarantees rough proportionality;

Step 3. Assign the achievable rate of SU, r^{SU}, for each SU to maximize the effective capacity while enforcing the proportionality;

Step 4. Using the Gale-Shapley algorithm make a match the power P_i for each FAP subject to the total power constraint P_{tot}.

5   Simulation Results

It is assumed that the simulation scenario consists one macrocell with one Primary Base Station, one Secondary Base Station, three femtocells with FAPs and randomly distributed SUs. The SINR threshold for PBS is set at 0.8, while the SINR threshold for FAPs is set at 0.3. Is assumed that the SUs are at the indoors and the outdoors of the femtocells. The simulation results are listed in Table 1.

According to the 3GPP-LTE standard the length of the cycle prefix is 4.69 μs and the symbol length is equal to 66.7 μs. Each subcarrier can carry maximally 15 Ksps (kilo-symbols per second). Assumed here modulation type, namely

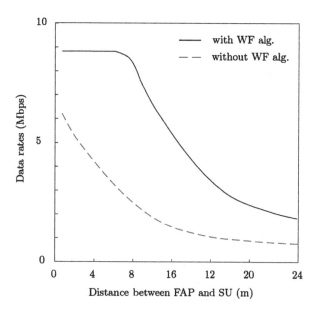

Fig. 2. Achieved data rate of SU versus the distance from their FAP.

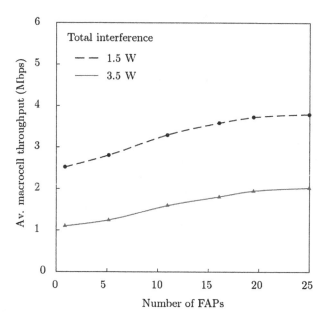

Fig. 3. Average throughput of system versus variable number of FAPs for constant number of SUs equal to 30.

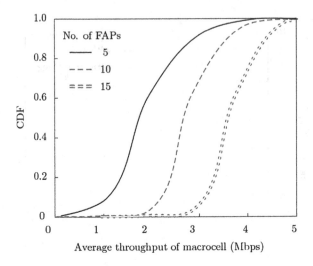

Fig. 4. Spectrum efficiency of the FAP can be treated here as the average spectrum efficiency of the subcarriers allocated to the SU.

64 QAM, represents 6 bits per symbol. Thus, the bandwidth is equal to 10 MHz and provides a symbol rate of 9 Msps or 54 Mbps.

In the simulation implemented in the Matlab language the effectiveness of the proposed method was studied in comparison with the waterfilling (WF) algorithm [5,13]. Figure 2 shows the achieved data rate of SU with respect to the distance from their serving FAP. It seems that with increasing distance from the FAP data rates decrease. It is evident that the Gale-Shapley algorithm gives slightly smaller value flow rates than is the case in the waterfilling algorithm.

Figure 3 shows the average throughput of studied cognitive LTE system versus variable number of FAPs for constant number of SUs. The total interference is varied from 1.5 - 3.5 W. As can be seen from the graphs in Fig. 3, the average throughput decreases with the increase in the total interference of the system.

Figure 4 shows the spectral efficiencies of the PUs described by CDF (cumulative distribution function). The spectrum efficiency of the FAP can be treated here as the average spectrum efficiency of the subcarriers allocated to the SU. It can be seen from the graphs in Fig. 4 that a dense femtocell LTE network with high value FAPs number ensures higher average throughput for the same value of CDF.

6   Conclusion

This paper focused on the problem of joint power control with QoS requirements and resource allocation in cognitive LTE-femtocell networks. The aim was to minimize the transmit power of each femtocell, while satisfying a maximum number of high-priority secondary users and servicing best-effort secondary users as well as possible. An algorithm based on the Gale-Shapley approach has been employed as a practical solution of this problem. The simulation results have shown that the presented solution of joint power control with QoS requirements and resource allocation considerably improves the number of high-priority secondary users in the network and provides the throughput satisfaction rate, as well as the spectrum spatial reuse.

References

1. Chandrasekhar, V., Andrews, J., Gatherer, A.: Femtocell networks: a survey. IEEE Commun. Mag. **46**(9), 59–67 (2008)
2. Chang, C.-S.: Stability, queue length, and delay of of deterministic and stochastic queueing networks. IEEE Trans. Autom. Control **39**(5), 913–931 (1994)
3. Courcoubetis, C., Weber, R.: Effective bandwidth for stationary sources. Probab. Eng. Inf. Sci. **9**(2), 285–294 (1995)
4. Gale, D., Shapley, L.S.: College admission and its stability of marriage. Am. Math. Mon. **69**, 9–15 (1962)
5. Gonter, J., Goertz, N.: An algorithm for highly efficient waterfilling with guaranteed convergence. In: 2012 19th IEEE International Conference on Systems, Signals and Image Processing (IWSSIP 12), pp. 126–129 (2012)
6. Gupta, N.K., Banerjee, A.: Power and subcarrier allocation for OFDMA femto-cell based underlay cognitive radio in a two-tier network. In: Proceedings of IMSAA 2011. IEEE, Bangalore, December 2011
7. Jang, J., Lee, K.B.: Transmit power adaptation for multiuser OFDM systems. IEEE J. Sel. Areas Comm. **21**(2), 171–178 (2003)
8. Kysti, P., et al.: WINNER II Channel Models. D1.1.2 V1.2 (2007). http://www.cept.org/files/1050/documents/winner2%20-%20final%20report.pdf. Accessed 30 Sept 2007
9. Li, Y.-Y., Sousa, E.S.: Cognitive femtocell: a cost-effective approach towards 4G autonomous infrastructure networks. Wirel. Pers. Comm. **64**(1), 65–78 (2012)
10. Lien, S.-Y., Tseng, C.-C., Chen, K.-C., Su, C.-W.: Cognitive radio resource management for QoS guarantees in autonomous femtocell networks. In: Communications (ICC), IEEE International Conference on Communication, pp. 1–6. IEEE (2010)
11. Mobile Broadband Access at Home, Informa Telecoms and Media, August 2008. http://www.informatandm.com/section/home-page/
12. Mitola, J., Maguire, G.Q.: Cognitive radio: making software radios more personal. IEEE Pers. Commun. **6**(4), 13–18 (1999)
13. Qi, Q., Minturn, A., Yang, Y.: An efficient water-filling algorithm for power allocation in OFDM-based cognitive radio systems. In: Systems and Informatics (ICSAI), 2012 IEEE International Conference on Systems and Informatics (ICSAI 2012), pp. 2069–2073 (2012)
14. Wu, D., Negi, R.: Effective capacity: a wireless link model for support of quality of service. IEEE Trans. Wirel. Commun. **12**(4), 630–643 (2003)
15. Wu, D., Negi, R.: Utilizing multiuser diversity for efficient support of quality of service over a fading channel. IEEE Trans. Veh. Technol. **54**(3), 1198–1206 (2005)
16. Xiang, J., Zhang, Y., Skeie, T.: Dynamic spectrum sharing in cognitive radio femtocell networks. In: Hei, X.J., Cheung, L. (eds.) AccessNets 2009. LNICSSITE, vol. 37, pp. 164–178. Springer, Heidelberg (2010). doi:10.1007/978-3-642-11664-3_13

17. Xiang, J., Zhang, Y., Skeie, T., Xie, L.: Downlink spectrum sharing for cognitive radio femtocell networks. IEEE Syst. J. **4**, 524–534 (2010)
18. Zhao, Z., Schellmann, M., Boulaaba, H., Schulz, E.: Interference study for cognitive LTE-Femtocell in TV white spaces. In: Proceedings of ITUWT 2011, Geneva, October 2011

Secure and Reliable Localization in Wireless Sensor Network Based on RSSI Mapping

Jakub Pyda[1], Wojciech Prokop[1], Damian Rusinek[1(✉)],
and Bogdan Ksiezopolski[1,2]

[1] Institute of Computer Science, Maria Curie-Sklodowska University,
Pl. M. Curie-Sklodowskiej 5, 20-031 Lublin, Poland
`damian.rusinek@gmail.com`
[2] Polish-Japanese Academy of Information Technology,
Koszykowa 86, 02-008 Warsaw, Poland

Abstract. The growing number of mobile devices and the popularity of Internet of Things (IoT) applications have caused an increase of local-area location-aware protocols and applications. Localization has become a popular topic in the literature and many localization algorithms in wireless sensor networks (WSN) have been proposed. However, the most of them are vulnerable to common wireless attacks. In this article, we introduce the secure localization protocol based on a RSSI-mapping, which uses symmetric cryptography. The security attributes of the protocol have been proved using the Scyther tool. We present the experimental results that demonstrate the ability to estimate sensor's location in the environment, where other networks are available. The results show, that the introduction of security mechanisms to localization protocols may significantly increase the execution time and exceed strict time restrictions in some real-time applications. Nevertheless, the existence of insecure protocol in hostile environment is unacceptable. To achieve a better efficiency, we suggest to use the hardware encryption provided in CC2420 radio chip.

Keywords: Localization · WSN · RSSI · Map-based algorithms · Localization algorithms · Security · Cryptography · Symmetric cryptography

1 Introduction

The topic of local-area localization, which assumes that GPS-based localization cannot be used, is well studied in the literature. The majority of localization process applications use routing algorithms with self-organizing nodes, mostly based on Received Signal Strength Indication (RSSI). There exist some practical implementations of this approach, such as position determination of nodes in construction sites [1] or inside a building [2,3].

Researchers have published a great number of new methods and improvements, which, contrary to the existing algorithms, are characterized by a wide range of possible approaches to deal with this problem. The most popular method

© Springer International Publishing AG 2017
P. Gaj et al. (Eds.): CN 2017, CCIS 718, pp. 55–69, 2017.
DOI: 10.1007/978-3-319-59767-6_5

of nodes positioning is the RSSI technique, because of its simplicity and low cost. The above advantages come from the fact, that most radios support measuring RSS. However, the downside of this method is the low accuracy, caused by the waves being reflected by the nearby objects or inferred. This leads to the radio frequency noise [4]. Furthermore, in a real world, the antenna radiation pattern is neither a circular nor a spherical shape and the path loss model is non-linear due to the environmental noises. This problem can be solved by including antenna patterns into localization algorithm [5] but it is impossible to exclude environmental noises completely.

One of the simplest approaches to deal with this problem, is to create a RSSI map, which is later used to localize sensor, using the k-nearest neighbour algorithm. Such an idea has been presented in [6]. By following the approach of map-based localization, we confirm the results of the proposed localization protocol on the test local-area, and propose a secure localization protocol. Finally, we compare the efficiency of both secure and insecure protocols. In the secure version of the protocol, we ensure the following charactetristics: confidentiality, integrity and authentication.

The contribution of this paper is the introduction of the secure version of the RSSI map-based localization protocol for wireless sensor networks, and efficiency comparison of secure and insecure protocols. It can be further enumerated as follows.

(1) We create mapping and localization algorithms in WSN, based on work presented in [6]. In contrast to the insecure protocol, where all the messages are send in a plaintext, we propose a secure version of this protocol. We do not include the description of how the new node joins the network, because we simply assume that this process is explained in [7].
(2) We formally verify proposed protocol with the use of the Scyther tool [8], which provides automatic verification of security attributes. Our secure protocol provides integrity, confidentiality and authentication (which are not provided by the insecure protocol, thus being vulnerable to many attacks). All the secure protocols were implemented with software implementation of the AES-CTR symmetric encryption and should work on all motes running the TinyOS system.
(3) We test our implementation on IRIS motes equipped with RF230 radio without the hardware support for encryption. Next, we compare the results of both secure and insecure protocol. We tested the protocol in a room 5 m long and 5 m wide, where we checked the accuracy of nodes localization.

The article is structured as follows: in Sect. 3 we describe the mapping (Sect. 3.1) and localization (Sect. 3.2) protocols. Then, in Sect. 3.3 we present the results of security verification of protocols. Further, we describe our experiment on IRIS nodes and compare mapping and localization protocols of both insecure and secure versions in Sect. 4, and finally we sum up with Sect. 5.

2   Related Work

Localization algorithms are crucial in Wireless Sensor Networks because in many cases the location of measured data is essential for further analysis (especially in dynamic networks). One can divide the proposed approaches into two groups: the range-free and range-based algorithms. Range-free algorithms use information about the topology and connectivity, while the range-based algorithms use range measurement techniques for localization. Many algorithms are hybrid and combine both range-free and range-based approaches.

In the article [9], the authors have tested and compared range-based approaches, such as Angle Of Arrival (AOA), Time of Arrival (TOA) and RSSI based. The results show that AOA outperforms both RSSI and TOA techniques. However, Angle Of Arrival approach requires directional antennas, which drastically increase the implementation cost, whereas one of the main goals in WSN is to reduce it. Therefore, in this article, we decided to analyse the RSSI approach.

Range-based algorithms examine received signal strength indicator (RSSI) to estimate the distance between nodes using techniques like trilateration, triangulation or statistical inference [10]. Unfortunately, the RSSI-based ranging is not accurate in many cases, due to the unpredictable radio propagation behavior [11]. There exists algorithms, that are characterized with higher accuracy. For instance, the region-based localization (RBL) algorithm based on trilateration (such as the Ring Overlapping), which uses the Comparison of Received Signal Strength Indicator (ROCRSSI) [1,2]. Futhermore, algorithms derived from this approach, such as the Weight Monte Carlo Localization (WMCL) [4] based on statistical inference and range-based Monte Carlo boxed (RMCB) [4]. However, the disadvantage of these solutions is greater complexity compared to the map-based approaches.

RBL algorithms use measurements to determine distances between anchors and the localized node. Within this process, one can distinguish two phases. In the first phase (RSSI propagation) anchors broadcast specific number of beacon messages and sample received signal strength indicator. In the second phase (location estimation) node uses information from the first phase, in order to calculate its own location. In the article [4], the authors compare range-free Weight Monte Carlo Localization (WMCL) and range-based Monte Carlo boxed (RMCB) localization algorithms, which are similar to ROCRSSI approach. RMCB gives better results in localization in terms of lower error and its variance.

In order to get a better accuracy, one needs more computing power. However, wireless sensor networks are limited by low-power hardware (for instance the sensor nodes). There exist approaches that use demanding algorithms such as Kalman filter to improve results of signal-to-noise ratio (SNR) methods. Using SNR gives better results but its cost and complexity is even higher than the cost of previously mentioned approaches. Effective-SNR estimation [12] uses multiple Kalman filter for non-linear input of received signal strength indicator (RSSI), noise indicator (NOI) and link quality indicator (LQI) to calculate received signal strenght (in dBm) and environmental noise. These values are the input to first Kalman filter, which returns SNR estimation result. Authors make it more precise using the second Kalman filter with SNR and signal quality degradation (SQD) as input and Effective-SNR estimation as the result. The root mean square (RMS) and standard deviation of the result is lower than in classic SNR approach.

In the literature, there are few works about localization algorithms using mapping techniques and their experimental results in WSN. The main problem of these approaches is highly time and storage consuming process of mapping of each small area, which we will describe later.

There are several existing indoor RSSI map-based localization systems, such as Microsoft's RADAR [13], RSSI Localization in buildings [1], Ekahau real-time location system (RTLS) [14] or LANDMARC [15] (which is the RFID based system). RADAR is the oldest approach of mapping the RSSI-based localization with an accuracy of 2–3 m, however, RTLS and RFID based LANDMARC give greater precision, equal to about 2 m (and less). Previous systems work on algorithms such as ROSSI (RTLS) or typical map-based approaches like RADAR. In [10], the authors analysed and compared different map-based algorithms, some of which are Maximum Likelihood algorithm (ML) (which is RSSI map based approach with Min-Max algorithm), Multilateration and ROCRSSI. They proved, that the ML algorithm is more accurate and yields better performance than the others when the number of anchor nodes is relatively high. The only drawback of this method, is that we need to map the area in which we want perform localization, which sometimes can be quite laborious.

None of the articles presented above addresses the problem of secure localization in WSN. Attacks and proposals for securing location algorithms have been described in the following publications: [16,17]. To the best of the authors' knowledge, there are no proposals for securing strictly RSSI map-based approaches.

3 RSSI Protocol

In this section, we present both secure and insecure mapping and localization protocols. We assume, that all the anchor nodes are stationary in all of the following experiments.

Full localization process consists of two phases. The first one is the mapping phase, in which the protocol collects samples from each area to be localized. It is presented on Fig. 1. The set up for the experiment was placed in students hall in our department, where few obstacles and radio communication (such as the Wi-Fi 2.4 GHz) are available. The goal of this phase is to estimate the location map for each area, which uses the k-nearest neighbour match algorithm. The map consist of characteristics of each field to be localized later. In the second localization phase, the real-time samples are received from mobile node and compared with the characteristics from the first phase, in order to find the best matching field.

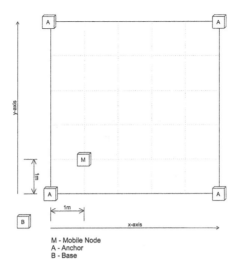

Fig. 1. The mapping phase.

3.1   The RSSI Mapping

In the mapping phase, we designed a protocol to gain RSSI from packets sent by the mobile node (denoted as M on Fig. 1) through anchors (denoted as A) to base station (denoted as B). In order to minimize the probability of collisions and make the data flow faster (since the algorithm needs a lot of packets to calculate characteristic of the field being mapped), we used packets with empty payloads. In the insecure version of the protocol, we do not use any cryptographic primitives, because we want to provide the fastest flow of packets within the network.

In our protocols, we use the HMAC function, random nonces and software AES-CTR symmetric encryption. In the insecure version, we do not use any securing mechanism. We assume, that such a protocol would be used in a safe environment, invulnerable to attacks.

To describe the proposed protocol in a formalized way, we use the notation presented in Table 1.

Below, the data flow diagram of insecure mapping protocol is presented:

Protocol shown on Fig. 2 is useful for quick mapping because of the fast response from Anchors. It needs about 5 or 6 s to receive and analyse 100 packets from one field. However, this protocol is vulnerable to many attacks on WSNs, and therefore is unacceptable to implement it in an existing production system in hostile environment. For this reason, we have designed and implemented the secure protocol, which is slower but ensures security attributes.

Below, on Fig. 3, we present the secure version of this protocol, using AES-CTR encryption, HMAC function and random nonce.

In this protocol, we secure the internal system (communication between the Anchor and Base, which is more critical than communication between the Anchor

Table 1. Utilized notation.

K, K_M - symmetric keys
ID_M - mobile node id
ID_A - anchor node id
N - nonce
$RSSI_{AVG}$ - average RSSI value
AKC - acknowledgement message
$\{M_1, M_2\}$ - tuple containing two messages
$\{M_1, M_2\}_K$ - symmetric encryption of tuple using key K
$HM(M, K)$ - HMAC function of message M and key K
$EoT = \{ID_M, ID_A\}$
$ResM = \{ID_M, ID_A, RSSI_{AVG}\}$
$SecResM = \{ResM, HM(ResM, K))$

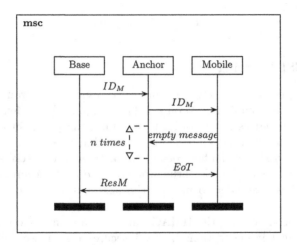

Fig. 2. The flow of the insecure mapping protocol.

and Mobile) from external attacks. We assume, that administrator inspects the mapping phase, and therefore we do not encrypt communication between Mobile node and Anchors. Key K used in communication between the Anchors and Base station can be established in many ways. In the introduced approach we use predefined, pre-installed keys. Another possible option is to use the TinyECC library in order to exchange keys.

We present Algorithms (1, 2 and 3), that were used in the protocol.

3.2   The RSSI Localization

In this section, we present both insecure and secure protocols for the second phase, namely the localization phase. Their construction is similar to the

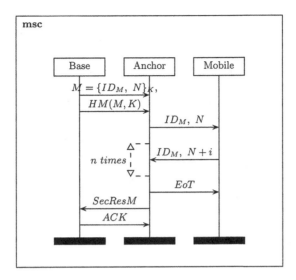

Fig. 3. The flow of the secure mapping protocol.

Algorithm 1. The algorithm of gathering anchors messages used by the BaseStation within the mapping phase.

1: AES.init()
2:
3: nonce = random();
4: plainText = [mobileId, nonce];
5: cipherText = AES.encrypt(plainText);
6:
7: send($\langle$ master AnchorId or broadcast $\rangle$, cipherText);

Algorithm 2. Algorithm used by Mobile node

1: testMsg = getMessageFromAnchor();
2: **while** !isEndOfTestMsg(testMsg) **do**
3:     testMsg.nonce++;
4:     send(broadcast, testMsg);
5: **end while**

Algorithm 3. The algorithm used by the BaseStation, when the BaseStation gets the response from anchors

1: response = getMessageFromAnchor();
2: responsePlainText = AES.decrypt(response);
3: **if** responsePlainText.storedNounce == nounce + 100 && responsePlainText.storedMobileId == mobileId **then**
4:     plainText = (response.source,
5:                     response.storedNonce +1);
6:     cipher = AES.encrypt(plainText);
7:     sendToAnchor(response.source, cipher);
8: **end if**

mapping algorithm apart from the messages sent from mobile node to the anchor. In this case, the anchors send queries more often to localize the Mobile node. This operation can be performed in a real-time. In the secure version, it is impossible to achieve the same performance results because of the encryption and other security mechanisms used.

One of the goals of this analysis, is to make this protocol work on every mote. Hence, it is more difficult to achieve short execution time because some radios are not equipped with hardware encryption (which is supported by, for instance, the CC2420 radio but not by the RF230 radio).

In order to develop the flow diagrams of the localization protocol, we utilize the notation presented in Table 1. In the insecure protocol, we put emphasis on the highest possible speed and therefore we do not add any security mechanisms. Insecure protocol is characterized by the longer life on battery power and fast node localization. This protocol can be used in safe environments. Its flow is presented on Fig. 4.

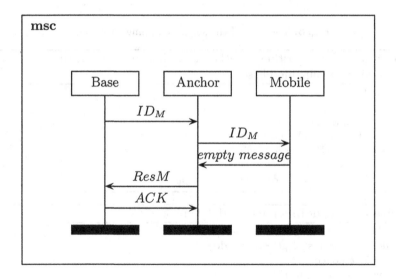

Fig. 4. The data flow in the insecure localization.

This data flow allows to collect information about one mobile sensor in a few milliseconds and send it to the base station, where it is analysed and saved in the system. The problem is that we can interrupt and tamper communication in every step using replay attacks or other injection of messages. Such situation is unacceptable in hostile environments because the attacker can join our network and conduct lots of different attacks. The protocol needs to be equipped with security mechanisms in order to use this protocol in hostile environment. Below, the data flow diagram of secure localization protocol is presented (Fig. 5).

In this protocol, we have two keys: the first one, to encrypt communication between the Base and Anchors (K), and the second one, to encrypt

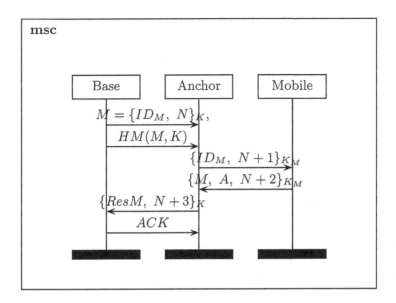

Fig. 5. The data flow in secure localization.

communication between the Anchors and Mobile (K_M). This solution forbids to communicate directly between Mobile and Base. Keys should be established when the mobile node joins our network, or simply pre-deployed in nodes.

In secure protocol mobile sensor has to encrypt messages, which degrades its lifetime [18] at the cost of security. The execution times extend as well, because every encryption and decryption operation, using software AES-CTR implementation, takes about 100 milliseconds for every 16 B data block. Therefore, the execution time of one full session takes about 1 s. Depending on the requirements, it can be enough in most cases, however it is too long for real-time applications with strict time restrictions.

3.3 Security Analysis

Insecure mapping and location protocol is exposed to a great number of attacks, some of which are: man in the middle, replay attack, wormhole attack, sybil attack or spoofing attack. In fact, a potential attacker is able to modify all packages or perform a classical denial of service attack (DDoS) [19] and block the whole network traffic. We are not able to defend our network against DoS attack but it is possible to defend the network against other attacks, by using cryptography inside the communication protocol.

To verify our protocol, we used the Scyther, which is automated security verification tool [8]. It provides its own modelling language, communication simulation and security functions, such as hash functions or symmetric/asymmetric cryptography.

We designed model corresponding to our protocol and tested it in Scyther, which confirmed our assumptions about security. We did not not find any possible attacks on our protocol (assuming that our keys were not compromised). We realize, that an automated tool is not able to test all the possible attacks which can be performed on WSN. Therefore, we present our analysis of the protocol and the services which have been provided.

(1) Confidentiality is provided with the symmetric encryption of packet's payload before sending. That is why the key must be stored by the Trust Center and must be securely established between nodes. We must assure that only authorized nodes are able to read the data from packet. Using symmetric cryptography, when every key is established, can secure us from spoofing attacks in the network.

(2) Integrity is assured by the keyed-Hash Message Authentication Code from the TinyECC Library. Only authorized nodes can generate the same HMAC because they know the secret key. HMAC prevents from the payload change attack, which can cause the system crash or errors (for instance, change RSS values).

(3) Authentication is also assured by the HMAC function. Keys for HMAC can be established using the Elliptic Curve Diffie-Hellman (ECDH) protocol, which is provided by the TinyECC Library [20]. However it may significantly slow the protocol because of time needed to generate points on elliptic curve. Authentication makes the protocol resistant to attacks such as the Sybil attack.

(4) Reliability is provided by sending ACK and resending messages if it is needed. We implemented time-outs and ACK messages. If the node does not receive any answer within a given time, we resend the packet.

(5) Freshness is provided by random nonces, in order to avoid attacks, such as replay attacks or flooding network with packets with random nonces (which is easy to detect).

In our case, the number of anchors is set at the beginning but in the dynamic ad-hoc networks adding new anchors should be verified by the additional authentication system.

In the case of physical attacks (destroying or stealing nodes), the only defence is the physical defence. ACK message has been added to prevent the retention of the entire system due to jamming. As the consequence, due to the the higher packet loss and retransmission, the time delivery of messages may be higher. Nevertheless, the message should be delivered to the recipient. Attacks on routing protocols depend on implementation of this protocol. In our case, the network of anchors is static and the routing protocol is simple and assures the exact security attributes enumerated above.

4 Experiment

4.1 Mapping

In order to conduct the experiment, we prepared a test environment, shown on Fig. 1 and described in detail previously. It is worth mentioning, that the test environment was exposed to the 2.4 GHz radio communication (due to the deployment of the Wi-Fi network), which affects the results of localization.

In our test we used the IRIS Memsic sensors with built-in RF230 radio, which, in the comparison to the popular CC2420 radio, does not have the hardware support for encryption (which is characterized by the shorter execution time, like presented in [21]). All tests were made using secure protocols. We created a map of the area and collected 500 samples from each place.

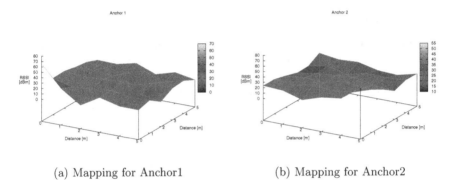

(a) Mapping for Anchor1   (b) Mapping for Anchor2

Fig. 6. Mapping results for anchor 1 and 2.

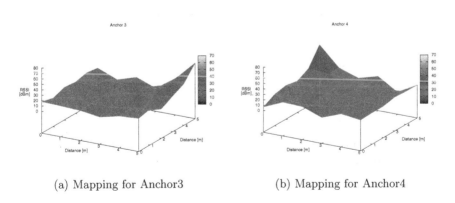

(a) Mapping for Anchor3   (b) Mapping for Anchor4

Fig. 7. Mapping results for anchor 3 and 4.

The results of our measurements for anchors 1–4 are shown on Figs. 6a, b, 7a and b, respectively. Anchor 1 is lower left on Fig. 1, anchor 2 is lower right,

Table 2. The comparison of times in mapping.

Packet no.	Insecure	Secure
100	6032,71 ms	7113,11 ms
200	10301,14 ms	12255,65 ms
500	25846,78 ms	26372,83 ms

anchor 3 is upper right and anchor 4 is upper left. By examining the results, we confirmed that there is no place on the map where the RSSI results of 4 anchors would be the same and that the signal propagates non-linearly, most likely due to the signal noise caused by Wi-Fi.

4.2 Localization

We performed two localization tests. In the first one, we localized the sensor every 1 m on the path presented on Fig. 8a. Black crosses represent the places, where sensor was placed during localization, the grid represents the map, while grey crosses are places, which were returned by the localization algorithm. We used the map built in Sect. 4.1. In this test, we used 5 packets from the mobile node (5 RSSI values) in order to locate the node.

Table 3. The comparison of times in localization.

Packet no.	Insecure	Secure
1	109,97 ms	1578,78 ms
10	614,63 ms	15319,02 ms
50	3014,35 ms	71729,1 ms

In the second test, we placed the mobile node on the straight path presented on Fig. 8b every 1 m. In both tests, we received the accuracy of approximately 1–2 m.

In Tables 2 and 3, we present time results for the mapping and localization stages, respectively. One can see that the cost of security in the case of localization is high (approximately 15–20 times longer). It comes from the fact, that the security mechanisms were software-based, which makes it more flexible but less time efficient.

The accuracy of localization is average due to the wireless networks available within the environment but it is more accurate than the one of the approaches without mapping. It is planned to conduct an experiment in target areas without any other wireless communication.

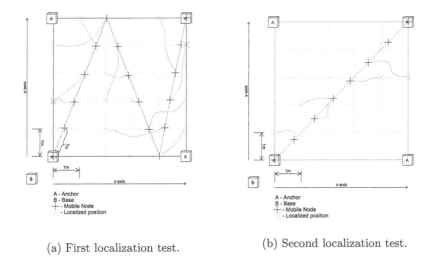

(a) First localization test.   (b) Second localization test.

Fig. 8. Paths localized during tests.

5 Conclusion

In this paper we designed, implemented and tested our approach of the map-based localization using the RSSI and compared it with the results given in [6]. We created an encrypted version of this approach and presented protocols of this solution. Finally, we conducted an experiment using the introduced protocol and the application designed for collecting the mapping results and localizing the nodes.

One can conclude, that localization protocols are commonly used, but they should be as well secured, since there exist a lot of attacks on Wireless Sensor Networks deployed in hostile environments. The secure localization should be used in critical systems such us power systems [22] or IoT architecture. However, to the best of authors' knowledge, there are no attempts to secure the map-based localization solutions using the RSSI.

We conducted the security analysis of our protocol using the Scyther tool, in order to check if it assures desired security attributes. The analysis indicated, that there are no attacks possible in the scope of our protocol. The experiment proved, that the secure version of the protocol can be slow because of encryption and other security mechanisms used.

In our analysis, we used the software encryption over the hardware encryption. The disadvantage of this solution is longer execution time, which exceed significantly the time difference between insecure and secure versions of our protocol. To achieve a better performance, one can use different radios, such as CC2420, which supports faster, hardware encryption. However, the advantage of using software encryption, is the possibility of applying the new concept of software-defined security, which allows us to gain the ability of changing the existing configuration of network security, during the application execution.

References

1. Luo, X., O'Brien, W.J., Julien, C.L.: Comparative evaluation of Received Signal-Strength Index (RSSI) based indoor localization techniques for construction job-sites. Adv. Eng. Inform. **25**(2), 355–363 (2010)
2. Liu, C.L.C., Wu, K.W.K., He, T.H.T.: Sensor localization with ring overlapping based on comparison of received signal strength indicator. In: 2004 IEEE International Conference on Mobile Ad-hoc and Sensor Systems (IEEE Cat. No.04EX975), pp. 516–518 (2004)
3. Dai, H., Zhu, Z.M., Gu, X.F.: Multi-target indoor localization and tracking on video monitoring system in a wireless sensor network. J. Netw. Comput. Appl. **36**(1), 228–234 (2013)
4. Adnan, T., Datta, S., MacLean, S.: Efficient and accurate sensor network localization. Pers. Ubiquit. Comput. **18**(4), 821–833 (2013)
5. Mwila, M.K., Djouani, K., Kurien, A.: Approach to sensor node calibration for efficient localisation in wireless sensor networks in realistic scenarios. Procedia Comput. Sci. **32**, 166–173 (2014). The 5th International Conference on Ambient Systems, Networks and Technologies (ANT-2014), the 4th International Conference on Sustainable Energy Information Technology(SEIT-2014)
6. Saxena, M., Gupta, P., Jain, B.N.: Experimental Analysis of RSSI-Based Location Estimation in Wireless Sensor Networks, pp. 503–510, January 2008
7. Mansour, I., Rusinek, D., Chalhoub, G., Lafourcade, P., Ksiezopolski, B.: Multihop node authentication mechanisms for wireless sensor networks. In: Guo, S., Lloret, J., Manzoni, P., Ruehrup, S. (eds.) ADHOC-NOW 2014. LNCS, vol. 8487, pp. 402–418. Springer, Cham (2014). doi:10.1007/978-3-319-07425-2_30
8. Cremers, C.: Scyther - Semantics and Verification of Security Protocols. Ph.D. Dissertation, Eindhoven University of Technology (2006)
9. Patwari, N., Ash, J., Kyperountas, S., Hero, A., Moses, R., Correal, N.: Locating the nodes: cooperative localization in wireless sensor networks. IEEE Sig. Process. Mag. **22**(4), 54–69 (2005)
10. Zanca, G., Zorzi, F., Zanella, A., Zorzi, M.: Experimental comparison of RSSI-based localization algorithms for indoor wireless sensor networks. In: Proceedings of the Workshop on Real-world Wireless Sensor Networks, REALWSN 2008, pp. 1–5. ACM, New York (2008)
11. Zhou, G., He, T., Krishnamurthy, S., Stankovic, J.A.: Impact of radio irregularity on wireless sensor networks. In: Proceedings of the 2nd International Conference on Mobile Systems, Applications, and Services, MobiSys 2004, pp. 125–138. ACM, New York (2004)
12. Qin, F., Dai, X., Mitchell, J.E.: Effective-SNR estimation for wireless sensor network using Kalman filter. Ad Hoc Netw. **11**(3), 944–958 (2013)
13. Bahl, P., Padmanabhan, V.: RADAR: an in-building RF-based user location and tracking system. In: Proceedings IEEE INFOCOM, vol. 2, pp. 775–784 (2000)
14. Ekahau page (2016). http://www.ekahau.com/. Accessed May 2016
15. Ni, L.M., Liu, Y., Lau, Y.C., Patil, A.P.: Landmarc: indoor location sensing using active RFID. Wireless Netw. **10**(6), 701–710 (2003)
16. Jiang, J., Han, G., Zhu, C., Dong, Y., Zhang, N.: Secure localization in wireless sensor networks: a survey (invited paper). J. Commun. **6**(6), 460–470 (2011)
17. Ammar, W., ElDawy, A., Youssef, M.: Secure localization in wireless sensor networks: a survey. CoRR abs/1004.3164 (2010)

18. Rusinek, D., Ksiezopolski, B., Wierzbicki, A.: Security trade-off and energy effi-ciency analysis in wireless sensor networks. Int. J. Distrib. Sens. Netw. **11**(6), 943475 (2015)
19. Katarzyna Mazur, B.K., Nielek, R.: Multilevel modeling of distributed denial of service attacks in wireless sensor networks. J. Sens. **2016**, 1–13 (2016)
20. Liu, A., Ning, P.: TinyECC: a configurable library for elliptic curve cryptography in wireless sensor networks. In: Proceedings of the 7th International Conference on Information Processing in Sensor Networks (IPSN 2008), SPOTS Track, pp. 245–256, April 2008
21. Rusinek, D., Ksiezopolski, B.: Influence of CCM, CBC-MAC, CTR and stand-alone encryption on the quality of transmitted data in the high-performance WSN based on imote2. Ann. UMCS Informatica **11**(3), 117–127 (2011)
22. Michal Wydra, P.K.: Power system state estimation accuracy enhancement using temperature measurements of overhead line conductors. Metrol. Meas. Syst. **23**(2), 183–192 (2016)

Application of Fault-Tolerant GQP Algorithm in Multihop AMI Networks

Sławomir Nowak, Mateusz P. Nowak, Krzysztof Grochla$^{(\boxtimes)}$, and Piotr Pecka

Institute of Theoretical and Applied Informatics,
Polish Academy of Sciences, Gliwice, Poland
{emanuel,mateusz,kgrochla,ppecka}@iitis.pl

Abstract. The paper presents the evaluation of the GQP algorithm in AMI networks using real map based network topologies (low, medium and high density of buildings). The algorithm was compared to the reference algorithms: CB, SBA and DP. It was shown that the GQP algorithm presents the good scalability, relatively optimal number of selected forwarders and by introducing the redundancy factor, good fault tolerance. An improvement to GQP algorithm to add a redundancy factor is proposed, what allows to execute fault tolerant broadcast transmission.

Keywords: AMI networks · Wireless networks · Broadcast · IoT

1 Introduction

Advanced Metering Infrastructure (AMI) is a popular architecture for automated, two-way communication between smart utility meters and an utility company. The AMI networks usually consist of thousands of relatively simple metering devices, creating a complex, multihop network. There is a need to develop methods for the management of such networks, as well as a need of development of reliable communication method. The unicast transmission is usually directed to the central node. Communication in opposite direction, from central node to meters, is often using the multicast or broadcast transmission. The AMI networks are heterogeneous, and may have different specific. The topology may be static or dynamic, multihop or direct. The number and location of nodes also may vary. The connection can be wireless or wired.

In this work we concentrate on an specific AMI network use case, with meters communicating by wireless interfaces. Meters are located within the buildings and they have a power supply (we don't focus on energy efficiency). Changes in placement of sensor nodes are rare and done under control of a network operator, so there is no need of automatic reconfiguration of network topology. There are no limitations of battery power, but it is the necessity of reliable communication and possibly optimal usage of network resources (bandwidth). We assume that a designated control node is distinguished, which typically has the access to a backhaul interface and forwards the traffic to and from the Internet to the AMI network.

© Springer International Publishing AG 2017
P. Gaj et al. (Eds.): CN 2017, CCIS 718, pp. 70–80, 2017.
DOI: 10.1007/978-3-319-59767-6_6

In presented evaluation we focus on broadcast solution for the low-dynamic, wireless, multihop network of meters with the designated central point. The physical connectivity is wireless, what implies limited range of communication and non-zero probability of transmission errors. The network has low dynamics, which means that the number of nodes, their positions and physical connectivity may change over time, although changes in nodes are relatively rare and done under control of network operator. The reconfiguration of network topology is possible but costly.

The multihop network requires the selection which nodes shall forward messages, forwarding packets coming from other nodes. Such node is referenced as a forwarder in the following part of the paper. The remaining nodes act as the communication endpoint.

This work is in part based on [9], which compares the selected set of possible broadcast solution and proposes the own solution, the Global Queue Pruning algorithm (GQP) in the random topologies of nodes. In practical AMI deployments the network topology is based on the physical locations of buildings, in which meters are installed. In this article we compare the set of reference methods to the GQP solution, based on the selected, map-based topologies of residential buildings. We also propose improvement to the algorithm to achieve better fault tolerance. The presented evaluations are based on simplifying assumption that the distance between nodes determines whenever they may be connected. The location of nodes and, in consequence, the connectivity, is now better correlated with the physical topology of buildings than random distribution, however the realistic radio signal propagation model is planned as a future work.

2   Related Work and Problem Formulation

The topology of physical connections in AMI network may be expressed as a graph. A connection is an edge of the graph. To create the logical structure for broadcast communication we have to create a spanning tree of logical connection with central node as a root. A node in the tree with degree higher than one act as a forwarder node. A forwarder may only forward a packet once (to avoid infinite loops) and all nodes shall receive the packet in no-failure conditions (the topology of connections don't change during the transmission and there are no transmission errors or node fails).

The RPL protocol [12] is a popular, general solution for IP level routing for wireless multihop networks. However RPL it not designed to solve the problem of broadcast, in particular does not specify dedicated method for performing data broadcasting [2]. The RPL broadcast results in many overlapping transmission, particularly problematic for larger networks, adequate to dense urban area where the overlap of nodes' range is high.

The straightforward solution for broadcast is flooding [11], in which every node retransmitts once every packet. The advantage of flooding is simplicity and reliability. The algorithm is simple to implement and guarantees 100% coverage, but has disqualifying disadvantages: can be costly in terms of wasted bandwidth and can impose a large number of redundant transmissions.

Various approaches have been proposed to solve the broadcast storm [10]. Some method are designed to reduce the redundant broadcast. The solution is to limit the broadcasting nodes only to limited set of forwarding nodes. The selection of forwarders can be dynamic or static [5] and, as a consequence, globally or locally. In the static approach a global algorithm determines the status (forwarder/non forwarder) of each node. In the dynamic approach the status is decided "on-the-fly" based on local node information, and the state can be different for every transmitted message. Some methods Counter Based select forward node based on probability, which cannot guarantee the reachability of the broadcast.

The assumption of AMI network structure being static also can be discussed, as node failures and changing propagation conditions (for example due to external interference) will result in structure changes.

The selection of nodes that will forward messages is a challenging task. Assuming that the vertexes are not weighted, the optimal solution is the Connected Dominant Set with the central node as a root. But it was proved [3] that the optimal selection of CDS is a NP hard problem, even if the network topology is known and in case of complex AMI networks cannot be used.

Two basic approaches are possible: local (only local, n hops neighbourhood is known) and global (the global information about topology is known). In the previous work [9] we have selected a set of basic, reference solutions: counter based (CB) [4] and two neighbour knowledge methods: scalable broadcast algorithm (SBA) [1] and dominant pruning (DP) [6]. We also proposed own solution, Global queue pruning, based on the global approach.

Counter Based (CB) is the probabilistic method example. Is executed locally on every node in the network. It has two parameters: T_{RAD} and C. When new packet is received, time $T = (0..T_{RAD}]$ is drawn. Within T the packet counter c is incremented when duplicates of the packet are received. Then, if $c < C$, the packet is retransmitted. If $C = \infty$ (practically "large enough") the algorithm works as flooding. The CB algorithm does not guarantee the 100% cover.

Scalable Broadcast Algorithm (SBA) also works locally and assumes that the 1-hop neighbour is known. When new broadcast packet is received, a period $T = (0..T_{RAD}]$ is drawn. Packet headers contain list of sender's neighbours. Within time T node analyses incoming packets. After T, if there are still nodes in the range that not received packets, the node forward a packet. 100% cover is guaranteed and the algorithm exhibit good scalability properties as the network size increases.

Dominant Pruning (DP) method utilises 2-hop neighbourhood information to reduce redundant transmissions. A forwarder selects the set of next forwarders among its 1 hop neighbours, to achieve the full cover of all nodes within 2-hop range. Then all designated forwarders repeat that step. The optimal solution is a NP-complete problem ($N!$ combinations to check), but the amount of nodes to analyse is usually relatively small. 100% network coverage is guaranteed.

3   The Global Queue Pruning Method

The new method was presented in details in [9]. The designation of forwarders is global (done e.g. by a server or central node) and is based on a queue of potential forwarders. Global approach was assumed as it was expected to have significantly more efficient topology at the expense of the communication cost in the initialisation phase.

In the initial phase every node sends to the known, central node the list of its 1hop neighbours. The central node represents the gateway to the network operator, having the connection to the servers. The more complex system can be assumed, in which the large network is divided by segments, and each segment can be managed by one gateway. From the network point of view the location of central point can be random, although the system administrator can potentially influence on that selection, to obtain e.g. better performance or better logical topology.

The algorithm is performed on the central node. The algorithm selects which nodes should act as forwarders. It takes list of neighbour per each node as input. The algorithm operates on a ordered list of nodes, which will be later on called queue. The algorithm uses also a pointer p which points to the current node. Nodes in front (before p) are designated as forwarders, and the remaining nodes are nodes, that are covered by another forwarders and can be considered as potential forwarders. The pointer p indicates the last selected forwarder. The algorithm starts by adding the central node to the queue as the first forwarder and setting $p = 1$ to point to this node.

The algorithm iterates until all nodes are added to the queue. At each step of the algorithm all the 1-hop neighbour nodes of the selected node are inserted into the queue, as it is shown on Fig. 1. Each node is inserted into a position that assures keeping the queue ordered by the weight. The weight is calculated as the function of cover and rank. The different methods of weight calculation are discussed in Sect. 4.3, but in general form it can be represented as:

$$weight = f(cover, rank), \tag{1}$$

where cover is the number of neighbours and the rank means the distance (number of hops) from the central node.

A node with highest weight value is designated as the new selected forwarder. Each selection of a forwarder influences on the nodes in the queue. The queue is

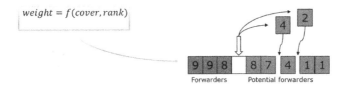

Fig. 1. Selection of forwarders from the global queue

rearranged. The weight function changes by reducing nodes' coverage according to the number of neighbours covered by the already selected forwarder (Fig. 1). The numbers on the figure represents the current weight, and each node can be inserted into the queue only once. The selected forwarder adds its neighbours (that are still unconnected e.g. by another forwarder) to the queue, and the operation is repeated. Each node has a flag to indicate whether it is covered by already designated forwarders or not. The algorithm step is repeated until all nodes are covered. The algorithm ends when all nodes were added to the queue, and than the set of forwarders is designated (nodes from the beginning of the queue up to node pointed by the pointer).

The algorithm is based on the concept of DP algorithm, but it utilises the global approach with the global queue of nodes. It also uses the weight function to designate the forwarder that influences e.g. to the length of paths.

In the evaluation presented in [9] we have compared the GQP to CD, SBA and DP algorithms. In current evaluation the MAGANET, a dedicated topology generator and analyzer [7] is used. The generator creates a random network topology based on square are on the map[1]. Subsequent nodes are located randomly within buildings and within the range of existing nodes to guarantee the connectivity between nodes). Parameters are: area size, number of nodes, minimum distance and radio range. According to selected algorithm the logical topology (spanning tree) is created. Finally, the broadcast communication is simulated to obtain a results for a single broadcast communication.

In the broadcast phase, it is possible for a node to receive duplicates when it was in the range of two or more forwarders. The number of duplicates was also evaluated. We also evaluated the cost of algorithm as the communication overhead to create the spanning tree.

In previous research the evaluation was conducted for random topologies only. They were significantly different than topologies of typical AMI networks, which depend on the size and density of buildings, because the meters that are network nodes in AMI networks are located within buildings, as it was shown in [8]. These differences can have a considerable impact on choosing the network protocols (in particular broadcast protocols). We decided to compare mentioned above algorithms but using the reliable map-based topologies. In this article we focus on topologies based on buildings location in different representative map areas.

The GQP algoritm can be the subject of improvements to increase the fault tolerance. It is important because the global approach produces the spanning tree with lower number of forwarders than in case of other algorithms. It makes the topology sensitive to the link or communication errors. To cope with that problem we propose the redundancy factor K.

In the original GQP algoritm every node held a flag, which was set when the node was covered by forwarder. We propose to use a counter k instead of the flag. If a node is covered by a forwarder, the k value is incremented (until $k >= K$). The node is treated as covered if at least K forwarders are in range

---

[1] Downloaded from http://www.openstreetmap.org.

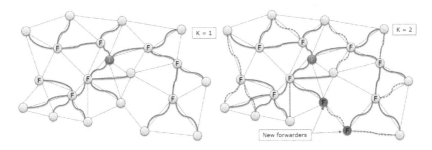

Fig. 2. New forwarders created with redundancy factor $K = 2$ to increase the fault tolerance

(in case if it is possible, because some nodes have just physical connectivity to number of forwarders that is less than K). Thus K is an algorithm parameter, chosen by network designer. If he or she would choose $K = 1$, the algorithm will work in a way equivalent to the original method without the redundancy factor (Fig. 2).

4   Performance Evaluation

In this work we focus on the evaluation of GQP algorithm in map-based topologies, using the generator described in [7]. We distinguish three types of topology: medium density town area, the residential area and the downtown area.

The assumed parameters were as follows: nodes count $n = 3000$, average number of neighbours for each node $a = 40, 45, 50, 55$. Minimum distance between nodes was 1 m.

The radio link range was adjusted automatically, according to the average number of neighbours. In case of a gap between buildings longer than calculated radio range the "long link" was created that reflects the additional cable (or equivalent, eg. microwave radio line) connection between distance nodes. In the real AMI systems, when it is impossible to connect nodes in distant locations, they have to be connected using some long distance technology (cable, directional antennas etc.). MAGANET simulator is able to generate "long link" when necessary. For the sake of simplification the free space radio signal propagation model is used and the decision whenever two nodes can communicate is based on the threshold on their distance. The central point is chosen randomly among all network nodes.

To evaluate the fault tolerance we simulated the certain probability of link failure and evaluated the scenario with $K = 1, 2, 3$. The following results present the average value for 20 simulations for each point on the graphs.

4.1   Number of Forwarders

The higher the number of forwarder is, the longer paths are created. The highest number of forwarders was created using the DP method, as it is shown on Fig. 3.

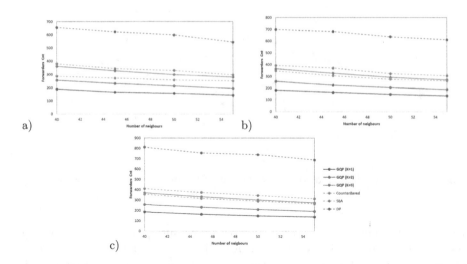

Fig. 3. The number of forwarders in the medium density (a), residential (b) and high density (c) areas

The lowest number was created using GQP with $K = 1$. The GQP with $K = 3$ had similar number of forwarders as the SBA and CB, but $K = 3$ means the better fault tolerance (as it is shown later in the article).

4.2   Broadcast Overhead

The broadcast overhead is defined as the redundant communication. It is assumed that each node must receive a message at least once. The overhead is calculated as the total sum of received messages beyond one. The results on Fig. 4 are expressed in %, where 100% is the case when there is no overhead (only 3000 messages were received, one per each node). The smallest overhead was observed in case of GQP with $K = 1$. The biggest is connected to DP local case. The interesting remark that in case of GQP the overhead is stable, independent from the average number of neighbours.

4.3   Influence of the Weight Formula

In this section we evaluate the influence of different formulas of weight calculation on the performance of the algorithm. The weight formula influences on the selection of forwarders. The simplest formula is:

$$weight = cover \qquad (2)$$

In that case the forwarder is selected according to best cover, what is dependent on the number of neighbours. This simplest formula was used in the previous

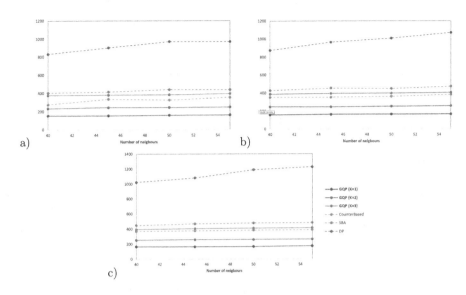

Fig. 4. The broadcast overhead in the medium density (a), residential (b) and high density (c) areas

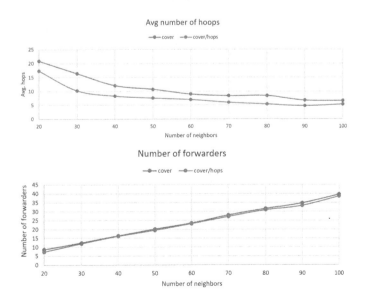

Fig. 5. Results of different weight functions calculations

calculations. The different approach takes into consideration the path length to the central point (rank):

$$weight = cover/rank \tag{3}$$

In such a case the forwarder with smaller cover but closer to the central point is preferred. This leads to the shorter paths. Better performance can be achieved for more clustered, lower density networks (as the large residential area). The influence of different weight function was is presented on Fig. 5.

5   Fault Tolerance

The introduction of the redundancy factor K created the additional forwarders to assure the redundant paths, possible if a node is in the range of two different nodes. The influence of K factor was evaluated by introducing the link error probability in the simulation phase. We assumed that in when a message is transmitted there is a probability $p = 0.1, 0.05, 0.02$ that the message is not received properly. The results are presented on Fig. 6.

By increasing the K factor the percent of delivery fails is decreased. The influence is significant. The broadcast communication assumes that a message is delivered to each node. Due to the link errors the successful scenario may use the unicast delivery after the broadcast phase to all remaining nodes. The larger K factor minimizes that unicast communication, while in case of $K = 1$ the fault tolerance is worse than in the reference algorithms (Fig. 7).

The CB also presents moderate fault tolerance (but still no guarantee of connecting all nodes to the spanning tree) and the SBA and DP has good tolerance but at the expense of greater number of forwarders.

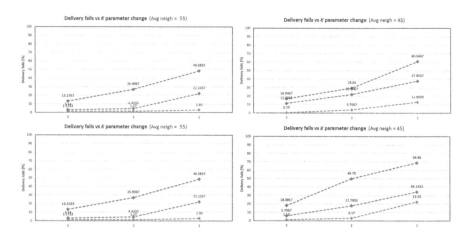

Fig. 6. The influence of the K factor on the delivery fails

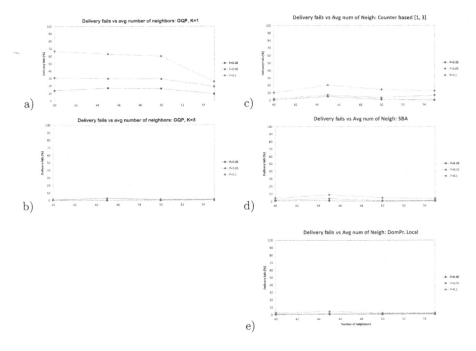

Fig. 7. The comparison of fault tolerance of GQP $K = 1$ (a), GQP $K = 3$ (b) and the reference algorithms (c-e).

6   Conclusions

The article presents the evaluation of the GQP algorithm (introduced in [9]) in comparison to set of reference algorithms. We used the map based network topologies that are more practical scenarios than in previous evaluations that used the random node distribution, as the real AMI networks usually depends on the topology of buildings. We assumed the semi-static topology, which does not change long enough to create the static designation of forwarder. In practice it is costly to obtain the complete knowledge of the network topology, however the AMI networks are very static, as the nodes don't move. We can assume that the central node has a partial network knowledge, based e.g. on the configuration. In the proposed GQP algorithm after the forwarders are selected, the algorithm decission can be used inspite of the interferences and temporal unavailability of nodes, thanks to the redundancy factor.

We considered the two simple cases of the weight function and proposed the improvement to the algorithm to increase the fault tolerance. It was shown that the GQP algorithm presents the good scalability, relatively optimal number of selected forwarders and by introducing the redundancy factor, good fault tolerance. It is also possible to adjust the GQP parameters and weight function to the specific of a certain case of AMI network.

References

1. Boukerche, A. (ed.): Algorithms and Protocols for Wireless and Mobile Ad Hoc Networks. Wiley, Hoboken (2009)
2. Clausen, T., Herberg, U.: Multipoint-to-point and broadcast in RPL. In: 2010 13th International Conference on Network-Based Information Systems, pp. 493–498, September 2010
3. Garey, M.R., Johnson, D.S.: Computers and Intractability: A Guide to the Theory of NP-Completeness. W. H. Freeman, San Francisco (1979)
4. Izumi, S., Matsuda, T., Kawaguchi, H., Ohta, C., Yoshimoto, M.: Improvement of counter-based broadcasting by random assessment delay extension for wireless sensor networks. In: 2007 International Conference on Sensor Technologies and Applications (SENSORCOMM 2007), pp. 76–81. IEEE, October 2007
5. Khabbazian, M., Blake, I.F., Bhargava, V.K.: Local broadcast algorithms in wireless ad hoc networks: reducing the number of transmissions. IEEE Trans. Mob. Comput. $11(3)$, 402–413 (2012)
6. Lim, H., Kim, C.: Flooding in wireless ad hoc networks. Comput. Commun. $24(3-4)$, 353–363 (2001)
7. Nowak, S., Nowak, M., Grochla, K.: MAGANET – On the need of realistic topologies for AMI network simulations. In: Kwiecień, A., Gaj, P., Stera, P. (eds.) CN 2014. CCIS, vol. 431, pp. 79–88. Springer, Cham (2014). doi:10.1007/978-3-319-07941-7_8
8. Nowak, S., Nowak, M., Grochla, K.: Properties of advanced metering infrastructure networks' topologies. In: 2014 IEEE Network Operations and Management Symposium (NOMS), pp. 1–6. IEEE (2014)
9. Nowak, S., Nowak, M., Grochla, K., Pecka, P.: Global queue pruning method for efficient broadcast in multihop wireless networks. In: Czachórski, T., Gelenbe, E., Grochla, K., Lent, R. (eds.) ISCIS 2016. CCIS, vol. 659, pp. 214–224. Springer, Cham (2016). doi:10.1007/978-3-319-47217-1_23
10. Shen, X., Zhao, R., Zhang, X.: Broadcast protocols for wireless sensor networks. In: Smart Wireless Sensor Networks, September 2010
11. Tanenbaum, A.S., Wetherall, D.: Computer Networks, 5th edn. Pearson Prentice Hall, Boston (2011)
12. Yi, J., Clausen, T., Igarashi, Y.: Evaluation of routing protocol for low power and lossy networks: LOADng and RPL. In: 2013 IEEE Conference on Wireless Sensor (ICWISE), pp. 19–24. IEEE, December 2013

A Comparative Analysis of N-Nearest Neighbors (N3) and Binned Nearest Neighbors (BNN) Algorithms for Indoor Localization

Serpil Ustebay[1(✉)], M. Ali Aydin[1], Ahmet Sertbas[1], and Tulin Atmaca[2]

[1] Department of Computer Engineering, Istanbul University, Istanbul, Turkey
{serpil.ustebay,aydinali,asertbas}@istanbul.edu.tr
[2] Laboratoire Samovar, Telecom SudParis, CNRS,
Université Paris-Saclay, Evry, France
tulin.atmaca@telecom-sudparis.eu

Abstract. In this study, performances of classification algorithms N-Nearest Neighbors (N3) and Binned Nearest Neighbor (BNN) are analyzed in terms of indoor localizations. Fingerprint method which is based on Received Signal Strength Indication (RSSI) is taken into consideration. RSSI is a measurement of the power present in a received radio signal from transmitter. In this method, the RSSI information is captured at the reference points and recorded for creating a signal map. The obtained signal map is knows as fingerprint signal map and in the second stage of algorithm is creating a positioning model to detect individual's position with the help of fingerprint signal map. In this work; N-Nearest Neighbors (N3) and Binned Nearest Neighbors (BNN) algorithms are used to create an indoor positioning model. For this purpose; two different signal maps are used to test the algorithms. UJIIndoorLoc includes multi-building and multi floor signal information while different from this RFKON includes a single-building single floor signal information. N-Nearest Neighbors (N3) and Binned Nearest Neighbors (BNN) algorithms are presented comparatively with respect to success of finding user position.

Keywords: N3 and BNN algorithms · Indoor localization · RSSI · Fingerprint · Accuracy

1 Introduction

With a common infrastructure such as the Internet, different applications have became available anywhere at any time. It is possible to obtain the instant status information of the person or objects with the positioning systems, Positioning systems are divided into two basic groups, indoor and outdoor. Outdoor satellite-based positioning systems achieve high success by processing more than one satellite information with triangulation. For this, some information is needed like the arrival angle of signals, arrival time of signals and so on. Also, the signals must be within line-of-sight.

© Springer International Publishing AG 2017
P. Gaj et al. (Eds.): CN 2017, CCIS 718, pp. 81–90, 2017.
DOI: 10.1007/978-3-319-59767-6_7

Many countries try to implement their own GPS applications. GLONASS [1], Galileo [2], Beidou [3] and IRNSS (NAVIC) [4] can be given as examples for developed applications. Although recommended indoor GPS [5] systems are realized for indoors, it is a disadvantage that the devices used are costly. The search for a high performance, low cost solution continues. There is no standard for any positioning systems. Offered solutions locate people inside a building using radio waves, Bluetooth and infrared signals, magnetic fields, acoustic signals or other sensory information. Indoor positioning is divided into two categories, active and passive. In active positioning, it is the direct transmission of information in positioning by means of a device carried by the users. RFID, Bluetooth, Infrared, ultra-wideband, IEEE 802.11 WLAN are examples of these. In passive systems they do not transfer information directly with any device they carry on them. Differential Air, Computer vision, Device-free Passive systems can be given as examples of passive systems [6]. Fingerprint, is a localization method based on obtaining signal map of place by power of signal of IEEE 802.WLAN. This technique is applied just other sensor data like magnetic field, RFID, Bluetooth.

The fingerprint consists two step that are called online and offline stage as shown on Fig. 1. Offline phase consists the radio map construction process. Although the offline phase takes a lot of time and effort, it provides high success rate and low cost implementation. Reference points are selected inside the building and Received Signal Strength (RSS) are measured from Wireless Access Points (WAP) with the help of a mobile device. Also Detailed information about

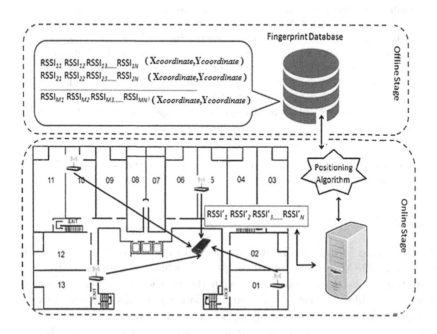

Fig. 1. Fingerprint algorithm stages

reference point as building id, floor id, space name...etc are recorded. RSS measurements can easily affected from noise (objects, people, wheatear.. etc.) to obtain high accuracy in the noisy wireless channel, different mobile devices are used to capture RSS. The signal map is used to generate a positioning model at second stage. The model can be grouped by mathematical formulas. Geometry based methods like triangulation. This method facilitates validation of data through cross verification from two or more sources. Probability based mathematical approaches can be used like Bayesian inference techniques. Machine learning algorithms especially classification methods [7–10], neural network [11–13], deep learning [14–16]...etc can be used to create high performed positioning model.

This study is aimed at finding position of a person in indoors by using fingerprint method. At positioning stage, N3 and BNN algorithms based on nearest neighbor approach are applied. Success and error distance of algorithms are tested on two different databases are presented.

2   Use of Wireless Signal Maps

Two different signal maps are used to test the algorithms. The first one is named RFKON and second one is named UJIIndoorLoc. The constructions from which both signal maps are obtained are shown in Fig. 2.

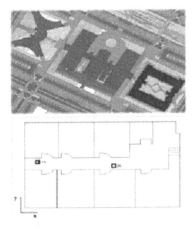

Fig. 2. Red, Green and Blue refers to the building of UJI Riu Sec Campus (top). KIOS Research Center floor plan (down) (Color figure online)

The Area was broken into 1.2×1.2 m grid squares and Received Signal Map Strength (RSS) from available Wireless Access Points (WAP) was measured at the center of each grid square to get RFKON [17]. Four different mobile devices are used to capture RSS and the number of reference points appearing in the

Table 1. Data samples in RFKON

ID	Time	Device ID	x	y	Floor	Battery	MAC1	MAC2	...	MACN
1	01.07.2015 09:10	1	1.2	1.2	1	100	−97	−65	...	100
2	02.07.2015 09:10	2	1.2	1.2	1	100	−95	−64	...	100
3	03.07.2015 09:10	3	1.2	1.2	1	100	100	−70	...	100
4	04.07.2015 09:10	4	1.2	1.2	1	100	100	100	...	100

Table 2. UJIIndoorLoc signal map attribute information.

Attribute	Information
Attribute 001 (WAP001)	Intensity value for WAP001. Negative integer values from −104 to 0 and +100. Positive value 100 used if WAP001 was not detected
...	...
Attribute 520 (WAP520)	Intensity value for WAP520. Negative integer values from −104 to 0 and +100. Positive Vvalue 100 used if WAP520 was not detected
Attribute 521 (Longitude)	Longitude. Negative real values from −7695.9387549299299000 to −7299.786516730871000
Attribute 522 (Latitude))	Latitude. Positive real values from 4864745.7450159714 to 4865017.3646842018
Attribute 523 (Floor)	Altitude in floors inside the building. Integer values from 0 to 4
Attribute 524 (BuildingID)	ID to identify the building. Measures were taken in three different buildings. Categorical integer values from 0 to 2
Attribute 525 (SpaceID)	Internal ID number to identify the Space (office, corridor, classroom) where the capture was taken. Categorical integer values
Attribute 526 (RelPosition)	Relative position with respect to the Space (1 - Inside, 2 - Outside in Front of the door). Categorical integer values
Attribute 527 (UserID)	User identifier (see below). Categorical integer values
Attribute 528 (PhoneID)	Android device identifier (see below). Categorical integer values
Attribute 529 (Timestamp)	UNIX Time when the capture was taken. Integer value

database is 54. Each recorded measurements attributes and sample values are given at Table 1. 100 dbm means that there is no WAP information at related reference point.

Second signal map is UJIIndoorLoc [18] which is the biggest multi-building and multi-floor database in the literature and first publicly available database that could be used to make comparisons among different methods shown. Totally 520 WAP data were recorded by more than 20 users using 25 different models of mobile devices. BuildingID is used to identify buildings which could be 3 different value. FloorID is used to identify floors of building. SpaceID is used to identify the particular space (offices, labs, etc.) where the capture was taken. All attribute information is listed at Table 2.

3   Algorithms

In this section, the structure of the N Nearest Neighbor (N3), Binned Neighbor Neighbors (BNN) algorithms [19] and the comparison parameters used to measure the success of the algorithms are presented.

3.1   N Nearest Neighbors (N3)

N3 algorithm is based on nearest neighbours which is take into account all neighbours except k-nearest neighbours. When a test sample is classified, all neighbours are sorted from most similar to the least similar with this a similarity rank vector is acquired. A weight matrix is W_{ig} calculated to classify the test sample. W_{ig} is defined as the ith object's g th class weight and calculated on the Formula 1.

$$W_{ig} = \frac{1}{\hat{n}_g} \cdot \sum_{j=1, j \neq i}^{(n-1)} \frac{s_{ij}}{r_{ij}^{\alpha}} \cdot \delta_j \qquad \delta_j = \begin{cases} 1 \ if \ C_j = g \wedge \frac{s_{ij}}{r_{ij}^{\alpha}} > \varepsilon \\ 0 \qquad otherwise \end{cases} \tag{1}$$

$\hat{n}_g$ is the number of neigbors that contribute to the class weight and calculated with the Formula 2. r_{ij} is a similarity rank and modulated by α which is a real-values parameter to be optimized in the range $[0.1, 2.5]$. c_i is the class of the jth object. ϵ (e.g., 10^{-7}) is a threshold parameter to define Dirac delta (δ_i).

$$\hat{n}_g = \sum_j \delta_j \tag{2}$$

S_{ij} is the similarity between ith and jth object and calculated from their distances with Formula 3. d_{ij} is a distance measure between ith object and jth object.

$$S_{ij} = \frac{1}{1 + d_{ij}} \tag{3}$$

Finally, the calculated W_{ig} matrix is normalized with Formula 4.

$$\acute{W}_{ig} = \frac{W_{ig}}{\sum_g W_{ig}} \tag{4}$$

All objects class definition is making with $\acute{W}_{ig}$ values. These values can be interpreted as fuzzy measures. Maximum $\acute{W}_{ig}$ value defines ith object class which is predefined at wireless signal map.

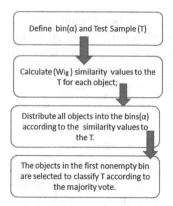

Fig. 3. The flow chart of the BNN algorithm

3.2 Binned Nearest Neighbors (BNN)

The BNN algorithm works similarly to the KNN algorithm. The test data are classified according to their nearest k neighbors. The BNN algorithm differs from the KNN algorithm in that it is the selection of neighbors used in classification. While k closest neigbors are selected in KNN algorithm, All neighbors are distributed into predefined similarity intervals (bins) and distributes into these intervals in BNN algorithm. Then, the first non-empty bin are chosen as the nearest neighbors in order to being used for prediction. The Prediction is taken as majority vote [19]. The flow diagram of the BNN algorithm is given in Fig. 3.

3.3 Comparison Parameters

Classification algorithm's performance are compared with different metrics. It affected by number of instances that are correctly classified, number of instances that are incorrectly classified, total number of instances, etc. The metrics which are used to compare performance of different classification algorithms are listed at Table 3.

Table 3. Algorithms comparison parameters

Name	Formula	Definition
Accuracy	$\frac{TP}{n}$	TP is the number of true predicted samples, n is the number of test instances
Error type 1	$\frac{\sum_i^n D_i}{n}$	D_i is ith class distance between predicted class and real class, n is the number of test samples
Error type 2	$\frac{\sum_i^n D_i}{FP}$	D_i is ith class distance between predicted class and real class. FP is the false predicted instance

4   Implementation and Results

In this study two different fingerprint map is used to test the BNN and N3 algorithms. The UJIndoorLoc dataset contains 21048 measurements (samples) taken from 520 KEN (attributes). However, 3329 measurements were used. The measurements used were re-labeled according to BuildingID, FloorID, and SpaceID, and 174 classes were distributed. RFKON data has 54 reference points recorded with coordinates values and 26 WAP measurements as listed at Table 4.

Table 4. Description of fingerprint signal map

Name	Number of sample	Number of variable	Number of class	Multi building	Multi floor
RFKON	18480	26	54	No	No
UJIIndoorLoc	33290	520	174	Yes	Yes

In machine learning algorithms; amount of data must be used in training process and amount of data must be used for to test accuracy of model. For this, 80% of data is separated for training, while 20% is for test is used both of fingerprint maps. For RFKON database, 14.784 records are used for training and 3696 records is used for test. This operation is repeated for 100 times so the average accuracy value is calculated. The optimal alpha value for N3 and BNN is determined 0.75. These values gave the highest accuracy according to the test results. Euclidean is used as the distance metric.

Table 5. Test results of BNN and N3 algorithms

	Algorithm	Error type 1	Error type 2	Accuracy
RFKON	BNN	0.3373	2.3949	85.9053
	N3	0.4031	2.3975	83.1872
UJIIndoorLoc	BNN	4.3420	20.6632	78.8905
	N3	2.546	14.2402	82.1514

Error distance shows the indicates how much error the position of a person is made. Error distance is a measurement may change related to reference point frequency. If frequent intervals of reference points are determined within the building, error distance will be reduced. Error distance of both signal maps is shown in Table 3. Obtained lower error types for RFKON could be a small grid squares as 1.2×1.2 m when measurements are taken at inside building.

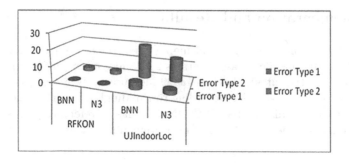

Fig. 4. Test results

5 Conclusion

In this study, two different indoor positioning models were created with finger-print method. For this; N3 and BNN algorithms were used to define user/objects position and tested with the two different signal maps. RFKON is a data set based on signal strengths consisting of a single building and a single floor. Conversely, the UJIIndoorLoc data set consisting of multi-floor and multi-building signal map has been preferred. Effect of adding different features (building and floor information) to the N3 and BNN algorithms have been examined. When the test results are examined, it is seen that the performance of the algorithms changes according to the signal map. For UJIIndoorLoc, N3 algorithm has higher accuracy than BNN algorithm as shown Table 5. Besides, error distance is lower. N3 algorithm could not show the same performance on RFKON dataset as shown Fig. 4. In single floor data, BNN has high performance and low error distance. Lower accuracy can be obtained at multi floor signal maps than one floor signal map. Floor information affect RSSI captures.

The measurement number of fingerprinting signal data, capturing method, the number of the floor and building of location effects the accomplishment of positioning. The results of these algorithms do not sufficient for localization. For obtaining a more accurate localization system, new classification algorithms and hybrid approaches are required. For this purpose, different sensor data may be included to the system. Measured RSS values when floor information is considered can lead to errors and reduce accuracy.

Acknowledgements. This work is also a part of the PhD thesis titled "Design of an Efficient User Localization System for Next Generation Wireless Networks" at Istanbul University, Institute of Physical Sciences.

References

1. Bykhanov, E.: Timing and Positioning with GLONASS and GPS. GPS Solutions **3**(1), 26–31 (1999)
2. O'Keefe, K., Julien, O., Cannon, M.E., Lachapelle, G.: Availability, accuracy, reliability, and carrier-phase ambiguity resolution with Galileo and GPS. Acta Astronaut. **58**(8), 422–434 (2006)
3. Zeng, Q.H., Liu, J.Y., Hu, Q.Q., Yang, D.: Research on Beidou and GNSS multi-constellation integrated navigation. GNSS World China **1**, 013 (2011)
4. Zaminpardaz, S., Teunissen, P.J.G., Nadarajah, N.: IRNSS/NavIC and GPS: a single-and dual-system L5 analysis. J. Geodesy, 1–17 (2017)
5. Van Diggelen, F., Abraham, C.: Indoor GPS technology. In: CTIA Wireless-Agenda, Dallas, vol. 89 (2001)
6. Pirzada, N., Nayan, M.Y., Subhan, F., Hassan, M.F., Khan, M.A.: Comparative analysis of active and passive indoor localization systems. AASRI Procedia **5**, 92–97 (2013)
7. Talvitie, J., Renfors, M., Lohan, E.S.: Distance-based interpolation and extrapolation methods for RSS-based localization with indoor wireless signals. IEEE Trans. Veh. Technol. **64**(4), 1340–1353 (2015)
8. Bozkurt, S., Elibol, G., Gunal, S., Yayan, U.: A comparative study on machine learning algorithms for indoor positioning. In: 2015 International Symposium on Innovations in Intelligent SysTems and Applications (INISTA), pp. 1–8. IEEE, September 2015
9. Khudhair, A.A., Jabbar, S.Q., Sulttan, M.Q., Wang, D.: Wireless indoor localization systems and techniques: survey and comparative study. Indonesian J. Electr. Eng. Comput. Sci. **3**(2), 392–409 (2016)
10. Obeidat, H.A., Dama, Y.A., Abd-Alhameed, R.A., Hu, Y.F., Qahwaji, R.S., Noras, J.M., Jones, S.M.: A Comparison Between Vector Algorithm and CRSS Algorithms for Indoor Localization Using Received Signal Strength. University of Bradford Engineering and Informatics Publications (2016)
11. Lin, T.N., Lin, P.C.: Performance comparison of indoor positioning techniques based on location fingerprinting in wireless networks. In: 2005 International Conference on Wireless Networks, Communications and Mobile Computing, vol. 2, pp. 1569–1574. IEEE, June 2005
12. Fang, S.H., Lin, T.N.: Indoor location system based on discriminant-adaptive neural network in IEEE 802.11 environments. IEEE Trans. Neural Networks **19**(11), 1973–1978 (2008)
13. Dai, H., Ying, W.H., Xu, J.: Multi-layer neural network for received signal strength-based indoor localisation. IET Commun. **10**(6), 717–723 (2016)
14. Wang, X., Gao, L., Mao, S., Pandey, S.: CSI-based & fingerprinting for indoor localization: a deep learning approach. IEEE Trans. Veh. Technol. **66**, 763–776 (2016)
15. Faragher, R., Harle, R.: SmartSLAM-an efficient smartphone indoor positioning system exploiting machine learning and opportunistic sensing. In: ION GNSS, vol. 13, pp. 1–14, September 2013
16. Gu, Y., Chen, Y., Liu, J., Jiang, X.: Semi-supervised deep extreme learning machine for Wi-Fi based localization. Neurocomputing **166**, 282–293 (2015)
17. Bozkurt, S., Yazıcı, A., Gunal, S., Yayan, U., Inan, F.: A novel multi-sensor and multi-topological database for indoor positioning on fingerprint techniques. In: 2015 International Symposium on Innovations in Intelligent SysTems and Applications (INISTA), pp. 1–7. IEEE, September 2015

18. Torres-Sospedra, J., Montoliu, R., Martínez-Usó, A., Avariento, J. P., Arnau, T.J., Benedito-Bordonau, M., Huerta, J.: Ujiindoorloc: a new multi-building and multi-floor database for wlan fingerprint-based indoor localization problems. In: 2014 International Conference on Indoor Positioning and Indoor Navigation (IPIN), pp. 261–270. IEEE, October 2014

19. Todeschini, R., Ballabio, D., Cassotti, M., Consonni, V.: N3 and BNN: Two new similarity based classification methods in comparison with other classifiers. J. Chem. Inf. Model. **55**(11), 2365–2374 (2015)

Evaluation of Connectivity Gaps Impact on TCP Transmissions in Maritime Communications

Michal Hoeft[(✉)] and Jozef Wozniak

Faculty of Electronics, Telecommunications and Informatics,
Gdańsk University of Technology, Gdańsk, Poland
michal.hoeft@pg.gda.pl, jowoz@eti.pg.gda.pl

Abstract. Many organizations and research working groups, including among others IEEE, ITU, ETSI and IMO are currently working towards improvements in communication of different types of vehicles (cars, trains, planes and vessels) and upgrading utilities and services offered to their crews and passengers travelling all over the world. The paper deals with selected aspects of the TCP protocol connectivity in maritime wireless networks. The authors show results of measurements performed in a real-world maritime environment consisting of on- and off-shore devices and transmission systems, partly located on vessels. Based on the collected measurements, the authors provide statistical analysis, including connectivity periods and connectivity gaps observed as a result of vessels motions. The obtained statistical results were used, in turn, in a simulation test-bed to verify efficiency of TCP connections, assuming the use of Reno algorithm, together with F-RTO enhancement (Forward RTO-Recovery), Cubic algorithm, and further adjustments proposed by the authors. The results show significant increase of the average throughput (up to 3 times) and decrease of the average download time (up to 53%), as compared to the pure Reno TCP solution. The proposed modifications seem to be especially important and attractive for highly unreliable wireless links - expected to be very common in a maritime environment.

Keywords: Maritime communications · TCP · Connectivity gaps · e-navigation services

1 Introduction

The need for effective and reliable communication between heterogeneous networking equipment installed on different kinds of vessels and on- and off-shore fixed network elements is becoming more and more important. The lack of reliable maritime solutions, offering high throughput and available to a wide audience (including crews and passengers travelling all over the world) for a relatively low price is one of the main barriers in effective implementation of e-navigation. The e-navigation services, defined by IMO (International Maritime Organization), should ensure: "harmonized collection, integration, exchange, presentation

© Springer International Publishing AG 2017
P. Gaj et al. (Eds.): CN 2017, CCIS 718, pp. 91–105, 2017.
DOI: 10.1007/978-3-319-59767-6_8

and analysis of marine information on board and ashore by electronic means to enhance berth to berth navigation and related services for safety and security at sea and protection of the marine environment" [1]. The concept of e-navigation includes, for example, integration of a multitude of navigational systems and aids that currently have to be separately monitored by a bridge crew, as well as, making increased use of inter-ship data exchange for purposes of safety and efficiency of maritime travel. Unfortunately there are no commonly accessible maritime communication systems, fulfilling all needs requested by e-navigation services - there is no affordable Internet access at the sea.

It does not mean the lack of systems designed and used for maritime communications. Nowadays, several of wireless technologies are considered to be used in maritime communications. Some of them offer very long distance links but with a very limited throughput and reliability [2]. This group of terrestrial solutions operates with HF and VHF bands. On the other hand, there is a set of solutions offering higher bandwidth but at the price of relatively short communication ranges. The preliminary investigations show, however, that standardized solutions like LTE, WiMAX, as well as some vendor-specific products like RADWIN Fiber-in-Motion or Kongsberg Maritime Broadband Radio can be considered for maritime wireless links. Additionally, popular high-orbit satellite systems can be employed, with their advantage of global coverage, and limitation of a relatively low data rates (single Mb/s) and very high transmission delay values (RTT apps. 0.5 s). However, in both low-orbit (like IRIDIUM) and high-orbit (e.g., INMARSAT or VSATs) satellite systems, it is necessary to cover relatively high costs of both installation and maintenance.

Each of the mentioned above transmission techniques is or can possibly be used for different maritime purposes, however separately. Integration of these diverse terrestrial technologies in a heterogeneous system is one of the main objectives of the netBaltic project [3], launched in order to meet the increasing needs for wider and faster deployment of e-navigation services and to alleviate the mentioned above issues.

The aim of the netBaltic project is to define and implement in a maritime environment - a fully heterogeneous self-organizing broadband mesh network consisting of moving vessels employing a set of wireless technologies integrated by an intelligent interface selection module and equipped with effective switching mechanisms, efficient routing protocol providing optimum transmission parameters for an existing network topology - all independently of the transmission technique being used. netBaltic will provide mechanisms for economically viable, wideband network connectivity over sea areas, as an alternative to costly satellite-based solutions. Much of research is focused on solutions located in higher layers of TCP/IP architecture. The designed mechanisms create an architecture integrating different wireless broad-band technologies. The system utilizes a concept of dynamical division of its operational region into three areas, presented in Fig. 1, where different principles of network organization and different communication mechanisms are employed.

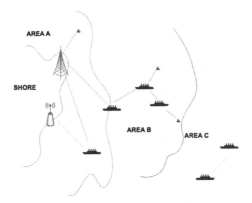

Fig. 1. Communication areas of the netBaltic system

The first one (area A), spread out along a coast, defines mobility management mechanisms for handling vessels' movements, to provide uninterrupted connections across a set of different wireless technologies. The one-hop communication in this area employs popular wireless technologies (such as WiMAX, LTE, Wi-Fi) or technologies dedicated to maritime usage (e.g. RADWIN Fiber-in-Motion). Advanced mobility management mechanisms are used to provide uninterrupted communication at the network layer with IPv6 protocol.

In the case of the second area (area B), nodes form a heterogeneous, self-organizing mesh network, capable of providing connectivity in both ship-to-ship and ship-to-shore scenarios. Additional information from the AIS system (Automatic Identification System) is utilized to provide an increased efficiency and reliability of the proposed solution, by giving the system an ability to predict ongoing changes in mesh network structure [4].

The last one (area C) is dedicated to nodes located far away from the rest of vessels, and as a result, being able to establish connections only very occasionally. The group of mechanisms needed in this area utilizes concepts of a dedicated, delay-tolerant network system [5].

Using the communications mechanisms described above, important data e.g. weather forecasts will be delivered to vessels. In practice Transmission Control Protocol (TCP) is used to provide reliable data segments dissemination in existing applications (e.g. NaviWeather developed by NavSim Technology [6]). In zones A and B we assume possibility of establishing TCP sessions. However due to different factors influencing propagation conditions at the sea (e.g., sea waving, motion of vessels, etc.) we have to accept the assumption that the link quality will be limited and the link synchronization can be broken quite often, thus causing communication gaps, seriously influencing TCP connection and its basic configuration parameters. In the paper we focus our attention on one of possible solutions of the above problem, quite essential, however for each TCP connection, realized under bad propagation conditions.

The main contributions of the paper are:

- Evaluation of connectivity gaps at the data-link layer of a wireless technique dedicated for maritime communications.
- Evaluation of TCP (common TCP Reno, F-RTO and TCP Cubic) data dissemination in a maritime environment.
- Proposal of TCP adjustment suitable for maritime communications and its comparable verification with above-mentioned solutions.

The rest of the paper is organized as follows. In Sect. 2 related works are discussed. Section 3 includes results of wireless link performance evaluations in real-world measurements (Subsect. 3.1) and details of our simulation environment used for evaluations of connectivity gaps impact on TCP connections efficiency (Subsect. 3.2). The proposition of TCP adjustments increasing transmission efficiency in a maritime environment is presented in Sect. 4. The paper is summarized with conclusions in Sect. 5.

2   Related Works

Maritime wireless communication systems represent a group of networks relatively close to the MANET solutions. However, the specifics of the maritime environment cause a number of differences, as compared to terrestrial networks. It is know that vessels experience two types of motions: displacement motions (including heaving, swaying or drifting and surging) and angular motions (yawing, pitching and rolling) due to fluctuation of sea waves. Horizontal and vertical movements, when considering wireless communications, influence connectivity between wireless terminals located on ships or base stations (see Fig. 2).

Due to changes of the angle between transmitting and receiving antennas, and as a result, changes in the signal power at the receiver, communication can be influenced by periodic connectivity gaps. This issue is more important for wide-band communication systems using higher frequencies and working with antennas offering narrowed beams. This problem is studied in [7,8]. In the simulations results shown there, variations of received power as great as 10 dB are observed for two dipole antennas (maximum gain of 2.1 dBi) located on the moving vessels.

Performance evaluations of wireless links implementing different communication standards (e.g. WiMAX, LTE, Wi-Fi) have been presented in [9,10]. These evaluations include values of signal level parameters (e.g. RSSI, SNR) and some data transmission parameters measurements (e.g. bandwidth) for stable connections. These works do not include verification of maritime communication efficiency in unstable environment. Such verification is presented in our paper.

The problem of TCP usage in a wireless system has been examined by the number of researchers during the last several years [11]. The results of this work bring us the group of solutions commonly named wireless TCP. Some of them [12–14] implement a mechanism in which connections are split at a base station where data buffering can be used. These solutions result in separated TCP flows and offer higher throughput for a sender, but they also break the end-to-end TCP architecture. Solutions in another group assume a tight integration between

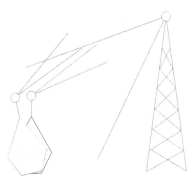

Fig. 2. Waving impacts on antenna characteristics

TCP mechanisms and the data-link layer mechanisms - they are so-called TCP-aware data-link layer mechanisms. In case of [15], mechanisms of the data-link layer are able to cache TCP segments, monitor ACK, drop duplicated ACK and retransmit data segment on duplicated ACK. In [16] authors propose a feedback generated by the data-link layer to the transport layer, where the decision classifying the reason of losses is made. If losses are result of a congestion, an appropriate congestion algorithm is used, otherwise it is assumed to be a result of poor wireless conditions. There are also solutions [17,18], in which data-link layer mechanisms (e.g. schedulers) are designed to prioritize TCP traffic [18] or allocate resources based on TCP flows parameters [17].

Most of these mechanisms are verified under conditions including losses and burst losses of data packets. However, the problem of connectivity gaps and its impact on the transmission efficiency is not investigated.

Aforementioned modifications of TCP protocol are, in general, designed for wireless networks. They address problems of packet losses and burst packet losses but are not suitable for periodic connectivity gaps being observed in maritime conditions. During a connectivity gap, there is no possibility to receive acknowledgments from a remote host. Thus, if Retransmission Timer (RTO) calculated for this connection, is shorter than a currently observed connectivity gap, the retransmission occurs. The number of retransmissions during the connectivity gap depends on the current value of RTO. The algorithm of TCP's Retransmission Timer computation is defined in RFC 6298 [19]. It uses two variables: sRTT (smoothed round-trip time, sometimes called short round-trip time) and RTTvar referring to round-trip time variation. Until the current value of RTT is calculated, RTO at the sender side is set to 1 second. With the first value of RTT measured, the host has to use the following equations to calculate its RTO:

$$sRTT = newRTT \tag{1}$$

$$RTTVAR = \frac{newRTT}{2} \tag{2}$$

$$RTO = sRTT + max(G, K * RTTvar) \tag{3}$$

where K = 4, and G refers to clock granularity in seconds.

Each subsequent measurement of new RTT value ($newRTT$) performed in accordance with Karn's algorithm (as it is stated in [19]) must result in RTO update in following manner:

$$RTTvar = (1 - \beta) * RTTvar + \beta * |sRTT - newRTT| \qquad (4)$$

$$sRTT = (1 - \alpha) * sRTT + \alpha * newRTT \qquad (5)$$

where $\alpha = 0.125$ and $\beta = 0.25$, following assumptions from [19]. Finally the current value of RTO should be calculated in accordance with (3).

As RTO values highly depend on RTT values, an incorrect RTT estimation due to e.g. periodic connectivity gaps leads to spurious RTO calculations. As a solution for this problem in [20,21], Forward RTO-Recovery (F-RTO) is proposed. It is a variant of Limited Transmit algorithm applied to RTO recovery. What is important, the mechanism relays on standard TCP header and does not require neither usage of TCP options fields nor additional bits in the header. In [21] authors investigate TCP communication efficiency under a few different scenarios resulting in RTO expiration: sudden delay in a network, lost retransmission, burst losses and packet reordering. The F-RTO configuration is available in nowadays Unix-based operating systems. The F-RTO algorithm has not been yet verified in periodically interrupted connectivity scenarios. Such evaluation is presented in this paper.

3   Problem of TCP Efficiency in Maritime Communications

3.1   Evaluation of Wireless Links in a Maritime Environment

In order to inspect the impact of vessels motions on performance of wireless communications some measurements in a real-world environment have been conducted in the netBaltic project. As a wireless technique the RADWIN Fiber-in-Motion system being a part of Laboratory of Mobile Wireless Technologies at Gdansk University of Technology [22] has been selected. RADWIN Fiber-in-Motion installations are successfully used in maritime communications systems over the world e.g. by cost-guards or search and rescue groups.

Due to the fact that small vessels are more sensitive to waving, in the measurement campaign a nine-meter yacht was used. The base station of our wireless system was located close to the Baltic Sea shore (on Sobieszewo Island - see Fig. 3) on the tower at the height of 45 m. Mobile terminals were mounted on the vessel's bow. The experiment purposes verification of two types of terminals used with directional (HMS1 - High capacity Mobile Subscriber unit 1) and omni-directional antennas (HMS2 - High capacity Mobile Subscriber unit 2). The details of the vessel node's architecture are depicted in Fig. 4. The onshore node is very similar, however there is no a GPS receiver connected to the main node and RADWIN terminals are replaced with a RADWIN base station.

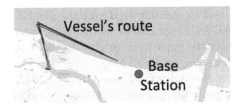

Fig. 3. Vessel's route during measurements

Dedicated measurement stations have been implemented to monitor signal parameters of wireless links (e.g. RSSI), as well as, connectivity at the network layer between terminals. For this purpose uplink and downlink UDP streams have been generated between nodes. All traffic on both terminals has been captured and stored in pcap files (Packet CAPture files, providing a high level interface to packet capture systems). Resulting datasets make it possible to analyze connection gap periods both at the wireless data-link layer and at the network layer. In Fig. 5, results of an example RSSI measurement for both antennas used in the campaign are presented. The values in range between -80 dBm and -60 dBm refer to acceptable connections conditions. The value of -120 dBm was reported by the terminal when the connection has been lost.

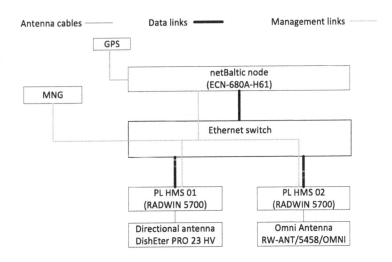

Fig. 4. Test-bed architecture of a vessel node

Collected data shows that regardless of the antenna setting used in tests, connection interruptions at the wireless link layer follow similar distributions. As RADWIN Fiber-in-Motion uses time slots assignments for end-users transmissions, its terminals have to be synchronized with RADWIN base station.

Similar solutions are used in standardized systems like LTE or WiMAX. Synchronization of the RADWIN terminal in the maritime environment took about 4.5–5 s. Shorter connectivity gaps were practically not observed. The average connectivity gap was 9.3 s long or 11.3 s long, respectively for HMS1 and HMS2. Whereas, the average connectivity time was 8.9 s for HMS1 and 3.4 s for HMS2.

Verification of the probability density function estimation is presented in Fig. 6. In the same plot, empirical cumulative distribution functions of connectivity periods and connectivity gaps for measured data (solid lines), as well as random samples (dashed lines) generated in accordance with estimated probability density functions using KDE method (see Fig. 6b) are presented.

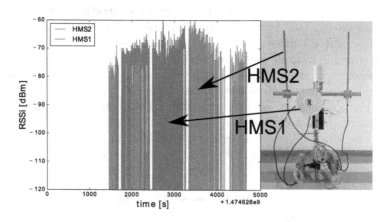

Fig. 5. Example results of RSSI measurement and client terminals used in experiments

Having probability density functions estimated based on data measured in real-world experiments, it was possible to evaluate and set up initial parameters for maritime communications in a simulation environment described further in this section.

3.2 Evaluation in the Simulation Environment

To verify the impact of periodic connectivity gaps in wireless links on TCP communication in a greater scale a dedicated simulation environment has been prepared. As this research is a part of the netBaltic project, implemented models are consistent with the netBaltic's node architecture. The netBaltic node has been modeled as a set of virtual machines (grey boxes in Fig. 7) with one referring to the main netBaltic node, and others acting as external CPEs (Customer-Premises Equipment). The test-bed with two netBaltic nodes is presented in Fig. 7. Each physical interface in the main node is connected to two modules proposed in the netBaltic node's architecture - Link Quality Evaluation Module (LQE module) and Mobility Management Module (MM module).

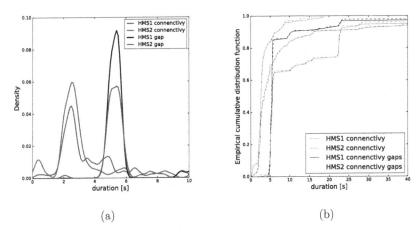

(a)   (b)

Fig. 6. Statistical analyzes of connectivity in the maritime environment: (a) Estimation of probability density function; (b) Empirical cumulative distribution function for measurement data (solid lines) and data generated based on estimated probability density function (dashed lines)

LQE module is responsible for collecting data from each wireless technique interface available in the node, and using it to comparably evaluate available connections resulting in calculation of a metric value, designated as Link Quality Indicator (LQI). Link parameters (LP) describing signals (e.g. RSSI, RSRP) allow to estimate an accurate data-transmission rate. They are strictly related to considered wireless connections, thus for each wireless technique, separate mechanisms of data gathering (e.g. SNMP, AT commends) must be used. Link parameters of the data-link layer - connectivity gaps periods, connectivity periods do not depend on particular wireless technique. They can be measured by means of link-up and link-down SNMP traps mechanisms. The results presented in the further part of the paper will be used by LQE module to obtain a metric describing a high-level throughput of TCP connections to estimate LQI metric and choose the best link for ongoing connections.

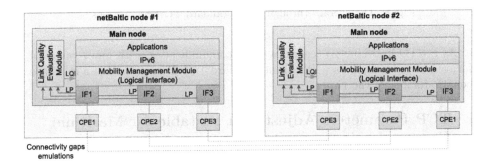

Fig. 7. Simulation test-bed architecture

Mobility Management module implements a dedicated scheme extending the Proxy Mobile IPv6 procedures [23]. In the proposed solution, all physical interfaces can be used to perform a data-link layer handover independently. As consequence, the interface providing better link quality (higher LQI value), can be used by the network layer in mobility management procedures to maintain the connectivity with corresponding nodes. To verify the impact of vessel's motions on TCP connections, and owing to the fact that time between topology changes due to vessels' movements in longer than duration of evaluated TCP connections, static topology with point-to-point connections (between a vessel and an on-shore base station) has been evaluated in described experiments. Thus, in this case mechanisms of Mobility Management module are not needed and were not used - the module is presented for the sake of completeness of the netBalitc node's architecture.

Temporary connection losses between two nodes were emulated by means of a dedicated script build on top of *ip* (it is a collection of utilities for controlling TCP/IP networking and traffic control in Linux) and *tc* (it is the user-space utility program used to configure the Linux kernel packet scheduler) tools, periodically blocking and unblocking links between connected CPEs. The values of link bandwidth between CPEs have been reduced to 4 Mbps at a hyper-visor level. In our experiments for a throughput evaluation the iperf tool was used. For each pair of up-time (a connectivity period) and down-time (a connectivity gap period), twenty connections lasting 60 s each were conducted. Based on these results the average throughput presented in Fig. 8 for up-time and down time in the range of 6 s to 15 s has been calculated.

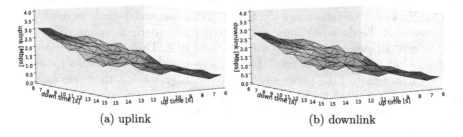

(a) uplink   (b) downlink

Fig. 8. The average throughput of standard TCP connections

As F-RTO in evaluated scenarios offers communication with lower values of the average throughput (see Fig. 9) the standard TCP Reno is used as a reference scenario further in the paper.

4   TCP Parameters Adjustment Suitable for Maritime Communications

A detail inspection of periodic connectivity gaps impact on TCP performance allows to conclude that with standard values of TCP parameters the

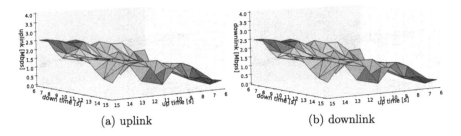

(a) uplink   (b) downlink

Fig. 9. The average throughput of TCP connections with F-RTO

communication efficiency is very poor. Due to this, it was decided to modify the standard TCP implementation and adjust some TCP parameters to reduce time between subsequent re-transmissions. In order to do that, the dedicated patch for Linux kernel has been adopted to the current 4.9 CentOS kernel [24]. It gives ability to change the default value of TCP_RTO_MAX variable. TCP_RTO_MAX variable is the upper boundary of RTO values calculated in accordance with the procedure described in Sect. 2. The default value of this parameter is 120 s.

As a consequence of a shorter time between retries of subsequent segments retransmissions the maximum number of allowed retransmissions has to be increased. This parameter can be adjusted by means of TCP_RETRIES2 variable in Linux kernels and its default value is 15. It might be simply estimated that time needed to recognized that the current connection is broken is no longer than TCP_RETRIES2 * TCP_RTO_MAX (for default values, it is 1800 s).

To increase the efficiency of TCP communication in a maritime environment, TCP parameters have been adjusted. In the simulation experiments, the value of TCP_RTO_MAX has been changed and set up to 1 s. Authors are aware that such proposition breaks the rules defined in [19]. Although, as it is discussed further, this solution does not break ability to communicate with hosts implementing the standard TCP configuration. It is evident that with such configuration the ability to react properly on network congestion is limited, however with wireless systems using a strict resource allocation mechanisms (like RADWIN Fiber-in-Motion) and especially, in area C of the netBaltic system, proposed solution will be beneficial. To keep constant time needed to recognize that the current connection is broken, the value of TCP_RETRIES2 was changed to 1800. Tools used in evaluation procedures do not support a TCP connection reestablishing during a communication session, thus an appropriate setting of the above mentioned parameters is important. In consequence, presented results include values describing parameters of once established connections (in practice TCP connection timeouts were not observed).

The results referring to the average throughput for adjusted TCP connections are depicted in Fig. 10. It can be seen that the proposed adjustment brings significant increases in TCP connections throughput, especially for scenarios with relatively low values of up-time and high values of down-time. For up-time equal to 6 s and down-time about 9 s, the average throughput offered by the

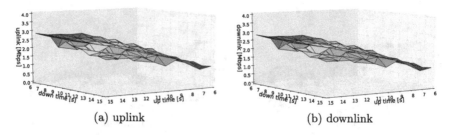

(a) uplink   (b) downlink

Fig. 10. The average throughput of adjusted TCP connections

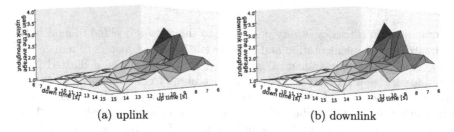

(a) uplink   (b) downlink

Fig. 11. Gain of the average throughput of the proposed solution over standard TCP connections

proposed adjustment is more than 3 times greater than the average throughput of standard TCP connections (with standard parameters) (see Fig. 11).

Reno is one of the first (and still popular) congestion control algorithms proposed for TCP. As such, it can be treated as a reference solution. However, in order to evaluate more recent mechanisms, additionally the Cubic algorithm was selected for comparison. The results of the average uplink throughput gain for several up/down time interval pairs have been presented in Table 1. In a maritime evaluation the throughput degradation is mainly a result of connectivity gaps thus the obtained gains are comparable for both evaluated congestion algorithms. However, the impact of connectivity gaps on a congestion window's size for different congestion algorithms is a topic of further research.

To verify the increase of TCP connections efficiency in a practical use-case scenario, the second set of tests has been conducted. In this case, times required to download files with sizes of 2 MB and 10 MB containing updates for e-navigation maps were measured. A file exchange has been performed by means of scp command (which copies files between hosts on a network via a secured SSH connection). Durations of connectivity gaps and connectivity periods have been generated randomly with the distribution estimated by means of KDE method based on real-world measurements (see Fig. 6). The results of these tests are presented in Table 2. The values in the table are the average values with 95% confidence intervals calculated based on 40 different connections.

As it can be seen, the proposed TCP enhancement results in a reduction of the average download time. The average value is reduced up to 53% for small files

Table 1. Cubic evaluation - gain of the average uplink throughput

Up/Down time [s]	TCP cubic	adj. TCP cubic	Gain
10/15	1.39	2.24	1.61
6/9	0.62	1.66	2.67
6/6	1.18	2.06	1.75
6/15	0.76	1.23	1.63
15/6	3.04	3.37	1.11
15/15	1.74	2.32	1.33

Table 2. Download times for evaluated scenarios

Scenario	TCP	adj. TCP	adj. TCP - TCP	TCP - adj. TCP
10 MB file	62.79 ± 25.75 s	44.83 ± 6.84 s	46.74 ± 9.98 s	48.58 ± 9.35 s
2 MB file	19.47 ± 7.63 s	9.03 ± 2.59 s	10.89 ± 4.17 s	10.82 ± 2.97 s

TCP column – standard sender and receiver.
adj. TCP column – both sender and receiver adjusted.
adj. TCP – TCP column – adjusted sender and standard receiver.
TCP– adj. TCP column – standard sender and adjusted receiver.

(2 MB) and up to 28% for files with size of 10 MB (as compared to the standards TCP Reno). It is also visible that there is no requirement for implementation of the proposed solution in both communicating nodes. The possibility of a cooperation between the standard and the described in this paper TCP configuration is not broken.

5   Conclusions

In the paper selected problems of TCP communication efficiency in a maritime environment are discussed. The authors present evaluations of connectivity gaps impact on data dissemination in communication between on-shore base stations and vessels. Based on real measurements they built a simulation environment - using results of these measurements to set up number of initial parameters. The simulation test-bed was in turn used for examining the usefulness of the standard TCP Reno protocol, together with its F-RTO enhancement and more recent TCP Cubic in maritime conditions. Both of them turned out to be non-acceptable in a context of maritime environment and e-navigation services. In order to improve the TCP performance under specific maritime conditions the authors proposed adjustments of TCP protocol configuration parameters leading to significant increase of the average throughput for TCP connections (more than 3 times), and to reduction of the average time needed to download a file with an e-navigation maps update up to 53% in comparison with the standard Reno TCP version. In the future work, the authors will propose a solution adopting a current TCP_RTO_MAX value to observed durations of connectivity gaps.

Acknowledgement. This work has been partially supported by the Applied Research Program under the Grant: ID PBS3/A3/20/2015, founded by the National Center for Research and Development.

The infrastructure was supported by "PL-LAB2020" project, contract POIG.02.03.01-00-104/13-00 and "Future Internet Engineering" project, contract POIG.01.01.02-00-045/09-00.

References

1. International Maritime Organization, e-Navigation Information - IMO (2017). http://www.imo.org/en/OurWork/safety/navigation/pages/enavigation.aspx
2. Hoeft, M., Gierłowski, K., Wozniak, J.: Heterogeneous wireless communications system over the baltic sea. Telecommun. Rev. + Telecommun. News **8–9**, 1196–1200 (2016)
3. Hoeft, M., Gierłowski, K., Nowicki, K., et al.: netBaltic: enabling non-satellite wireless communications over the baltic sea. IEEE Commun. Mag. (2016)
4. Gierłowski, K.: Characteristics of self-organizing, multihop wireless network for netbaltic system's zone b. Telecommun. Rev. + Telecommun. News **12**, 1316–1321 (2016)
5. Guminski, W.: The concept of providing access to services in network of discontinuos and sporadic communication. Telecommun. Rev. + Telecommun. News **12**, 1339–1342 (2016)
6. NavSim Poland, NaviWeather (2017). http://www.navsim.pl/for_professionals_and_pilots/naviweather
7. Huang, F., Bai, Y., Du, W.: Maritime radio propagation with the effects of ship motions. J. Commun. **10**, 345–351 (2015)
8. Hubert, W., Roux, Y.-M.L., Ney, M., Flamand, A.: Impact of ship motions on maritime radio links. Int. J. Antennas Propag. **2012**, 1–6 (2012)
9. Bronk, K., Lipka, A., Niski, R., et al.: Hybrid communication network for the purpose of maritime application. Int. J. Maritime Eng. **A2** (2016). (in print)
10. Bronk, K., Lipka, A., Niski, R., et al.: Weryfikacja pomiarowa zasięw systemów komórkowych uzyskiwanych w warunkach morskich. Telecommun. Rev. + Telecommun. News **6**, 463–466 (2016)
11. Liu, K., Lee, J.Y.B.: On improving tcp performance over mobile data networks. IEEE Trans. Mob. Comput. **15**(10), 2522–2536 (2016)
12. Jain, R., Ott, T.J.: Design and implementation of split TCP in the linux kernel. In: IEEE Globecom 2006, pp. 1–6 (2006)
13. Kim, B.H., Calin, D., Lee, I.: Advanced split-TCP with end-to-end protocol semantics over wireless networks. In: 2016 IEEE GLOBECOM, pp. 1–7, December 2016
14. Pathak, A., Wang, Y.A., Huang, C., Greenberg, A., Hu, Y.C., Kern, R., Li, J., Ross, K.W.: Measuring and evaluating TCP splitting for cloud services. In: Krishnamurthy, A., Plattner, B. (eds.) PAM 2010. LNCS, vol. 6032, pp. 41–50. Springer, Heidelberg (2010). doi:10.1007/978-3-642-12334-4_5
15. Li, Y., Jacob, L.: Proactive-WTCP: an end-to-end mechanism to improve TCP performance over wireless links. In: Proceedings of the 28th Annual IEEE International Conference on Local Computer Networks, LCN 2003, pp. 449–457, October 2003
16. Molia, H.K., Agrawal, R.: A comprehensive study of cross - layer approaches for improving TCP performance in wireless networks. In: 2015 ICCCT, pp. 362–367, February 2015

17. Rath, H.K., Karandikar, A.: On TCP-aware uplink scheduling in IEEE 802.16 networks. In: 3rd International Conference on Communication System Software and Middleware and Workshops, COMSWARE 2008, pp. 349–355, January 2008

18. Shojaedin, N., Ghaderi, M., Sridharan, A.: TCP-aware scheduling in LTE networks. In: Proceedings of IEEE International Symposium on a World of Wireless, Mobile and Multimedia Networks 2014, pp. 1–9, June 2014

19. Paxson, V., Allman, M., Chu, J., Sargent, M.: Computing TCP's retransmission timer. RFC 6298, RFC Editor, June 2011. http://www.rfc-editor.org/rfc/rfc6298.txt

20. Sarolahti, P., Kojo, M., et al.: Forward RTO-recovery (F-RTO): an algorithm for detecting spurious retransmission timeouts with TCP. RFC 5682, RFC Editor, September 2009

21. Sarolahti, P., Kojo, M., et al.: F-RTO: an enhanced recovery algorithm for TCP retransmission timeouts. SIGCOMM Comput. Commun. Rev. **33**(2), 51–63 (2003)

22. Gierłowski, K., Hoeft, M., Gierszewski, T., et al.: Laboratory of mobile wireless technologies. Telecommun. Rev. + Telecommun. News **12**, 1422–1431 (2015)

23. Hoeft, M., Kaminski, P., Wozniak, J.: Logical interface for soft handover - an effective scheme of handovers in proxy mobile IPv6. In: 2015 8th IFIP WMNC, pp. 72–79, October 2015

24. Noboru, O.: Make TCP-RTO-MAX a variable (2007). http://lists.openwall.net/netdev/2007/06/25/34

Path Loss Model for a Wireless Sensor Network in Different Weather Conditions

Dariusz Czerwinski[1(✉)], Slawomir Przylucki[1], Piotr Wojcicki[1], and Jaroslaw Sitkiewicz[2]

[1] Lublin University of Technology, 38A Nadbystrzycka Str, 20-618 Lublin, Poland
{d.czerwinski,s.przylucki,p.wojcicki}@pollub.pl
[2] SuperDrob S.A., 05-480 Karczew, Poland
j.b.sitkiewicz@gmail.com

Abstract. The paper presents path loss model for a WSN in an open space on the basis of measurements. The measurements were performed in different weather conditions i.e. temperature and humidity in agricultural area. Theoretical path loss models are always flawed because of their exponential relationship. This causes an increase in error during the localization process. Thus, the determination of the path loss exponent constant based on measurement results become so important. The study focused on designation of the path loss model for different weather conditions. The influence of the temperature and the humidity on the RSSI distribution was analysed.

Keywords: Path loss model · RSSI · ZigBee · Weather conditions

1 Introduction

Wireless sensor networks (WSNs) are used in a wide range of real-life systems. Widely spread sensors in WSNs can cover both indoor and outdoor environments [1,2]. The design and implementation of WSN require same form of an assessment of sensors position or the relative distances among the network nodes [3]. In most solutions the path loss models have been used to predict the attenuation and distortion of the radio frequency [4]. Localisation and deployment of the nodes are becoming even more complex in the case of different area types (i.e. urban, forest, agriculture area) and environment conditions (rain, snow etc.). One of the promising approaches in solution of this issue is the use of the Received Signal Strength (RSS). Recently the wireless sensor nodes communication modules have implemented built-in feature able to provide an estimation of RSS by means of the dependable RSS Indicator (RSSI). Nowadays it can be observed, that many test and commercial solutions are based on the processing of RSSI indicator values [5–7].

A path loss versatile theoretical model for predicting the feasible node connectivity for WSNs was developed in [8]. The authors verified the model by comparing the results with the measurements carried out by other researchers

© Springer International Publishing AG 2017
P. Gaj et al. (Eds.): CN 2017, CCIS 718, pp. 106–117, 2017.
DOI: 10.1007/978-3-319-59767-6_9

in outdoor open areas. In this model authors considered among others the effects of frequency of operation and terrain electrical and geometrical properties on the connectivity.

Signal propagation between wireless sensor nodes deployed in a concrete surface environments and adequate path loss model was presented in [4]. This has been accomplished by in-field measurements of radio waves propagation and on the basis of the results an empirical path loss model was elaborated. The authors also made the measurements for long grass and sparse tree environments. The path loss exponent was equal to 2.55 for long grass, 3.21 for concrete surface and 3.34 for sparse tree.

Similar approach of developing an empirical path loss models for WSN in grassy environments was described in [9]. The authors use XBee Digi modules for measurements. Proposed by the authors models were compared with the theoretical one to demonstrate their inaccuracy. The results of that research show values obtained form theoretical models differs from measurements by 12%–42%.

The methods of measuring path loss exponent in various environments have been studied in [10]. The focus has been put on the allocation and arrangement of transmitters and receivers. Based on the experimental results the authors propose properly arrange WSN nodes which provides better estimation of the path loss exponent.

Our contribution, presented in the article consists of the analysis of the path loss exponent value in path loss model based on the RSSI distribution measurements made in different weather conditions. On the basis of measurement results it can be seen that humidity has a big impact on the path loss exponent value. The remainder of the paper is organised as follows: Sect. 2 describes the hardware configuration of the WSN measurement test stand and scenarios. Section 3 describes the experiments and models elaboration. Conclusions and some further research ideas are presented in Sect. 4.

2   WSN Measurement Test Stand

The proposed system for measurements of RSSI values between WSN ZigBee nodes consists of the following main elements:

- radio XBee XB24-Z7WIT-004 modules from Digi, which are XBee 2mW Series 2 wire antenna modules [12],
- programming platform, the Arduino Uno board [13] used for data collection, acts as node together with XBee module,
- temperature and humidity sensor DHT22, which measures the temperature in range -40–$80\,°C$ and humidity in range 0–100 % [11]
- SD card module and LCD display.

The Arduino Uno board is placed inside the plastic box, as well as other elements due to the weather protection. The XBee module is fitted on the top cover (Fig. 1). At the back wall of box the temperature sensor and power supply (9V DC battery) are fitted (Fig. 2).

Fig. 1. The architecture of the WSN node

Fig. 2. Completely assembled WSN nodes: router and coordinator

The programming of the Arduino UNO board is performed in open-source development IDE. Arduino boards uses a simplified C/C++ language for developing the software necessary for storage and measurement purposes [14].

2.1   Hardware Configuration of Wireless Nodes

In order to establish the wireless connection, two Digi XBee series 2 modules were used [12]. In both radios the firmware was set to the XB24-ZB family, which allow to work with ZigBee protocol stack. This wireless modules operate in the 2.4 GHz band at 250 kbps baud rate. The summary of the XBee modules parameters is shown in the Table 1.

The personal area network ID in all modules was the same, therefore these settings allow to form the mesh wireless network. The XBee modules were connected with adequate pins of Arduino UNO boards, i.e.: supply 3.3 V, GND, XBee DOUT (TX) to pin 2 (Arduino's Software RX), XBee's DIN (RX) to pin 3 (Arduino's Software TX). Arduino platform with ATmega 328 AVR was used as a main element for programming, as also as sending and collecting data. It also allows to obtain the RSSI values from XBee modules thanks to the library for

Table 1. Summary of XBee modules parameters

Model	XB24-Z7WIT-004
Standard	802.15.4
Protocol	ZigBee
Transmission speed	250 kbps
Inner range	up to 40 m
Outer range	up to 120 m
Frequency	2.4 GHz
Receiver sensitivity	92 dBm
Communication interface	UART
Configuration methods	API, AT
Output power	2 mW
Antenna	Omnidirectional wire antenna
Module operating temperature	−40 °C to +85 °C
Power consumption	40 mA
Supplying voltage	3.3 V

communicating with XBee radios in API mode. For the measurement purposes each WSN node was programmed by dedicated software with the functionalities:

- sending packets and receiving acknowledgements,
- delays measurement,
- measurement of packet loss,
- RSSI measurement,
- measurement of temperature and humidity,
- display the measured values,
- record measurements on the SD card in a txt file.

2.2 Measurement Scenarios

The experiment relied on checking the influence of the weather conditions on the attenuation between two nodes of wireless sensor network. Measurements were conducted in the LOS conditions in the open agricultural area. For collecting the RSSI freqency distribution the devices were placed at a distance of 25 m and a height of 47 cm.

The first measurement was carried out in winter, in January 2016, at below zero temperatures (Fig. 3). During the test the following conditions prevailed:

- average temperature: −5.8 °C,
- average humidity: 70.4%.

The next measurement was made during the high humidity at the end of winter. Prevailing conditions:

Fig. 3. Measurements at below zero temperatures

Fig. 4. Measurements during summer

– average temperature 1.5 °C,
– average humidity: 86.3%.

The third measurement was made during the summer time in July 2016 (Fig. 4). At that time weather conditions were very good:

– average temperature 29.3 °C,
– average humidity: 34.1%.

The last fourth measurement was made during the winter in January 2017. At that time the humidity had the lowest value:

– average temperature −7.7 °C,
– average humidity: 28.5%.

The analysis of the measurements was performed based on 100 samples for each scenario during the LOS tests. There were no packet's losses all through measurements. The average values of selected parameters are summarized in the Table 2.

Table 2. Results of measurements, average values

Scenario	Temperature °C	Humidity %	RSSI dBm	Delay ms	Packet loss %
1	−5.8	70.4	−83.8	33.3	0
2	1.5	84.2	−84.2	29.4	0
3	29,3	34.1	−76.4	32.1	0
4	−7.7	28.5	−66.6	31.3	0

3   Path Loss Model in Different Weather Conditions

A common methodology for the study of radio wave propagation and path loss modelling is an empirical research. The presented research follows this methodology. Measurements described in above scenarios were conducted to investigate outdoor path loss characteristics. Based on the results, the authors propose the model of path loss in different weather conditions for WSN ZigBee nodes. Firstly the variations of the RSSI values were studied, due to the fact that they can fluctuate even in steady state conditions.

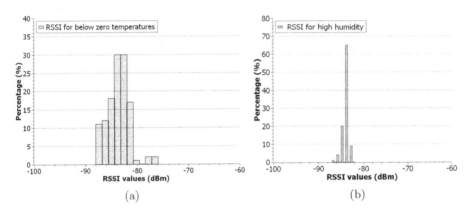

(a)   (b)

Fig. 5. Frequency distribution of RSSI: (a) below zero temperature (scenario 1), (b) high humidity (scenario 2)

The histograms of frequency distribution, for tests performed in winter season, are shown in Fig. 5. Standard deviation for results collected in scenario 1 shown in Fig. 5a was equal to 2.11 dBm and minimum and maximum values of RSSI were equal to −88 dBm and −76 dBm respectively. For that scenario the mean value of RSSI was equal to −83.8 dBm. For results collected in scenario 2 the histogram is shown in Fig. 5b. Standard deviation was equal to 0.74 dBm and minimum and maximum values of RSSI were equal to −87 dBm and −72 dBm respectively. The mean value was equal to −84.2 dBm.

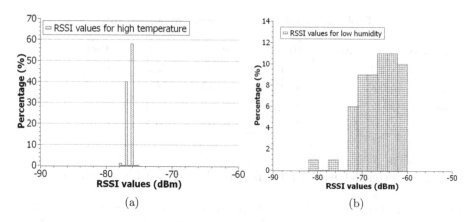

Fig. 6. Frequency distribution of RSSI: (a) high temperature in summer time (scenario 3), (b) lowest humidity in winter (scenario 4)

For low humidity and high temperature (scenario 3) the histogram of frequency distribution is shown in Fig. 6a. Standard deviation for results collected in this scenario was equal to 0.53 dBm and minimum and maximum values of RSSI were equal to -78 dBm and -75 dBm respectively, while the mean value was equal to -76.41 dBm. The lowest value of humidity was in scenario 4 (Fig. 6b). Mean value of RSSI was equal to -66.56 dBm, while the minimum and maximum values were equal to -82 dBm and -60 dBm respectively. Standard deviation collected in this measurements was equal to 4.33 dBm. Taking into account the RSSI frequency distribution shown in Figs. 5 and 6 the normal distribution of RSSI values were assumed.

In a free space the power of received signal is determined by the Friis's law (1) and depends on the distance [15].

$$P_r(d) = P_t \cdot G_t \cdot G_r \left(\frac{\lambda}{4\pi \cdot d} \right)^2 \tag{1}$$

where: $P_r(d)$ is the signal power, P_t is the transmitted signal power, G_t is the gain of the transmitter antenna, G_r is the gain of the receiver antenna, λ is the wave length, d is the distance.

In various environments, the signal strength is attenuated in different ways. In such cases the most commonly used outdoor path loss model is the classical log-distance model (2). In this model the attenuation factor, called a path loss exponent n is used. The attenuation factor is environmental depended variable in contrast to the square in Eq. (1) [4,9,16].

$$P_r(d) = P_r(d_0) - 10 \cdot n \cdot log_{10} \left(\frac{d}{d_0} \right) \tag{2}$$

where: $P_r(d_0)$ is the received signal power at the reference distance d_0, n is the path loss exponent. If there are obstacles on the line of sight or the signal is going by multiple paths then expression (2) becomes (3) [9,16]:

$$P_r(d) = P_r(d_0) - 10 \cdot n \cdot log_{10}\left(\frac{d}{d_0}\right) + X_\gamma \tag{3}$$

where: X_γ is the normal (or Gaussian) random variable which represents the effect of shadowing phenomenon.

In further analysis the authors took into account the model corresponding to the open free space in the conditions of LOS (Eq. (2)). Basing on the measurement results from scenario 3 the mean of the RSSI values and standard deviation at each position were calculated and presented in Fig. 7.

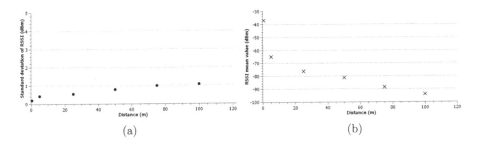

(a)   (b)

Fig. 7. Results of scenario 3: (a) standard deviation, (b) mean values of RSSI

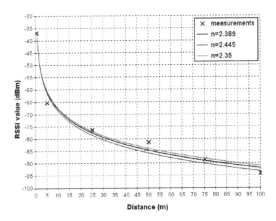

Fig. 8. Calculated RSSI versus distance for different values of path loss exponent n, measurement results from scenario 3

The variations of the signal values are not high, what is shown in Fig. 7a. It can be noticed that variations increase with the distance, however the maximum value is still low and equal to 1.06 dBm. The mean values of the RSSI versus the distance showed, that the signal decreases according to the logarithmic function (Fig. 7b). The next step was to determine path loss exponent based on the reference value of the signal obtained at the distance $d_0 = 1$ m and other measurements (Fig. 8). The value of n was determined in the three different ways. The first way was the calculation of exponent with the use of Scilab computation package and approximation of measurement points. Value obtained in that way was equal to $n = 2.389$. The second approach was the calculation of the mean value of n for distance range 5–100 m. The value thus obtained was equal to $n = 2.445$ due to the higher value for 5 m ($n = 2.82$). The last step was the calculation of the mean value of exponent for the distance range 25–100 m. The obtained path loss exponent was equal to $n = 2.35$.

The results of calculations were presented in Fig. 8. It might seem that the differences between n values are small, however if we consider the localization purposes using RSSI parameter, than it can be noticed that for chosen RSSI value differences in distance determining can vary from few meters to tens of meters. For the scenario 4 the path loss model (log normal approximation method) and measurement results were presented in Fig. 9. It can be noticed that path loss model follows the log-distance relationship however the exponent value is smaller and much closer to that one which is observed in closed area LOS conditions [17]. This can be explained due to the specific weather conditions. During the measurements, for several days prevailed freezing temperatures (below −6), and there was no snow. This had an impact on the humidity, which at that time was low. In addition, the snow that fell earlier became powder snow, which increased its reflectivity. To compare different path loss models additionally the two-ray ground reflection model was build. The model takes into account the impact of the reflected from the ground radio wave. According to the [18–21] path loss in this model can be defined as expression (4). In this model the: δ_d is additional reflected path Eq. (5), d is the distance between transmitter and receiver, h_t, h_r are heights on which the transmitter and receiver are mounted,

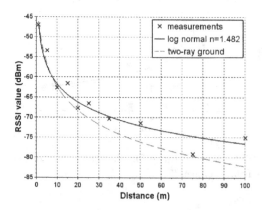

Fig. 9. RSSI measurement results from scenario 4 and calculated path loss models versus distance

Γ is Fresnel reflection coefficient for the vertical polarized signal Eq. (6), θ is the angle between the ground and reflected path, ε_r is ground relative permittivity and λ is wave length.

$$P_d(d) = P_r(d_0) - 10 \cdot log_{10} \left| \frac{1}{d} + \Gamma \cdot \frac{e^{j2\pi \frac{\delta_d}{\lambda}}}{d + \delta_d} \right| \tag{4}$$

$$\delta_d = \sqrt{(h_t + h_r)^2 + d^2} - \sqrt{(h_t - h_r)^2 + d^2} \tag{5}$$

$$\Gamma = \frac{sin(\theta) - \sqrt{\varepsilon_r - cos(\theta)^2}}{sin(\theta) + \sqrt{\varepsilon_r - cos(\theta)^2}} \tag{6}$$

The results of both models log normal and two-ray ground reflection were compared and shown in Fig. 9. It can be noticed that at the beginning distance (up to 10 m) both models behave almost the same. After this signal in two-ray ground model fades faster than in the log normal one. Taking into consideration measurement results in open space and LOS conditions, it looks like the log normal path loss model is better fitted in that case.

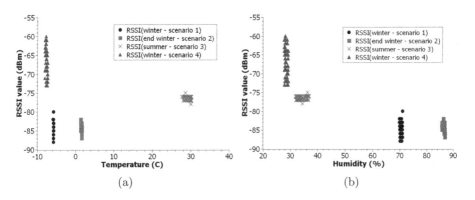

Fig. 10. RSSI distribution at the distance of 25 m in scenarios 1–4: (a) temperature dependence, (b) humidity dependence

To analyse the influence of the temperature and humidity on the distribution of RSSI values was worked out and shown in Fig. 10. The results were presented for measurements, for which the distance between devices was equal to 25 m and different weather conditions described in scenarios 1–4. It can be noticed that for increasing temperature (Fig. 10a) the values of RSSI have smaller dispersion (scenario 3 versus others). More clearly one can see the effect of humidity on the distribution of RSSI (Fig. 10b). When humidity decreases the signal propagation is better and RSSI value increases - scenario 3 and 4 versus scenario 1 and 2.

4   Conclusions and Future Work

The article presents the path loss model for different weather conditions, which was elaborated on the basis of measurement results. The measurements were made using Digi XBee S2 radio modules and RSSI values were recorded in a different weather conditions, described in scenarios 1–4, the same agricultural area, and different distance between devices. The compact design of the WSN nodes as well as recording capabilities were achieved with the use of Arduino programming platform.

In all experiment scenarios the path loss model follow the log-distance rule. It looks that in described measurement conditions log-normal model is better fitted comparing to the two-ray ground reflection one. Moreover it can be noticed that path loss exponent value is changing according to the weather conditions. It looks like the impact of the temperature is not such big as humidity, because the values of RSSI can significantly vary even in similar temperatures (scenario 1, 2, 4 in Fig. 10a). When the humidity increases values of RSSI decreases, which has a direct impact on the path loss exponent. With the growth of humidity the n factor also increases (from 1.48 in scenario 4 to 4.6 in scenario 2).

In the future work the authors plan to extend measurements for non line of sight (NLOS) propagation in open agricultural area during different weather conditions. Proper path loss model is very important for localization purposes and it is used in such application as tracking or ranging.

References

1. Gaj, P., Jasperneite, J., Felser, M.: Computer communication within industrial distributed environment-a survey. IEEE Trans. Industr. Inf. **9**, 182–189 (2013)
2. Kouril, J.: Using RSSI parameter in tracking methods – practical test. In: 2011 34th International Conference on Telecommunications and Signal Processing (TSP), pp. 248–251, 18–20 August 2011
3. Han, G., Xu, H., Duong, T.Q., Jiang, J., Hara, T.: Localization algorithms of Wireless Sensor Networks: a survey. Telecommun. Syst. **52**(4), 2419–2436 (2013)
4. Alsayyari, A., Kostanic, I., Otero, C.E.: An empirical path loss model for wireless sensor network deployment in a concrete surface environment. In: 2015 IEEE 16th Annual Wireless and Microwave Technology Conference (WAMICON), Cocoa Beach, pp. 1–6 (2015)
5. Mistry, H.P., Mistry, N.H.: RSSI based localization scheme in wireless sensor networks: a survey. In: Fifth International Conference on Advanced Computing and Communication Technologies, pp. 647–652, 21–22 February 2015
6. Kwiecień, A., Maćkowski, M., Kojder, M., Manczyk, M.: Reliability of bluetooth smart technology for indoor localization system. In: Gaj, P., Kwiecień, A., Stera, P. (eds.) CN 2015. CCIS, vol. 522, pp. 444–454. Springer, Cham (2015). doi:10.1007/978-3-319-19419-6_42
7. Czerwinski, D., Przylucki, S., Mukharsky, D.: RSSI-based localisation of the radio waves source by mobile agents. In: Gaj, P., Kwiecień, A., Stera, P. (eds.) CN 2016. CCIS, vol. 608, pp. 370–383. Springer, Cham (2016). doi:10.1007/978-3-319-39207-3_32

8. Torabi, A., Zekavat, S.A.: A rigorous model for predicting the path loss in near-ground wireless sensor networks. In: 2015 IEEE 82nd Vehicular Technology Conference (VTC 2015-Fall), Boston, MA, pp. 1–5 (2015)

9. Olasupo, T.O., Otero, C.E., Olasupo, K.O., Kostanic, I.: Empirical path loss models for wireless sensor network deployments in short and tall natural grass environments. IEEE Trans. Antennas Propag. **64**(9), 4012–4021 (2016)

10. Chuan Chin, P., Lim, S.Y., Ooi, P.C.: Measurement arrangement for the estimation of path loss exponent in wireless sensor network. In: 2012 7th International Conference on Computing and Convergence Technology (ICCCT), Seoul, pp. 807–812 (2012)

11. Aosong Electronics Co., Ltd., Digital-output relative humidity and temperature sensor/module DHT22, December 2016. https://www.sparkfun.com/datasheets/Sensors/Temperature/DHT22.pdf

12. Digi International, XBee and XBee-PRO ZB Modules - Datasheet (2011). https://cdn.sparkfun.com/datasheets/Wireless/Zigbee/ds_xbeezbmodules.pdf

13. Arduino, Arduino Uno board, September 2014. https://www.arduino.cc/en/Main/ArduinoBoardUno

14. Arduino, Arduino Software (IDE), September 2014. https://www.arduino.cc/en/Guide/Environment

15. Rappaport, S.T.: Wireless Commnications: Principles and Practice, 2nd edn. Prentice Hall, Upper Saddle River (2002)

16. Alsayyari, A., Kostanic, I., Otero, C., Almeer, M., Rukieh, K.: An empirical path loss model for wireless sensor network deployment in a sand terrain environment. In: Proceedings of the IEEE World Forum Internet Things (WF-IoT), pp. 218–223 (2014)

17. Zaarour, N., Kandil, N., Affes, S., Hakem, N.: Path loss exponent estimation using connectivity information in wireless sensor network. In: 2016 IEEE International Symposium on Antennas and Propagation (APSURSI), Fajardo, pp. 2069–2070 (2016)

18. Ahmed, N., Kanhere, S.S., Jha, S.: Link characterization for aerial wireless sensor networks. In: 2011 IEEE GLOBECOM Workshops (GC Workshops), pp. 1274–1279 (2011)

19. Thompson, R., Cetin, E., Dempster, A.: Unknown source localization using RSS in open areas in the presence of ground reflections. In: 2012 IEEE/ION Position Location and Navigation Symposium (PLANS), pp. 1018–1027 (2012)

20. Malajner, M., Benkic, K., Planinsic, P.: A new study regarding the comparison of calculated and measured RSSI values under different experimental conditions. Przeglad Elektrotechniczny **89**, 214–219 (2013)

21. Piyare, R., Lee, S.: Performance analysis of XBee ZB module based wireless sensor networks. Int. J. Sci. Eng. Res. **4**, 1615–1621 (2013)

Behavioral Analysis of Bot Activity in Infected Systems Using Honeypots

Matej Zuzcak[✉] and Tomas Sochor

University of Ostrava, Ostrava, Czech Republic
matej.zuzcak@osu.cz
http://www1.osu.cz/home/sochor/en/

Abstract. New Internet threats emerge on daily basis and honeypots have become widely used for capturing them in order to investigate their activities. The paper focuses on a detailed analysis of the behavior of various attacks agains 7 Linux–based honeypots. The attacks were analyzed according to the threat type, session duration, AS, country and RIR of the attack origin. Clusters of similar objects were formed accordingly and certain typical attack patterns for potential detection automation as well as some aspects of threat dissemination were identified.

Keywords: Honeypot medium–interaction · Internet threat · Malware · Trojan · AS · RIR · Country of origin · Session duration · Threat detection · Threat dissemination · IPv4

1 Motivation of the Study and Related Works

Recent publications on honeypots [1,2] describe a detailed procedure of attacks against systems with emulated Linux shell. Existing research included mainly used passwords and login names and command executed by attackers. Only few older works applied clustering to data obtained from honeypots. While [3] presents rather outdated conclusions, the newer [4] focuses on finding similarities in sandbox–executed malware. Clustering analysis is applied here, however, it is limited only to malware outputs. Other aspects of honeypot applications have also been studied recently from the point of legal issues [5] or malware behavior monitoring [6] as well as security techniques related to honeypots [7–9]. However, proper up-to-date analysis of attacker behavior is still missing. Therefore, this paper focuses to finding similar characteristics of countries and/or autonomous systems where individual attacks came from. This is different from [10] where similar approach is applied to finding similarities in numbers of connections, succesful logins etc. while specific threats as categorial data similarities are analyzed here. This paper broadens our perspective by focusing to the analysis of attackers; specific activities or behavior. The main emphasis was given to attacked systems seemingly being real but, in fact, were honeypot–emulated. The paper also focuses on the similarity of countries, regional internet registries (RIR) and autonomous systems (ASes) according to the incidence of specific threats as

© Springer International Publishing AG 2017
P. Gaj et al. (Eds.): CN 2017, CCIS 718, pp. 118–133, 2017.
DOI: 10.1007/978-3-319-59767-6_10

detected during the research period. The primary goal of the study was to learn details about behavior of attackers in infected systems. Another goal was to find what threats are disseminated in specific geographic territories and/or ASes. In fact, this study broadens the perspective of [10] with primarily numeric analysis to the real behavioral analysis. The study results will serve as a base for automated recognition of malicious activities coming from the Internet of any other networked nodes (including e.g. infected LAN node).

2   Introduction to Honeypots and Honeynets

A honeypot represents a system "whose value lies in being probed, attacked, or compromised" [11]. Consequently, the attacker's activity can be analyzed in detail so that the information can be applied in later constructing or tuning protecting systems (e.g. firewalls, IDS and/or IPS). A honeypot allows to obtain an efficient overview about currently spreading threats and possible new threats as well. Server honeypots represent a passive operating software offering a certain service on a specific transport–layer port (e.g. SSH on 22). Server honeypots are the most frequently used and they were used for our research, too. There is another classification criterion consisting in the level of interaction (i.e., roughly speaking, the level of similarity between the honeypot and the real operating system). Usually, low–interaction, medium–interaction, and high–interaction honeypots are distinguished. In this research, a medium–interaction honeypot emulating a Linux SSH shell on port 22 was used. Nevertheless, because a medium level of interaction is applied, any more sophisticated attacker easily and quickly recognizes that not all outputs correspond to a real system. Therefore, such a honeypot is suitable mainly to analyze the behavior of botnets and script–kiddies.

Multiple honeypots connected into different networks form logical networks called honeynets in order to obtain more relevant results by combining data from all the honeypots. The research honeynet used for this study has been collecting logs into a central site with MySQL database for further processing and analysis.

3   Specification and Topology of Research Honeynet

The honeynet composed of multiple Kippo [Kippo], honeypots with minor modifications.

The version developed by CZ-NIC[1], an organization operating CSIRT.CZ[2], was used. Kippo is a well-known medium–interaction server honeypot emulating Linux system by cloning the directory structure of a real system. In our case, Linux Debian 8 clone was used.

Kippo operates certain system utilities like netstat and records executed commands, downloaded files (via wget or SFTP), used login names, passwords,

---

[1] Available at https://github.com/CZ-NIC/kippo.

[2] https://csirt.cz.

clients etc. The aim of the study was to obtain the data on attackers' behavior as
detailed as possible. Bearing this aim in mind, the sensors (honeypots) have been
deployed into various networks of several types. The honeynet logical topology
is shown in Fig. 1. Measurements were done in the period from February 2015
till February 2016.

Whilst sensors HP1 and HP2 were connected to academic networks – in the
Czech academic network CESNET and the Slovak academic network SANET,
sensor HP3 was deployed on a virtual private server (VPS) by a Czech provider
offering also "adult services". Sensor HP4 was deployed on a server of the
provider of common VPS services. Sensor HP5 was connected to a Czech ISP
network and HP5-B was in the same network but the dynamic port 2222 was
used here for providing the SSH service instead of the well-known one (22) as
used in all other sensors. HP6 was deployed in the Slovak ISP network with an IP
address assigned dynamically (while all other sensor have static IP addresses).

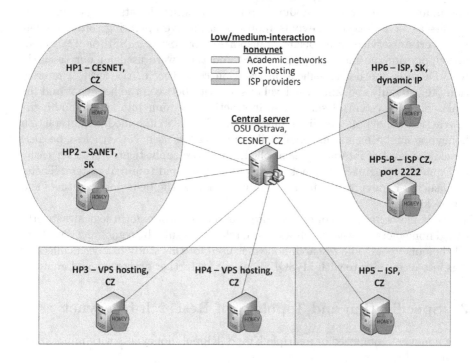

Fig. 1. Logical topology of the research honeynet

4   Observed Session Duration and Related Anomalies

At a glance, session duration seems uninteresting but a detailed analysis of older
and present results showed that certain sessions persisted even more than one
day (24 h). In practice, it means that the attacker – usually bot or a script–kiddy

but even a more advanced human user cannot be foreclosed – connected to the honeypot system and the connection that was established was not terminated properly after the attacker activity was completed, either due to a technical issue, or a negligence, or even no activity was done. The diagram in Fig. 2 shows the average session duration on each sensor (excluding very long sessions). After a detailed investigation of a significant portion of long sessions, a decision has been made to exclude sessions longer than one hour from the calculation of the average session duration. The analysis of long session showed that there was not even a single session where the real duration of attacker activity (entering inputs) was longer than one hour. As an example, a session that persisted for several days was selected; it contained the following commands only:

```
id;id;help;help;?;?;/bin/sh;/bin/sh
```

Despite the fact that the session persisted for more than 24 h, all the above commands were entered within the first minute after establishing the connection. After the attacker activity ended, they did not terminate the connection, however. Therefore, only sessions shorter than 1 h has been taken into account for the average calculation in Figs. 2 and 3. Just to show how big the influence was to average calculation, the averages calculated from all sessions are shown in Table 1, which can be compared to data in Fig. 3.

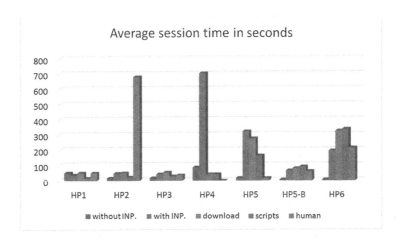

Fig. 2. Average session duration per activity type and per sensor

Also a comparison between average session duration without any subsequent attacker activity (see Fig. 2 "without INP." indication), and the average duration of session with any input after login was detected (see Fig. 2 "with INP." columns). In addition, the diagram shows the average duration of sessions with file download ("download" columns), and sessions where a file or script was run ("scripts" columns). The last case analyzed in this context was the average duration of sessions where such commands were used that are related to a human activity ("human" indication). When commands specific to human interaction occurred, the session duration usually did not exceed 1 min, with the exception of HP2 sensor. However, a more detailed analysis showed that this was due to a minor technical issue in termination connections from script–kiddies so the above conclusion can be generalized.

Most sessions with any subsequent activity after login (e.g. command execution) exist for approx. 2 min as one can observe in Fig. 2. This duration results from an execution of a predefined sequence of commands entered almost immediately after login. Sessions with no input – that means that they can terminate even on entering an invalid login (this does not happen in our honeynet where all login data are considered to be valid) – usually exist for less than a minute. Sessions with a file download took approx. 30 s with an exception of HP5 and HP6 sensors where such a duration is longer. When an attacker executes a file ("scripts") during the session, its duration is varied heavily ranging from several seconds till approx. 5 min.

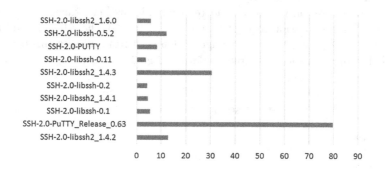

Fig. 3. Average session duration per SSH client (in seconds)

The session duration according to SSH client used was analyzed, too. The diagram in Fig. 3 shows this relation. Ten most frequently used SSH clients for the research period are shown. As one can see that some attackers using PUTTY client (see "SSH-2.0-PUTTY" in the diagram), which could indicate a "human" activity (because of PUTTY inconvenience for automated attacks), do not spend more than approx. 20 s in a honeypot. When various derivatives of LibSSH library was used (presumably by a bot), the command sequence is done usually within 30 s. The recorded data confirm the already mentioned prevalence of automated attacks that are done quickly and in an analogous manner and occur repeatedly.

5   Disseminated Malicious Activities

A malicious activity in the context of this study means a sequence of commands that was executed during a session when an attacker was performing

Table 1. Selected average SSH session duration per SSH client

Client name	Session duration
SSH-2.0-libssh2_1.6.0	90.2
SSH-2.0-libssh-0.5.2	2253.9
SSH-2.0-PUTTY	103.7
SSH-2.0-libssh-0.11	3.7
SSH-2.0-libssh2_1.4.3	4673.5

their activities. As a starting point for the analysis, a number of identical command sequences entered by an arbitrary attacker during their session should be determined. A 100% match of activity is required here – this serves as an indication of a machine-made attack. The condition for incorporating the sequence into the research was minimum 2 occurrences of the sequence. In such a case, it is highly unlikely that any "human activities" are analyzed because it is unlikely that different human users can enter the same commend sequence multiple times in a completely identical way. Altogether, 1,805 input sequences repeated two or more times was recorded.

The most frequently occurring input sequence was the following command sequence executed in total 1,424 times. It occurred at 5 out of 7 sensors in the honeynet and was distributed by 153 unique IP addresses from various countries.

```
uname;uname;free -m;free -m;ps x;ps x
```

This sequence seems interesting due to the following reasons. First, each command was entered twice. This could indicate an attacking software error (either originated by its design or its initiation), but it could be author's intention (to ensure its execution) or to verify the output data. In addition, the used commands are rather uncommon: uname command is used without arguments resulting in single–word output indication system type (e.g. Linux) and other details about the system (e.g. core version) are not shown. The request for available RAM size and ps x command displaying only running processes with TTY follow. The average duration of sessions with this sequence was approx. 18 s. For this sequence, the 2 SSH clients were used, namely SSH-2.0-libssh2_1.4.1 and SSH-2.0-libssh2_1.6.0. The sequence was occurring throughout the period of research so a conclusion could be drawn that the attacker likely wanted to gather specific information before a real executive attack begins; such an attack would be performed in the next phase after a new connection. Such an attack can be classified among so-called 'reconnaissance attacks'. This seems to be an attempt to identify that the target is Kippo honeypot[3]). The attacker then decides only if the output is analyzed to be in accordance with expectations.

---

[3] Details about Kippo detection using the ps x command see in https://github.com/desaster/kippo/issues/39.

The following second most frequently used sequence was used 1,289 times against our honeynet. However, this sequence does not require any sophisticated analysis because of its simplicity:

```
echo -n test;echo -n test
```

This sequence was distributed by 1,261 unique IP addresses with the average session duration of 7 s. The likely purpose of this seemingly reasonless sequence can be explained easily as follows. Kippo honeypot had suffered a certain defect causing that the fact of emulation was indicated by an incorrect response to the above command.[4] This example implies that even many administrators and/or designers of unsophisticated bots can implement simple mechanisms to detect honeypots (it means they are up-to-date in IT security field).

The third most frequently used sequence used in 920 sessions (3 sensors, 21 unique IP addresses) was a sequence that is rather typical representative of most bot behavior. The sequence consisted of the following commands:

```
cd /tmp; wget -q http://173.242.117.*/Sharky/gb.sh
sh gb.sh;rm -rf gb.sh
```

A noteworthy aspect of the individual commands was that each of the commands was entered within a single second that can hardly be done by any human. The downloaded file gb.sh contains another command sequence using the shell to download and execute another file. The average duration of such a session was approx. 14.5 s. The sequence had strictly confined timeframe: it occurred solely from Sept till Nov. 2015. Almost all remaining inputs corresponded to the above procedure – downloading a file, its execution and deletion, frequently combined with steps to deactivate security tools (e.g. iptables), changing file attributes to allow its execution etc.

However, another approach than individual (manual) analysis was necessary to analyze thousands of input types. At present, due to the existing infrastructure, the detection of specific threats using an antivirus service inside the VirusTotal[5] seemed as the most promising approach. Subsequently, such data was the basis for clustering analysis whose results are described below.

6   Analysis Based on Dissemination of Specific Threats

The analysis according to individual threat or malware dissemination can be utilized in practice efficiently both in research and primarily by national CSIRT[6] teams as well as CSIRT teams with constituency[7] for a specific single autonomous system. Thus, they can get an awareness on distributors of specific

---

[4] Possibilities of Kippo detection – see https://github.com/desaster/kippo/issues/190.

[5] VirusTotal service – see http://virustotal.com/.

[6] CSIRT = Computer Security Incident Response Team.

[7] Constituency means a part of the Internet where the CSIRT operates as an authority.

malware types in the scope of their authority with subsequent making necessary precautions (frequently just advising the affected ISPs).

Because most sessions captured in the research honeynet involved a file download, subsequently analyzed via submission to VirusTotal, the resulting thousands of records from the honeynet can be analyzed by clustering so that to find similarities among various objects in the data. Four n-tuples were used as input of the clustering algorithms. The first one, called "Sensors", represents whether a specific malware has been detected by a sensor or not. Thus, the first attribute is the sensor's name, with every subsequent attribute being a 1 or 0, representing detection, or lack thereof, respectively, of a specific malware by the given sensor. The second one, called "Countries", points out which malware originated in a specific country. The first attribute is the country's name, while all the other attributes are 1 or 0 representing whether or not has each malware originated in the given country. Third n-tuple called "ASN's", and the fourth one, "RIR's", are analogous to the second one, representing the existence or lack thereof of a specific malware originated in the specific ASN or RIR. This approach can distinguish differences among various objects (by assigning them into different clusters). It should be noted that downloaded files were compared according to SHA1 hash[8]. Before clustering analysis started, the selection of a suitable antivirus had to be decided. The sole requirement was the best possible detection and classification of a downloaded code specimen – threat in Linux environment. Based on our research among available antivirus engines in VirusTotal service, Avast antivirus was chosen due to its best performance in the number of identified malware specimens. Clustering was done using data from all sensors except HP4 (here the analysis could not be done due to a technical issue. The quantities according clustering was done were the classified downloaded malware, more specifically their names as identified by Avast. They are categorial (nominal, not ordinal) so the clustering method selection taken this fact into account. If a specimen was detected as a malware just based on a heuristic rule, it was identified as unknown. On the other hand, if the downloaded file was not detected as a threat, the source was not considered to be a malware source. Objects to classify were either honeypots – sensors, or source ASes identified with their ASN, or RIRs, or source countries. Therefore, not only trends in distributing AS but also trends in distributing threats from a specific country or global trends etc. can be analyzed.

6.1 Clustering Analysis of Captured Threats

Clustering Algorithms. The occurrence of specific malware at a specific sensor or distribution from a specific AS, country or RIR has just two possible values: either it happened (True/1) or not (False/0). Moreover, asymmetric approach was chosen because the significance of both values is not the same – the occurrence (value 1) is much more significant for assigning the object into a cluster/category. This decision is essential because it allows to determine the way

---

[8] Only files with identical SHA1 hash were considered identical.

how to measure the similarity between objects. According to [12], Jaccard similarity index was chosen for a hierarchic approach resulting in dendrograms [13] while ROCK [14] and Proximus [15] algorithms were chosen for a non-hierarchic approach. The clustering was implemented using R statistical language[9].

Binary Clustering Based on Captured Malware per Sensor. Results of hierarchic clustering are shown in a dendrogram in Fig. 4. In the dendrogram, a single cluster (involving HP1 and HP3) was formed at the dissimilarity level 0.3 so HP1 and HP3 are considered to be the most similar from the point of view of detected malware. With the dissimilarity level increasing, other sensors join the first cluster, no new cluster was formed. The last cluster to join, HP5, was considered the most different from others in the honeynet. For a comparison, non-hierarchic algorithms were applied, too. The expected number of clusters had to be set a-priori so 4 clusters were estimated (for academic, VSP, ISP, and HP5-B with dynamic port, respectively). All parameters were left on their defaults. ROCK algorithm resulted in the same sensor clustering as the above hierarchic dendrogram that means that HP3, HP1 and HP2 were assigned to the same cluster while the remaining sensors were left alone in their clusters. Also Proximus algorithm produced the same clusters.

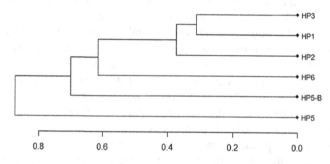

Fig. 4. Clusters of sensors formed using Jaccard index and single–linkage approach

To visualize the similarity between individual sensors regarding specific threat dissemination, the Multiple Correspondence Analysis (MCA) technique [16] was applied. MCA is similar to Principal Components Analysis (PCA) [17] technique routinely applied to numerical (ordinal) quantities. The diagram in Fig. 5 shows the sensor clustering using 2 new (artificial) variables generated so that they explain almost 68% of the original variability. Sensors HP5 and HP5-B seem much closer and much more similar here comparing to results from ROCK and Proximus. It is noteworthy that the influence of multiple quantities (specific threat) is neglected here. The neglected minor quantities (threats) could have an impact on final sensor clustering but due to the fact that almost 70% of original information was preserved after MCA application, the difference is not considered significant.

---

[9] Details are available at http://r-project.org.

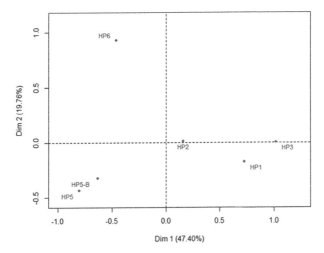

Fig. 5. Sensor clusters based on MAC technique with 2 artificial variables covering the majority of variability of original variables

Binary Clustering of Captured Malware per AS. The similarity between autonomous systems regarding the malware distributed from them is important, too. Based on such data, one can observe whether specific threats are disseminated from a limited range of ASes only etc. Avast antivirus was used for threat classification again. The dendrogram was made using Jaccard index, too, and the single–linkage technique was applied. Due to the high number of ASes, it was necessary make the dendrogram more readable as it is shown in Fig. 6. Here one can observe that ASes formed several smaller clusters that later joined in a single cluster (except certain exceptions like AS4134 and AS3462 forming separate cluster, as well as ASes 46844, 4812 and 4837 that made a separate cluster, too, but joined later with the above cluster and formed a large cluster at the same dissimilarity level as the "large cluster" of remaining ASes. A bit later, another cluster of AS: 35908 and 54600 joined while AS134764 and AS31400 are characterized as completely different.

On the contrary, Proximus algorithm classified the ASes among 21 clusters in a highly similar way as the dendrogram in Fig. 6. Among notable differences, one can see that AS8560, AS20857, and AS31400 formed separate clusters while they joined in the dendrogram at a high dissimilarity level. The ROCK algorithm classified ASes into 10 clusters. In the case of low–population clusters, ROCK algorithm usually agreed with Proximus (e.g. for AS 2527, 59491, 59743). However, multiple ASes that were clustered in small clusters (usually having 3 ASes) by ROCK were left alone by Proximus. It happened vice versa in certain cases, too. When the clustering results were analyzed in detail, it seemed obvious that ASes standing separate in their clusters used to disseminate just a single threat or a small number thereof. For instance, AS31400 standing alone in the dendrogram (in the right) disseminated the threat ELF:HideProcB. Similarly,

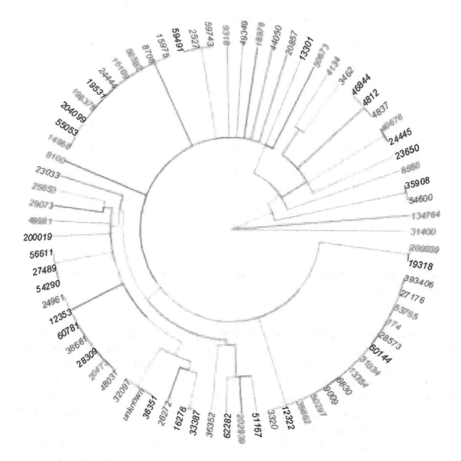

Fig. 6. Dendrogram of ASes. Colors assigned to AS are random, for better visibility

AS134764 joins the big cluster at the very end (an the highest dissimilarity level) because this was the only AS not distributing any malware.

Binary Clustering of Captured Malware per RIR. When source RIRs were considered as objects for clustering, the resulting dendrogram looks as shown in Fig. 7. Again, the Jaccard index was applied here. Here one can see that LACNIC and "unknown"[10] are the most similar. On the other hand, APNIC differs completely from remaining RIRs while RIPENCC and ARIN formed a cluster but the dissimilarity level is quite high (around 0.5) that means that threats disseminated by them are quite different. Even Proximus result fully confirms those conclusions.

---

[10] The unknown category includes ASes with no RIR data available. Usually this is the case for private ASes. Details can be found in IETF RFC 6996 – Autonomous System (AS) Reservation for Private Use available at https://tools.ietf.org/html/rfc6996.

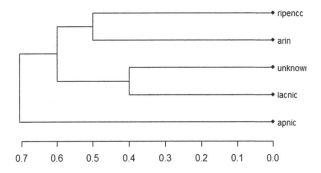

Fig. 7. Dendrogram of RIR clusters using Jaccard index and single–linkage

Quite expectedly, the least threats were disseminated by "unknown" ASes. This is given primarily by the fact that their share on total traffic in the honeynet was rather marginal. Also LACNIC RIR disseminated quite small number of threats – like Trojan Tsunami, Trojan Gafgyt, Trojan Perl:Shellbot O and Dropper BV:Powbot B. As well as other RIRs, also LACNIC ASes distributed "unknown" threats (i.e. heuristic–only detected malware). On the other hand, ARIN RIR distributed almost all detected threats with minor exceptions (Trojan Xorddos, Trojan ELF:YangjiC, and ELF:HideProcB tool). RIR APNIC was specific by the fact that also threats distributed uniquely from their ASes were detected in the honeynet, namely Trojan ELF:YangjiC and Trojan Xorddos. This means that these threats were distributed solely from Far East and Australia. Other unusual combinations could also be found, e.g. there were threats disseminated only from APNIC and ARIN ASes like ELF:ChinazN Trojan.

Binary Clustering of Captured Malware per Countries. The dendrogram in Fig. 8 as well as the ROCK clusters show that the most populous cluster (top frame – red) distinguished at the dissimilarity level of approx. 0.3 contains various countries ranging from Brazil through France, Canada, Lithuania to Russia and others. Also communication identified as "unknown" originating from private ASes belongs here. Another large cluster was formed from attackers from Portugal, Korea, Poland, Japan, Turkey and others, at the dissimilarity level of approx. 0.5 (see the second frame from the top with green border). The following cluster of the USA and Panama (2 yellow frames) was formed but the dissimilarity level is quite high here so that the ROCK algorithm classifies each of the countries in a different cluster. The last cluster of China and Taiwan (blue frame) shows quite an acceptable dissimilarity level less than 0.5. Despite this, ROCK classifies them again in separate clusters. Nevertheless, the dendrogram and clusters resulting from ROCK algorithm provide enough information so that the analyst could concentrate on detailed investigation into threats distributed by countries in a specific cluster.

When clustering is analyzed in details, ROCK algorithm seems to classify China and Taiwan properly into separate clusters because they distributed only

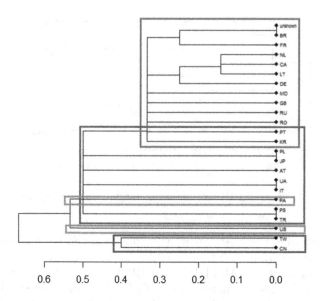

Fig. 8. Dendrogram of countries clustered using Jaccard index and single–linkage (Color figure online)

3 common threats. What is notable, China as a sole country distributed the threat called ELF:YangjiC Trojan, and together with USA, only those 2 countries disseminated the threat ELF:ChinazN. This malware was developed in China and its primary intention is to be incorporated into a botnet for DDoS attacks. Its infection is done either via wget download, or drive-by download[11] After being infected into a system, the malware connects to its control server that is advised about the data on the infected system (e.g. uname output) and remains ready to obtain instructions for DDoS attack performing. According to sources from MalwareMustDie[12], this malware targets primarily (or even exclusively) against the IT infrastructure in the U.S. Our measurements confirmed that except China (likely country of origin) and U.S. as a target country, is not disseminated by any system in the world. This example demonstrates what findings can be discovered from indices extracted from properly applied cluster analysis.

The most populous cluster (red line in the dendrogram) joined primarily countries where attackers distributing Tsunami Trojan came from. This threat has been prevailing for a very long time – this is a backdoor active in various versions since 2002. This Trojan listens to commands from a remote control server via an IRC channel. Its primary usage is, again, to incorporate the victim into DDoS attacks. The Trojan uses several simple commands, e.g. to choose among the DDoS attack type (e.g. TCP/UDP flooding). Since 2011, there has also been

---

[11] Details can be found in https://securingtomorrow.mcafee.com/consumer/family-safety/drive-by-download/.

[12] http://blog.malwaremustdie.org/2016/10/mmd-0060-2016-linuxudpfker-and-chinaz.html.

a Mac OSX version. In addition, various countries in this cluster distributed various other malware with no remarkable trend. The second most populous cluster (indicated by the green line in the dendrogram) was characteristic primarily by Gafgyt Trojan (also known as Mirai) dissemination. This Trojan does not attack solely against Linux servers and workstations but also against routers and devices that can be classified as IoT (Internet of Things) like IP cameras or other devices based on a Linux firmware. Its primary aim is similar to the above described one. Tsunami, i.e. incorporating the victims into a botnet for performing DDoS attacks. Its dissemination uses password attacks, either frequently used combinations, or dictionary attacks, and/or brute–force attacks. It infected more than a million devices[13] and maintains a network of bots counting approx. 120,000 of active stations (computers). While individual countries in this cluster also disseminated other threats, there is no indisputable trend among them again. Just to confirm the proper assignment into the cluster, it can be seen that the countries here disseminated the same malware or did not differ much in them. In addition, ROCK algorithm identified that Ukraine and Italy disseminated solely Tsunami Trojan while Palestine and Turkey only threats identified as unknown. Nevertheless, despite the fact that certain RIRs or ASes seem to be sources of significant part of malware for the period of research, it is highly likely that the situation is not permanent. Therefore it is vital for any network administrator to monitor threats and their sources continuously.

7   Conclusions

The research results presented above indicate that attackers utilizing botnets prepare tools that repeat the same activity with a limited or missing feedback towards its success or failure. Nevertheless, even such a seemingly extremely simple code includes elementary detection mechanism for discovering honeypots. This means that attackers observe events in the security field actively and prepare countermeasures. Vice versa, this forces honeypot designers to propose various modifications continuously. The presented analysis of attack provides an enormous space for subsequent specific threat dissemination in specific AS or countries – i.e. from the global view as well as in mutual interactions. A proper setup of clustering to find similarities between specific countries and AS and its application to various data can indicate prevailing trends for specific areas efficiently. This is useful both for research and for CSIRT teams, too. For instance, a CSIRT can identify IP addresses that disseminate similar threats and apply preventive measures subsequently. In addition, they can monitor number of clients distributing similar malware using the presented method (e.g. to find that just small group of IP addresses was infected with a specific infiltration and other is spread massively. Similarly, potential identification of a methodology for efficient automation of multiple clustering algorithm outputs seems

---

[13] According to Softpedia, see http://news.softpedia.com/news/mirai-ddos-trojan-is-the-next-big-threat-for-iot-devices-and-linux-servers-507964.shtml.

beneficial because finding such relationships usually requires empirical analyst knowledge, thus eliminating the human labor need.

Acknowledgment. The paper was supported by the project No. SGS08/PrF/2017 *Network Services Security* of the Student Grant Competition of the University of Ostrava.

References

1. Sochor, T., Zuzcak, M.: Study of internet threats and attack methods using honeypots and honeynets. In: Kwiecień, A., Gaj, P., Stera, P. (eds.) CN 2014. CCIS, vol. 431, pp. 118–127. Springer, Cham (2014). doi:10.1007/978-3-319-07941-7_12
2. Sochor, T., Zuzcak, M.: Attractiveness study of honeypots and honeynets in internet threat detection. In: Gaj, P., Kwiecień, A., Stera, P. (eds.) CN 2015. CCIS, vol. 522, pp. 69–81. Springer, Cham (2015). doi:10.1007/978-3-319-19419-6_7
3. Almotairi, S., Clark, A., Mohay, G., Zimmermann, J.: Characterization of attackers' activities in honeypot traffic using principal component analysis. In: 2008 Network and Parallel Computing, pp. 147–154. IEEE (2008). doi:10.1109/NPC.2008.82
4. Rieck, K., et al.: Automatic analysis of malware behavior using machine learning. J. Comput. Secur. **19**(4), 639–668 (2011)
5. Sokol, P., Andrejko, M.: Deploying honeypots and honeynets: issues of liability. In: Gaj, P., Kwiecień, A., Stera, P. (eds.) CN 2015. CCIS, vol. 522, pp. 92–101. Springer, Cham (2015). doi:10.1007/978-3-319-19419-6_9
6. Skrzewski, M.: About the efficiency of malware monitoring via server-side honeypots. In: Gaj, P., Kwiecień, A., Stera, P. (eds.) CN 2016. CCIS, vol. 608, pp. 132–140. Springer, Cham (2016). doi:10.1007/978-3-319-39207-3_12
7. Skrzewski, M.: System network activity monitoring for malware threats detection. In: Kwiecień, A., Gaj, P., Stera, P. (eds.) CN 2014. CCIS, vol. 431, pp. 138–146. Springer, Cham (2014). doi:10.1007/978-3-319-07941-7_14
8. Savenko, O., Lysenko, S., Kryshchuk, A., Klots, Y.: Botnet detection technique for corporate area network. In: 2013 Intelligent Data Acquisition and Advanced Computing Systems (IDAACS), pp. 363–368. IEEE (2013). doi:10.1109/IDAACS.2013.6662707
9. Pomorova, O., Savenko, O., Lysenko, S., Kryshchuk, A., Bobrovnikova, K.: A technique for the botnet detection based on DNS-traffic analysis. In: Gaj, P., Kwiecień, A., Stera, P. (eds.) CN 2015. CCIS, vol. 522, pp. 127–138. Springer, Cham (2015). doi:10.1007/978-3-319-19419-6_12
10. Sochor, T., Zuzcak, M., Bujok, P.: Statistical analysis of attacking autonomous systems. In: International Conference on Cyber Security and Protection of Digital Services, pp. 1–6. IEEE (2016). doi:10.1109/ICUFN.2016.7537159
11. Spitzner, L.: Honeypots: Tracking Hackers, vol. 1. Addison-Wesley, Reading (2003)
12. Fichet, B.: Distances and Euclidean distances for presence-absence characters and their application to factor analysis. In: Proceedings of a Workshop Multidimensional Data Analysis 1985, pp. 23–46. DSWO Press, Cambridge (1986)
13. Jaccard, P.: Etude Comparative de la Distribution dans une Portion des Alpes et du Jura. Bulletin de la Societe Vaudoise des Sciences Naturelle **4** (1901)

14. Guha, S., Rastogi, R., Shim, K.: ROCK: a robust clustering algorithm for categorical attributes. In: Proceedings of the 15th International Conference on Data Engineering (Cat. No. 99CB36337), pp. 512–521. IEEE (1999). doi:10.1109/ICDE. 1999.754967
15. Koyuturk, M., Grama, A., Ramakrishnan, N.: Compression, clustering, and pattern discovery in very high-dimensional discrete-attribute data sets. IEEE Trans. Knowl. Data Eng. **17**(4), 447–461 (2005). doi:10.1109/TKDE.2005.55. http://ieeexplore.ieee.org/document/1401886/
16. Abdi H., Valentin D.: Multiple Correspondence Analysis. University of Texas at Dallas, Texas (2007). utdallas.edu, http://www.utdallas.edu/~herve/Abdi-MCA2007-pretty.pdf
17. Jolliffe, I.T.: Principal component analysis and factor analysis. In: Principal Component Analysis. Springer Series in Statistics, pp. 150–166. Springer, New York (2002)

Enhancements of Encryption Method Used in SDEx

Artur Hłobaż[(✉)], Krzysztof Podlaski, and Piotr Milczarski

Faculty of Physics and Applied Informatics, University of Lodz,
Pomorska 149/153, 90-236 Lodz, Poland
{artur.hlobaz,krzysztof.podlaski,piotr.milczarski}@uni.lodz.pl

Abstract. In the paper, we present enhancements of the encryption method for secure data transmission used in Secure Data Exchange method (SDEx) [1–3]. Potential weaknesses and possible attacks have been previously described in [4]. The new version of the encryption method is immune to indicated in [4] vulnerabilities connected with the potential knowledge of the part of the original message and uneven distribution of bits in the message blocks. In the new solution, we use also Davies-Meyer scheme in order to enhance immunity to meet-in-the-middle attacks.

Keywords: Data security · Secure transmission · Encryption algorithm · Cryptography · Cryptanalysis · Security analysis · Meet-in-the-middle

1 Introduction

The concern about the data security is still growing. The applications we use still lack basic security or they use deprecated solutions. In previous works [1–3] authors presented a mobile secure method of data exchange Secure Data Exchange (SDEx). The cryptographic algorithm implemented in SDEx method used for data transmission is based on hash functions. The method is easily scalable and allows to use any new hash functions. The cryptanalysis of the algorithm was published in [4], where the author pointed some weaknesses of the algorithm. In this paper we present how to enhance the original cryptographic algorithm used in SDEx method.

A secure hash function is a collision-resistant, one-way function. Collision resistance means that it should be impossible to find two different messages/set of bits that will produce the same hash value in both cases [5]. One way means it is extremely difficult to reproduce the input from the calculated hash value, or to find another input that will produce the same hash value. Hash functions are also used to determine whether or not data has been changed.

Hash functions are often used with digital signature algorithms and processes that provide a security service, including:

© Springer International Publishing AG 2017
P. Gaj et al. (Eds.): CN 2017, CCIS 718, pp. 134–143, 2017.
DOI: 10.1007/978-3-319-59767-6_11

- keyed-hash message authentication code (HMAC) [6],
- digital signatures [5],
- key derivation functions (KDFs) [7],
- random number generators (RNGs) [8].

There are several families of hash function algorithms e.g. Message Digest (MD) or Secure Hash Algorithm (SHA). Some of them like MD5 are vulnerable to collision search attacks [5,9,10]. There are theoretical weaknesses of SHA-1 but no collision (or near-collision) has been found yet [11–13]. However, to avoid these problems, it is suggested that new applications should use later members of the SHA family, such as SHA-2, SHA-3, or implement techniques such as randomized hashing that do not require collision resistance.

Therefore, the authors in the paper propose a new enhancement to cryptographic algorithm, based on Davies-Mayers schema (DM). Davies-Mayers schema is pointed in [14] as one of 12 the best collision resistant algorithms/methods. This DM enhancement will make the Secure Data Exchange (SDEx) method [1–3] more collision resistant e.g. immune to meet-in-the-middle type attacks.

The paper is organized into five sections. In Sect. 2 the description of data encryption algorithm used in SDEx method [1–3] is given. Section 3 shows cryptanalysis of the basic version of the encryption algorithm that use hash functions. Statistical cryptanalysis attack is also discussed. In the next Section a new strengthening method as well as its cryptanalysis are given. Conclusions are drawn in Sect. 5.

2   Data Encryption Method Used in SDEx

The algorithm of encryption used in SDEx method [1–3] is based on hash functions. In a standard application, hash functions are used to check integrity of data. These functions take as an input any length string of bits and return the result of a fixed size called a hash or a message digest. The length of hash usually is in the range of 128 to 512 bits, while the whole message can contain thousands or even millions of bits.

All hash functions are constructed iterative and divide the input data into a sequence of fixed size blocks $M_1, \ldots, M_t$ (Fig. 1). Message blocks are sequentially processed using a hash function to the intermediate state of constant size. The process starts with the predetermined value h_0, while successive states $h_1, \ldots, h_t$ are defined as $h_t = \text{hash}(h_{t-1}; M_t)$. The h_t - result of the last iteration of the process - is the searched message digest.

The hash function used in SDEx encryption method (Fig. 2) performs a role of dynamic pseudorandom string of bits generator.

The common symbols used on diagrams 2, 3, 4 and 5 and Eqs. (1)–(15) are:

- $M_1, M_2, \ldots M_i$ - plaintext blocks,
- $C_1, C_2, \ldots C_i$ - ciphertext blocks,
- IV - initialization vector (session key),
- $h_1, h_2, \ldots h_k$ - particular iterations of hash computation,
- H_{IV} - hash from the initialization vector,

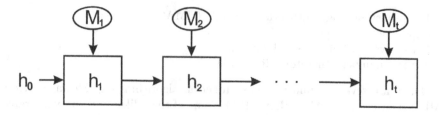

Fig. 1. Message digest counting for any length string of bits.

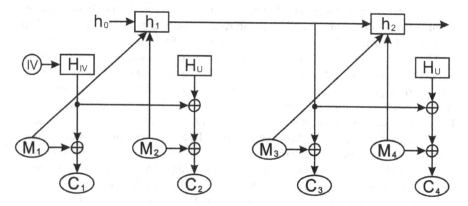

Fig. 2. Encryption for even number of blocks.

- H_U - hash from user password,
- $\oplus$ - XOR operation.

In order to make communication more secure we should use encryption keys instead of passwords. Hence, in the paper "password" and "encryption key" are used as synonyms.

Corresponding decryption process is presented in the Fig. 3.

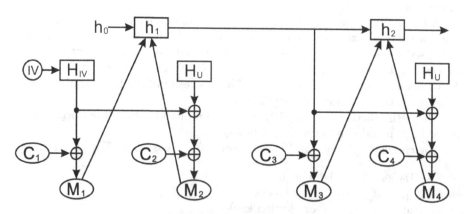

Fig. 3. Decryption for even number of blocks.

The next figures (Figs. 4 and 5) show the encryption and decryption for an odd number of blocks of the explicit message M_i.

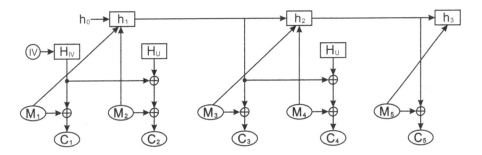

Fig. 4. Encryption for odd number of blocks.

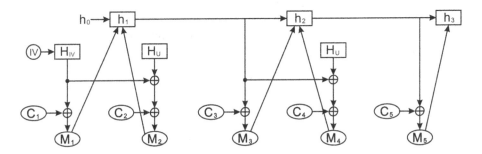

Fig. 5. Decryption for odd number of blocks.

Generated blocks of pseudorandom string of bits are XORed with plaintext blocks. The results of that operation are ciphertext C_i blocks. The encryption method is described in details in [1, 15, 16].

The encryption algorithm (Figs. 2, 3, 4 and 5) can be also defined in algebraic operations as follows:

$$C_1 = M_1 \oplus H_{IV} \qquad \text{(where } \oplus \text{ - XOR operation),} \quad (1)$$
$$C_2 = M_2 \oplus H_U \oplus H_{IV}, \qquad (2)$$
$$C_{2k+1} = M_{2k+1} \oplus h_k \qquad \text{(where } k \geq 1, \text{ for odd blocks),} \quad (3)$$
$$C_{2k+2} = M_{2k} \oplus H_U \oplus h_k \qquad \text{(where } k \geq 1, \text{ for even blocks),} \quad (4)$$
$$h_1 = \text{hash}(h_0; M_1 + M_2) \qquad + \text{ string concatenation operator,} \quad (5)$$
$$h_k = \text{hash}(h_{k-1}; M_{2k-1} + M_{2k}) \qquad \text{(where } k \geq 2), \quad (6)$$

where $\text{hash}(h; M)$ is a hash value, i.e. the result of the function hash that takes a parameter h. The parameter h represents an initialization vector of a hash function and input M is a message [10].

And the appropriate decryption algorithm is defined as follows:

$$M_1 = C_1 \oplus H_{IV}, \tag{7}$$

$$M_2 = C_2 \oplus H_U \oplus H_{IV}, \tag{8}$$

$$M_{2k+1} = C_{2k+1} \oplus h_k \qquad \text{(where } k \geq 1, \text{ for odd blocks)}, \tag{9}$$

$$M_{2k+2} = C_{2k} \oplus H_U \oplus h_k \qquad \text{(where } k \geq 1, \text{ for even blocks)}. \tag{10}$$

As we can see from Eqs. (3) and (4) the method allows to encrypt messages with any number of blocks. If the length of the last block of the message is shorter then block size, than thanks to the XOR operation the ciphertext is complemented. The overall size of ciphertext is always a multiplicity of the block size.

3   Cryptanalysis of the Basic Version of the Encryption Algorithm

The presented cryptanalysis is a recollection of analysis presented in [4]. The proposed algorithm [1] is vulnerable to two types of cryptographic attacks: known plaintext attack and uneven bits distribution. Taking into account Eqs. (1) and (2) we can show that:

$$C_1 \oplus C_2 = M_1 \oplus H_{IV} \oplus M_2 \oplus H_{IV} \oplus H_u = M_1 \oplus M_2 \oplus H_u, \tag{11}$$

$$C_3 \oplus C_4 = M_2 \oplus h_1 \oplus M_3 \oplus h_1 \oplus H_u = M_3 \oplus M_4 \oplus H_u. \tag{12}$$

We can observe that using XOR operations on two neighboring blocks the result depends only on the message blocks and a password:

$$C_{2k+1} \oplus C_{2k+2} = M_{2k+1} \oplus M_{2k+2} \oplus H_u, \text{ where } k \geq 1. \tag{13}$$

The same password is used for all communication of a user and so the analysis of the expressions $C_{2k+1} \oplus C_{2k+2}$ allows an attacker to gather some knowledge about the user password H_u. This also means that all future communication with the same password is no longer secure.

3.1   Plain Text Type Attack

If an attacker knows first two blocks of original message M_1, M_2 he can derive the user password H_u, using Eqs. (11) and (12) we obtain:

$$H_{IV} = C_1 \oplus M_1, \tag{14}$$

$$H_u = C_2 \oplus M_2 \oplus H_{IV}. \tag{15}$$

In case of hash functions SHA-256 and SHA-512 the size of blocks M_1, M_2 is accordingly 256 and 512 bits long. The knowledge about only this two first blocks allows an attacker to decrypt the rest of the message. It has to be stressed that most of the known file types have standard header blocks that do not depend on their content.

3.2   Statistical Cryptanalysis Attack

An attacker can count any number of XOR operations of neighboring blocks $C_1 \oplus C_2$, $C_3 \oplus C_4$ etc. Then the intruder obtains XORs of appropriate message blocks with H_u. If on some position in a block we have 0 bit, then it is highly probable that there is also 0 on the appropriate position of password H_u. The intruder has to search for blocks where a distribution of 0 and 1 is highly uneven, then it is possible do recreate some parts of the user's password (see details in [4]).

As we can see in both of presented attack methods the basic vulnerabilities are connected with the fact that the result of $C_{2k+1} \oplus C_{2k+2}$ depends only on message blocks and a password H_u in Eq. (13). This can lead to the discovery of the whole password or a part of it.

4   Enhanced Secure Data Exchange (SDEx) Method

The security level of hash functions has been estimated at half of the generated hash length because of the possibility of collision. To avoid this type of attack, we can modify the previously described method of calculating the message digest to hinder the reversal of the various stages of calculation. One of the possibilities is to use Davies-Meyer schema (Fig. 6) which allows hash function collision resistance [17]. In [14] the analysis of 64 possible modifications of black box ciphers with use of Davies-Meyer type schemas from [17] is presented. As the result, authors identified a group of 12 good solutions (called in the paper $H1, \ldots, H12$) which are regarded to be the most collision-resistant from all 64 schemes. The "security bound is identical for all of these 12 schemes" in relation to collision-resistance. In order to improve our encryption method we have chosen Davies-Meyer scheme marked in the article as $H5$, because:

- it is one of 12 good solutions pointed in [14],
- it is the most proven and audited scheme,
- it is faster; it has only one XOR operation (most of schemes have two XOR operations).

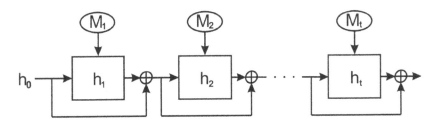

Fig. 6. Davies-Meyer schema.

4.1 Description of the Enhanced SDEx Method

The common symbols used on diagrams 7 and 8 and Eqs. (16)–(28):

- $M_1, M_2, \ldots, M_i$ - plaintext blocks,
- $H_0 = H_{IV + h_0}$.

In order to make an encryption algorithm resistant to the attack described in Sect. 3 we change the method into a new one. The new enhanced schemes for encryption and decryption methods are shown in Figs. 7 and 8. As it is shown on that diagrams there are additional XOR operations needed to obtain C_1 and C_4. This time the equations that describe encryption operations take the following form:

$$C_1 = M_1 \oplus H_{IV} \oplus H_{IV + h_0}, \tag{16}$$

$$C_2 = M_2 \oplus H_U \oplus H_{IV}, \tag{17}$$

$$C_{2k+1} = M_{2k+1} \oplus h_k \oplus h_{k-1} \qquad \text{(where } k \geq 1), \tag{18}$$

$$C_{2k+2} = M_{2k} \oplus H_U \oplus h_k \qquad \text{(where } k \geq 1), \tag{19}$$

$$h_1 = \text{hash}(H_{IV + h_0}; M_1 + M_2), \tag{20}$$

$$h_2 = \text{hash}((h_1 \oplus H_{IV + h_0}); M_3 + M_4), \tag{21}$$

$$h_k = \text{hash}((h_{k-1} \oplus h_{k-2}); M_{2k-1} + M_{2k}) \qquad \text{(where } k \geq 3). \tag{22}$$

Hence, the new decryption algorithm is defined as follows:

$$M_1 = C_1 \oplus H_{IV} \oplus H_{IV + h_0}, \tag{23}$$

$$M_2 = C_2 \oplus (H_U \oplus H_{IV}), \tag{24}$$

$$M_{2k+1} = C_{2k+1} \oplus h_k \oplus h_{k-1} \qquad \text{(where } k \geq 1), \tag{25}$$

$$M_{2k+2} = C_{2k} \oplus H_U \oplus h_k \qquad \text{(where } k \geq 1). \tag{26}$$

The element $H_{IV + h_0}$ in Eqs. (16) and (23) is just a hash from the string created as a concatenation of IV and h_0. The enhanced encryption/decryption methods are presented of Figs. 7 and 8.

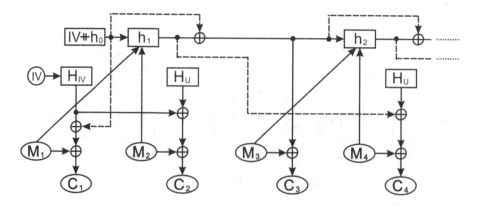

Fig. 7. Strengthened encryption method.

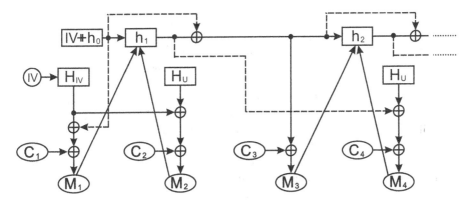

Fig. 8. Strengthened decryption method.

4.2 Cryptanalysis of the Enhanced Version of SDEx Method

In order to prove that the strengthened version is immune to attacks described in Sect. 3 we should analyze the expressions for $C_1 \oplus C_2$ and $C_{2k+1} \oplus C_{2k+2}$:

$$
\begin{aligned}
C_1 \oplus C_2 &= M_1 \oplus H_{IV} \oplus H_{IV + h_0} \oplus M_2 \oplus H_U \oplus H_{IV} \\
&= M_1 \oplus M_2 \oplus H_{IV + h_0} \oplus H_U.
\end{aligned} \tag{27}
$$

$$
\begin{aligned}
C_{2k+1} \oplus C_{2k+2} &= M_{2k+1} \oplus h_k \oplus h_{k-1} \oplus M_{2k} \oplus H_U \oplus h_k \\
&= M_{2k+1} \oplus M_{2k} \oplus h_{k-1} \oplus H_U.
\end{aligned} \tag{28}
$$

As we can see the expression $C_{2k+1} \oplus C_{2k+2}$ depends not only on the password H_u but also on h_{k-1}. The randomness of the sequence h_{k-1} implies that any knowledge over $h_{k-1} \oplus H_U$ cannot lead to any useful information on H_u. That means that whatever knowledge an attacker has on message blocks M_{2k+1}, M_{2k+2} or cipher blocks C_{2k+1}, C_{2k+2} no information can be gained on the password H_u. Moreover the usage of Davies-Mayer scheme guaranteed that this sequence h_{k-1} is a better pseudorandom sequence than in the original version of the encryption algorithm [14]. Additionally, it should be stressed that in [14] the DM type schemas where applied to black box blocksciphers. In our paper, DM is applied only to the hash function in order to increase randomness of its results. What is more important, except the DM-schema in cryptographic algorithm, we use additional XOR operator with message i.e. $C_3 = M_3 + (h_1 \oplus h(IV + h_0))$. As a result, it is very similar to the $H4$ scheme (in [14]) that belongs to the group of 4 schemas with the best second-preimage resistance [14].

5   Conclusions

The main security vulnerabilities of the mobile applications are connected with often usage of that mobile applications in the non-secure environment e.g. airport or city hotspots. This is the main reason to research and develop more secure cryptographic algorithm used in (SDEx) method (see [1–3]).

In the paper, we describe the hash functions and their use in secure encryption algorithm. The families of hash functions are often implemented in applications. However, some of the hash functions e.g. MD5 and SHA1, show vulnerabilities and it is recommended not to use them. Some authors e.g. [9–13] present that the encryption algorithms that use hash functions show the potential weaknesses to possible collision attacks, like meet-in-the-middle one.

In the paper, we present enhancements of the encryption method for secure data transmission used in (SDEx) method. Potential weaknesses and possible attacks have been previously described in [4]. The new version of the encryption method is immune to indicated in [4] vulnerabilities connected with the potential knowledge of the part of the original message and uneven distribution of bits in the message blocks (Sect. 4.2). In the new solution, we use also Davies-Meyer scheme in order to enhance immunity to meet-in-the-middle attacks. We discuss statistical cryptanalysis of the attack and the collision resistance of the new encryption SDEx method.

In the mobile applications of Secure Data Exchange method we propose to use the enhanced version of cryptographic method.

References

1. Hłobaż, A., Podlaski, K., Milczarski, P.: Applications of QR codes in secure mobile data exchange. In: Kwiecień, A., Gaj, P., Stera, P. (eds.) CN 2014. CCIS, vol. 431, pp. 277–286. Springer, Cham (2014). doi:10.1007/978-3-319-07941-7_28
2. Milczarski, P., Podlaski, K., Hłobaż, A.: Applications of secure data exchange method using social media to distribute public keys. In: Gaj, P., Kwiecień, A., Stera, P. (eds.) CN 2015. CCIS, vol. 522, pp. 389–399. Springer, Cham (2015). doi:10.1007/978-3-319-19419-6_37
3. Podlaski, K., Hłobaż, A., Milczarski, P.: Secure data exchange based on social networks public key distribution. In: Internet of Things. IoT Infrastructures - Second International Summit, IoT 360° 2015, Rome, Italy, 27–29 October 2015, Revised Selected Papers, Part I, pp. 52–63 (2015)
4. Rzońca, D.: Cryptoanalysis of HPM14 encryption algorithm. Studia Informatica 36(3), 5–10 (2015)
5. Katz, J., Lindell, Y.: Introduction to Modern Cryptography, 2nd edn. Chapman & Hall/CRC, Boca Raton (2014)
6. Bellare, M., Canetti, R., Krawczyk, H.: Keying hash functions for message authentication. In: Koblitz, N. (ed.) CRYPTO 1996. LNCS, vol. 1109, pp. 1–15. Springer, Heidelberg (1996). doi:10.1007/3-540-68697-5_1
7. Bezzi, M., Capitani, D., di Vimercati, S., Foresti, S., Livraga, G., Paraboschi, S., Samarati, P.: Data Privacy, pp. 157–179. Springer, Berlin (2011)
8. Goldreich, O., Goldwasser, S., Micali, S.: How to construct random functions. J. ACM 33(4), 792–807 (1986)
9. Wang, X., Yu, H.: How to Break MD5 and Other Hash Functions, pp. 19–35. Springer, Berlin (2005)
10. National Institute of Standards and Technology: Secure Hash Standard (SHS). Technical report, NIST (2008)

11. Wang, X., Yin, Y.L., Yu, H.: Finding Collisions in the Full SHA-1, pp. 17–36. Springer, Berlin (2005)
12. Szydlo, M., Yin, Y.L.: Collision-Resistant Usage of MD5 and SHA-1 Via Message Preprocessing, pp. 99–114. Springer, Berlin (2006)
13. Rijmen, V., Oswald, E.: Update on SHA-1, pp. 58–71. Springer, Berlin (2005)
14. Black, J., Rogaway, P., Shrimpton, T., Stam, M.: An analysis of the blockcipher-based hash functions from PGV. J. Cryptology **23**, 519–545 (2010)
15. Hłobaż, A.: Security of measurement data transmission - message encryption method with concurrent hash counting. SEP 1, 13–15 (2007)
16. Hłobaż, A.: Security of measurement data transmission - modifications of the message encryption method along with concurrent hash counting. FSNT NOT **1**, 39–42 (2008)
17. Preneel, B.: Davies-meyer hash function. In: van Tilborg, H.C.A. (ed.) Encyclopedia of Cryptography and Security, p. 136. Springer, Boston (2005)

The Possibilities of System's Self-defense Against Malicious Software

Mirosław Skrzewski[1]($\boxtimes$) and Paweł Rybka[2]

[1] Institute of Informatics, Silesian University of Technology,
Akademicka 16, 44-100 Gliwice, Poland
miroslaw.skrzewski@polsl.pl
[2] Faculty of Automation Control, Electronics and Computer Science,
Silesian University of Technology, Gliwice, Poland
pawel.rybka@interia.eu

Abstract. For many years the detection of malware and preparation of the ways of preventing them have been treated as two distinct issues. Malware monitoring should provide information on how to detect the presence of malware and attempts to infect the system. The protecting systems using this information should identify and stop malware operation. This paradigm led to current solutions, where protecting systems focus on detection of incoming threats and do not pay attention on the presence of not previously detected malware in the system. Malware authors have developed various methods of circumventing the defense lines of protecting systems, what results in a growing stream of information of systems security breaches. This indicates the need for additional line of defense, focused on detection of the malware, which penetrated defenses of the system. The paper presents the concept of such additional defense line, discusses the sources of necessary informations, method for detection of unknown malware and possible method of blocking malware operation.

Keywords: Malware detection · System self-defense · Monitoring of outbound communication · Blocking malware operation

1 Introduction

Malware monitoring and system protection are generally treated as two sepparated tasks. Threats monitoring is generally aimed on inspection of malware's distribution channels, acquiring of malware code and patterns of infection attempts, often simulating vulnerabilities in the operation of the various system services running on standard and non-standard system ports. In order to achieve its goal, the monitoring system must be infected by malicious software to obtain evidence on its operation and for the development of its signature but rarely can monitor its operation in the longer term.

Protection systems attempt to block malware propagation and system infection, based on its descriptions of malware properties. Most protecting systems

© Springer International Publishing AG 2017
P. Gaj et al. (Eds.): CN 2017, CCIS 718, pp. 144–153, 2017.
DOI: 10.1007/978-3-319-59767-6_12

checks the programs before they are executed, rarely paying attention to their continued operation in the system. When the behavior of the program file does not quite corresponds to patterns of "bad" program actions (patterns do not match), the program is considered harmless and remains active in the system.

Most of malware protection systems is not interested in a long monitoring the operation of programs that do not fit the patterns of malware activity and have not been caught by the traps systems. These programs may demonstrate malicious activity deviating slightly from the schemes, on which is based operation of protecting systems or they can significantly delay its start of activity.

As a result the detection of the fact that system was infected takes place with great delay, often by some chance. Some monitoring systems (eg. Honeypots) may longer analyze program operation after its start and classify it accordingly as a threat, but alone they are not able to stop its operation nor to remove it from the system.

Such situation indicate the need to develop a supporting protection systems based on different principle of operation. Since contemporary malware is able to pass through security systems, the main task of this additional line of defense should be monitoring of the system operation to detect signs of suspicious or malicious activity of running programs, blocking their operations and signaling their presence in the system.

This additional line of defense should operate continuously and consist of two parts: a detecting section, identifying suspicious running processes and the executive section, controlling the activity of suspicious system processes and implementing self-defense against detected active threats.

Proposed detection process merges the functionality of host-ids systems (detection of changes in the list of running processes) and network-node ids (monitoring of outgoing network communication and detecting of processes responsible for it). To minimize the errors in interpretation of the list of running processes it is assumed the use of the list of processes of installed applications as the "white list".

Discussed system retains features of malware monitoring system – it records information about the place and manner of operation of a suspicious process and attempts to stop its further action, while the full removal of infection leaves other tools and intervention of user or system administrator. In authors intention the proposed solution should supplement the operation of security systems and prevent the possibility of prolonged exposure on undetected malware.

The description of proposed solution is organized as follows: Sect. 2 discusses methods of malware presence detection used in monitoring systems, details of proposed method of system's self-defense against operation of malware are presented in Sect. 3, Sect. 4 discusses chosen method of stopping the operation of selected processes in the system and its tests with malware samples. Section 5 discusses the remains in the system after stopping operation of malware with conclusions closing the paper as Sect. 6.

2   Methods of Malware Detection

Most methods described in the literature for the detection of malware operation
is intended for support of anti-virus systems and proposes methods to generate signatures that identify specific types of of malware based on code analysis,
analysis of program actions (behavioral) and mixed (heuristics), adjusting malware detection criteria to the observed symptoms of their action. For detection
of malware operation are used different methods of collecting malware performance statistics and various algorithms their classification eg. machine learning,
neural networks. Due to the volumes of data processed and the goals of the algorithms generally they are not suitable for direct monitoring the safety status of
the system.

Another approach to malware detection represent system traps (honeypots)
[4]. To better reveal details of malware operation, whole users activity were eliminated of the system. The traps are waiting for the infection of emerging malware
programs, record their actions during infection, and often record their behavior
after infection to determine the types and purposes of their communication with
the environment.

They were created and still are created various types of honeypots to monitor
different types of communication channels and access to services in the system.
As server side honeypots was created dionaea [15] (to capture windows malware),
Kippo - to capture attacks on ssh, glastopf, honeyweb [16] - for monitoring
attacks on web service. Client side honeypots are looking for malware infected
servers, pretending to be a Web browser like HoneyC, Thug, PhoneyC [6]. Also
are created a trap systems dedicated to other services [5].

Detection of the fact of system infection allows to identify responsible
process(es) in the system. Detection methods range from the searches of changes
in the file system (file integrity monitoring eg. tripware, afick [20]) or changes in
the system registry (comparison of snapshots eg. regshot [18]), by recording the
program operation in the system (logs of the service, recording low-level operation eg. process monitor [19], capturebat [14]) to audit of the system network
communication (eg. argus [13], tdilog [7]).

Most of the solutions intended for detecting malware is dedicated for specialized protection systems (IDS and IPS), and not for computer users. Not all
threats activity monitoring solutions [10] are suitable for continuous operation,
due to the amount of generated data (eg. process monitor) or lack of access to
data during monitoring.

3   Concept of System Self-defense

The concept of self-defense (self-protection) systems was presented in the literature as proposals for solutions modeled on the human immune system [1],
introducing a mechanism for identifying as self and non-self software system
components. The basis of the proposals were the introduction of various forms
of classification of system elements on self (safe) and non-self (at least suspected)

which detection started some corrective actions [2]. The concept were generally discussed on systems management level, with correcting actions based on systems topology reconfiguration.

One of presented implementation of the concept was related to experimental multi-agent cluster system technology [3]. Self-management cluster system uses knowledge of software component architecture to reconfigure application as a form of active counter-measure to observed events in distributed application operation.

The concept of self-defense mechanisms presented in this work concerns an solution for a single user system. It comes down to the use of malware monitoring methods for the identification of new, unknown or suspicious processes running on the system and then to fast stopping their actions. Mechanisms for self-defense should ensure the prompt reaction to the appearance of an unknown threats and protection of system resources against malware, on the other hand the effective detection of hidden threats.

Informations about suspicious processes may be derived from various analyzing method, monitoring data from different time ranges. To confirm them is planned process for correlating data from various sources, to verify the legality of operation of previously unknown processes, to get proofs that they are not malware (Fig. 1).

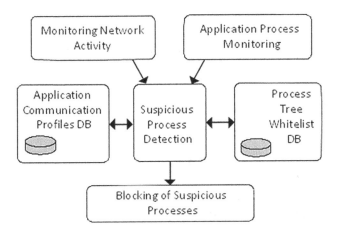

Fig. 1. General concept of system self-defense

To implement the fast detection of threat related processes it is planned the use of the direct monitoring of changes in the list of system processes, eg. by comparing the lists of active system processes in successive time slots. Detection of actuation of new process starts checking of its legality (a comparison with the list of processes that may be started by installed applications - test of the type of white list). Such method of controlling list of running processes should be especially effective in case of embedded systems with fixed list of active application like [9,12]. Long-term verification of the behavior of the running processes will be based on the analysis of system outgoing network communication using communication profiles of installed application, identifying suspicious processes running on the system.

Detection of malware process should activate the mechanism of stopping its operation, to deter the damage that the operation of such a process may cause and should indicate the presence of such process the user of the system.

3.1   Detection of Suspicious System Processes

Detection of processes related to the operation of hidden malware corresponds to the identification of processes responsible for suspicious outgoing network communication of the system. The usage of network activity profiles [11] of installed applications allows for elimination of transmissions initiated by individual application and system programs. Remaining transmissions, not matching the profiles of installed applications, that often generate a lot of connection errors can be attributed to the operating in the system malware programs.

Construction of network activity profiles based on a statistical analysis of network communication takes some time so detection of the presence of malware needs some time of its activity. Such possibility of malware detection is based on the use of locally operating network communication monitoring programs, that have access to necessary information about active processes, like tdilog or ndislog.

A quicker path of detection of appearing suspicious processes is to analyze the changes in the amount of running processes in the system. Monitoring changes of the list of running system processes, comparing the lists in the next time windows allows for detection of newly started processes. Verification of them with lists of processes started by active system applications (whitelisting) allows to detect unknown processes, possible malware.

3.2   Execution of Self-defense Action

Identified malware process should be stopped as soon as possible using eg. the kill process operation. Because modern malware often have embedded their own mechanisms of self-defense [8], that control the presence of their processes in the system and restarting the missing process in the event of its removal, more effective method seems to be the suspension of the operation of such a process, which should not start its defensive action.

Suspension of the process operation disables its further activities (conducting of possible destructive action) and allows to signal the user of the event, leaving him also the final solution of the problem.

4   Tests of Self-defense Executive Actions

Generally concept of self-defense looks quite simple – system stop the operation of every suspicious process and inform user/system administrator about the event. Earlier tests with captured malware samples on virtual machines shows that having identified malware responsible process it is not difficult to stop its operation using proper tools (at least looking on its network activity).

In case of "typical" malware there are no time constraints - the system was infected, we want to identify intruder and stop it activity to prevent further losses (eg. leaks of proprietary information). In recent years, increasingly appears the type of malicious software that aggressively encrypts documents, files, system, or even a part of the hard disk trying to extort ransom on the system's user for their recovery. This behavior changes the rules - the timing of self-defense action is becoming increasingly important the same as the effectiveness of its action. Suspension of selected process should reliably stop malware action to protect as much as possible from user resources.

To check the efficiency of proposed method of stopping the activity of malicious programs has been prepared a simplified version of self-defense algorithm as the s-def.bat script. The script uses a few simple text tools [19, 21] and consist from few steps. It starts by generating a current list of running system processes, compares this list with the initial list of started processes (of clean system) and identifies new started processes. To focus our tests on possibilities of stopping malware action, newly launched processes were treated as suspicious ones (there are no checking whether the new processes are on the list of allowed processes or not). Then the script calls simple program, which in loop attempts to suspend each of suspicious processes.

Pseudocode of self-defense script s-def.bat main steps

```
@echo off
:loop
get_process_list > new_list;
get_new_processes init_list new_list > plist;
REM  verify_process_list plist;  // now empty
get_PID_list plist > pid_list;
ps_suspend pid_list;
sleep 80;
goto loop;
```

4.1   The Test Environment

Test environment have been prepared on a system running Centos 6 with Xen. For testing were prepared virtual machines (VM) running Windows 7 system configured as a honeypot to monitor malware behavior. On Windows system were installed sample user applications like ms word, pdf reader, web browser and monitoring tools like process monitor, process manager, capturebat, regshot which were started before malware runs and were included on initial list (init_list) of running processes. For visualizing VM network communication on Centos system were started etherape monitor [17].

Tests were conducted according to a fixed pattern. Before each test on VM was started monitoring tools (procexplorer, capturebat, procmonitor) and was recorded actual initial processes list with current PID numbers, then was launched s-def.bat script. Next selected sample of malware was launched and was observed the results of its operations.

On process explorer screen were visible start of the original malware program and often also run by him his child programs. The first tests were carried out on registered by the dionaea honeypot malware samples, identified by AV program as a trojan pepex.F, worms allaple, peobot.

The results were as expected, malware started processes were stopped (suspended) on first run of s-def script and later there was no malware activity. Stopping of malware activity does not mean, of course, removal of the effects of his prior operation. A more detailed analysis of recorded activities of malware detected its copies stored in different system folders and various registry changes, providing malware start after system reboot.

For further testing was used examples of ransomware programs, malware modifying files and documents on the infected system. The aim of the test was to check whether launched periodically every tens of seconds script can stop such malware and further damage done in the system. The Table 1 presents data about tested malware samples.

Table 1. Malware samples tested

MD5 Hash	Malware	Type
31d0b80cf2b20e024c2176541a594a7b	Kryptik	Trojan horse
506cf4d78b44bc51b0ebd474b69dd611	Pepex.F	Trojan worm
7de33a523906522533cc65793f01ab3e	Allaple	Network worm
83befa17d31d68aadad93ef202c2a96b	Peobot	Worm
b06d9dd17c69ed2ae75d9e40b2631b42	Locky	Ransomware
2773e3dc59472296cb0024ba7715a64e	Jigsaw	Ransomware
108756f41d114eb93e136ba2feb838d0	Satana	Ransomware
0d1cb79feeb13d9ff4afcbc86e7b4a6a	Cerber	Ransomware
46bfd4f1d581d7c0121d2b19a005d3df	Tesla_crypt	Ransomware
6152709e741c4d5a5d793d35817b4c3d	Radaman	Ransomware
a92f13f3a1b3b39833d3cc336301b713	Petya	Ransomware

For tests were used programs called satana, tesla_crypt, jigsaw, Locky, cerber, Petya and others quite randomly selected from the GitHub [22] repository.

The course of the tests was twofold - some samples for the launch required "support" from the user - system windows open dialogue box and asking user if he has run a program of the given name (often almost random). After clicking on yes, system run the process with the given name, which often run another process (already without asking), then turn off itself (eg. tesla_crypt) or as the version of the file invoice.doc opened in a word processor completely unreadable file and suggested "enable macros". Most programs start after clicking on the sample file name supplemented with the .exe extension, and appeared on the list of processes visible in process manager, often running the subsequent processes with different names. These visible processes are practically always stopped by the script.

5   The Remains of Malware Operation

Virtually all ransomware programs were interested mainly in documents files (with extensions .txt, .doc, .pdf and the like) and those files attempted to encrypt or overwrite. Practically they did not try to modify any executable files. Of course, ransomware behaved also as a normal malware and to ensure own start after a system reboot, they wrote down their copies in different parts of the system, often using the names of legitimate system processes (eg. as c:/Users/user/AppData/Local/Temp/svchost.exe) and made modification of registry entries.

Changes made by various ransomware in the system before his stopping were different. The names of documents were changed in various ways on completely random, replaced part of the file name by fixed prefix or appended to the file name constant extension. Practically in every folder ransomware left his business card – information about the amount of ransom and payment method.

The amount of the changes depend on the time, which malware had for its activities. For modification of the documents, the program takes after safeguarding his start in the registry and after any other preparatory operations. At the period of approx. 60 seconds of the script operation it gave a different amount of changed files. eg. cerber encrypted 28 small files, giving them random names and standard a5a0 extension, jigsaw encrypt 3 files, storing them under original names with added .fun extension.

In one case malware (satana) acted like a boot sector virus, overwritten with its own code MBR and system loader (together almost 4 kB), and does not care about the registry, so had more time for system modifications. It stores his manifest message in about 580 folders during 51 second of its activity.

6   Conclusions

Possible results of self-defense actions depends on the amount of time in which identified as suspicious process is allowed to continue its operation. In experiment case this window was up to 60–80 s between checks, and in one case (Petya ransomware) it was too long.

Malware initiated system restart and started encryption process as the boot time file system errors correction process making the system unusable, in other cases it was enough time to stop damaging effects.

In normal mode of operation the monitoring of system processes should be event driven and should start investigation after every start of new system process. In general, the concept of suspension of suspicious processes as a method of securing the system works, allows more time for thoroughly checking and also allows for fast retraction in case of a wrong decision.

Acknowledgements. This work was supported by the European Union from the FP7-PEOPLE-2013-IAPP AutoUniMo project "Automotive Production Engineering Unified Perspective based on Data Mining Methods and Virtual Factory Model" (grant agreement no. 612207) and research work financed from the funds for science in the years 2016–2017, which are allocated to an international co-financed project (grant agreement no. 3491/7.PR/15/2016/2).

References

1. Dasgupta, D.: Immuno-inspired Autonomic System for Cyber Defense. http://citeseerx.ist.psu.edu/viewdoc/download?doi=10.1.1.372.1644&rep=rep.1&type=pdf
2. Sterritt, R., Hinchey, M.: Biologically-Inspired Concepts for Autonomic Self-protection in Multi-agent Systems. http://uir.ulster.ac.uk/2579/1/sasemas-sterritt-hinchey-book-finalv3.pdf
3. Hagimont, D., Boyer, F., Broto, L., De Palma, N.: Self-protection in a clustered distributed system. IEEE Trans. Parallel Distrib. Syst. **23**, 330–336 (2012)
4. Riden, J., Seifert, C.: A Guide to Different Kinds of Honeypots. https://www.symantec.com/connect/articles/guide-different-kinds-honeypots
5. Sochor, T., Zuzcak, M.: High-interaction linux honeypot architecture in recent perspective. In: Gaj, P., Kwiecień, A., Stera, P. (eds.) CN 2016. CCIS, vol. 608, pp. 118–131. Springer, Cham (2016). doi:10.1007/978-3-319-39207-3_11
6. Zeltser, L.: Specialized Honeypots for SSH, Web and Malware Attacks. https://zeltser.com/honeypots-for-malware-ssh-web-attacks/
7. Skrzewski, M.: Flow based algorithm for malware traffic detection. In: Kwiecień, A., Gaj, P., Stera, P. (eds.) CN 2011. CCIS, vol. 160, pp. 271–280. Springer, Heidelberg (2011). doi:10.1007/978-3-642-21771-5_29
8. Shevchenko, A.: The evolution of self-defense technologies in malware. https://securelist.com/analysis/publications/36156/the-evolution-of-self-defense-technologies-in-malware/
9. Cupek, R., Ziebinski, A., Franek, M.: FPGA based OPC UA embedded industrial data server implementation. J. Circ. Syst. Comput. **22**, 1350070 (2013)
10. Sochor, T., Zuzcak, M.: Attractiveness study of honeypots and honeynets in internet threat detection. In: Gaj, P., Kwiecień, A., Stera, P. (eds.) CN 2015. CCIS, vol. 522, pp. 69–81. Springer, Cham (2015). doi:10.1007/978-3-319-19419-6_7
11. Skrzewski, M.: System network activity monitoring for malware threats detection. In: Kwiecień, A., Gaj, P., Stera, P. (eds.) CN 2014. CCIS, vol. 431, pp. 138–146. Springer, Cham (2014). doi:10.1007/978-3-319-07941-7_14
12. Ziebinski, A., Cupek, R., Erdogan, H., Waechter, S.: A survey of ADAS technologies for the future perspective of sensor fusion. In: Nguyen, N.-T., Manolopoulos, Y., Iliadis, L., Trawiński, B. (eds.) ICCCI 2016. LNCS, vol. 9876, pp. 135–146. Springer, Cham (2016). doi:10.1007/978-3-319-45246-3_13
13. ARGUS- Auditing Network Activity. http://qosient.com/argus/
14. Capture BAT. https://www.honeynet.org/project/CaptureBAT
15. Dionaea capture malware. http://dionaea.carnivore.it/
16. https://www.honeynet.org/project

17. Etherape, a graphical network monitor. http://etherape.sourceforge.net/
18. Regshot. https://sourceforge.net/projects/regshot/files/regshot/
19. Windows Sysinternals. https://technet.microsoft.com/en-us/sysinternals
20. AFICK (Another File Integrity ChecKer). http://afick.sourceforge.net/
21. GNU utilities for Win32. http://unxutils.sourceforge.net/
22. GitHub ytisf/theZoo repository. https://github.com/ytisf/theZoo/tree/master/malwares/Binaries

Impact of Histogram Construction Techniques on Information - Theoretic Anomaly Detection

Christian Callegari[1,2], Stefano Giordano[2], and Michele Pagano[2(✉)]

[1] RaSS National Laboratory, CNIT, Galleria Gerace 18, 56100 Pisa, Italy
christian.callegari@cnit.it
[2] Department of Information Engineering,
University of Pisa, Via Caruso 16, 56122 Pisa, Italy
{s.giordano,m.pagano}@iet.unipi.it

Abstract. Thanks to its ability to face unknown attacks, anomaly-based intrusion detection is a key research topic in network security. In this paper anomalies are addressed from an Information theory perspective: in a nutshell, it is assumed that attacks determine a significant change in the distribution of relevant traffic descriptors and this change is measured in terms of Shannon entropy. In more detail, the traffic is first aggregated by means of random data structures (namely three-dimensional reversible sketches) and then the entropy associated to different traffic descriptors (for sake of brevity, we focus on the numbers of flows and bytes) is computed by using two alternative constructions of the corresponding empirical distributions, one based on the flows destination address and the other on their volume. The experimental results obtained over the MAWILab dataset validate the system and demonstrate the relevance of the way in which the histogram is built.

Keywords: Network security · Anomaly detection · Information theory · Entropy

1 Introduction

Network security and Information Theory are relevant research topics in the frameworks of applied statistics and networking. Although the latter was originally related to compact representations of data and error correcting codes, its basic ideas are easily applied to cryptography: for instance, the notion of entropy allows us to provide a mathematical definition of perfect secrecy. In this paper we point out the applicability of Information Theory in the framework of anomaly-based intrusion detection.

It is well-known that the universal use of the Internet has determined the continuous development of more and more sophisticated network attacks. Such attacks, at least in the initial phase (namely, until the corresponding "rules" are identified and the users have properly updated their software tools), cannot be detected by traditional signature-based Intrusion Detection Systems (IDSs). As new attacks might be extremely dangerous, the capability of capturing unknown

© Springer International Publishing AG 2017
P. Gaj et al. (Eds.): CN 2017, CCIS 718, pp. 154–165, 2017.
DOI: 10.1007/978-3-319-59767-6_13

attacks is the key motivation for research in the field of anomaly-based IDSs. The underlying rationale is quite simple and intuitive: the normal behavior of the network traffic is defined (starting from statistics collected from attack-free network data) and significant deviations from it are tagged as attacks.

Nevertheless, the design of efficient IDSs is an open research issue for two main reasons, which will be addressed in the paper: the identification of suitable traffic descriptors and the measure of the deviation from the normal behavior. In more detail, we consider two different traffic descriptors, number of flows and number of bytes, for random node aggregates. Note that, in order to guarantee the applicability of anomaly-based techniques to backbone traffic, some kind of aggregation is needed to ensure scalability; our focus on random aggregation via sketches is well-justified in the literature (see, for instance, [1]) since it outperforms standard deterministic approaches based on the network prefix and input/output routers.

An Information theoretic approach is employed to deal with the second issue. For each bucket of the sketch (i.e., for each random aggregate) a histogram is built during each time-bin and its distance from the histogram corresponding to the normal behavior for the same bucket is used to detect anomalies, according to a threshold-based voting mechanism. The level of similarity between the histograms can be measured with a simple geometric approach, based on the traditional Euclidean distance between the corresponding points in the multidimensional space. In this work, we assume that an attack determines a change in the information content of the corresponding (normalized) histogram and use (Shannon) entropy as the related measure. As highlighted in the next section, the use of Shannon entropy (as well as the related Kullback-Leibler and Jensen-Shannon divergences) is not completely new in this framework, but previous works do not investigate in detail the way in which the histograms are built. The main contribution of this paper consists in comparing two different constructions of the histograms, one based on the distribution of the relevant parameter (as mentioned above, number of flows or number of bytes) and the other focusing on the destination address of the flow (the relevant parameter in this case is just the increment for the corresponding bin of the histogram).

The ability of the different traffic descriptors and histogram definitions in capturing anomalies (note that the structure of our IDS is flexible and other traffic descriptors, histogram definitions and distances could be used) is evaluated in terms of the Receiver Operating Characteristic (ROC) curve for the well-known MAWILab traffic traces, taking into account the different labels that describe the attacks in the original dataset.

The remainder of this paper is organized as follows: Sect. 2 discusses related work, while Sect. 3 provides an overview of the theoretical background. Then, Sect. 4 describes the architecture of the proposed system. The dataset used for testing and validating our proposal is sketched in Sect. 5 and in Sect. 6 we describe the experimental results. Finally, in Sect. 7 we conclude the paper with some final remarks.

2   Related Work

The idea to use some entropy measurements in anomaly detection is not new, but a deep analysis of the impact of the different ways of computing entropy over a given dataset has not been presented yet.

For instance, entropy has been applied in [2] to detect fast Internet worms taking into account the entropy contents (more precisely, the Kolmogorov complexity) of traffic parameters, such as IP addresses, and in [3] to detect anomalies in the network traffic running over TCP. In both works an upper bound of Shannon entropy has been estimated through the use of different state-of-the-art compressors. Some more recent works include [4–6], where entropy-based methods have been applied to specific domains like cloud computing, android devices, and vehicular networks.

A different approach has been considered in [7], where Shannon entropy was used to "summarize" the distribution of specific traffic features to detect unusual traffic patterns. In [8] several information theoretic measures (including Shannon entropy, conditional entropy and Kullback-Leibler divergence) have been considered and their specific use is discussed defining a general formal framework for intrusion detection. Moreover, some recent works (see [9] and references therein) take into account also alternative entropy definitions, such as Tsallis and Rényi entropies, and their applicability to anomaly detection. It is also worth mentioning that some general weaknesses of entropy-based approaches are highlighted in [10], where "optimal camouflage" strategies are described. In our case, the combined effect of random aggregation and different kinds of entropy adds robustness to the method.

Finally, regarding sketches, even if they cannot be considered as a detection method, they have been used as a building block of several IDSs (e.g., [11]). Indeed, the use of sketches corresponds to a random aggregation that "efficiently" reduces the dimension of the data (wrt other deterministic aggregations, such as according to input/output routers [1]).

To the best of our knowledge, this paper is the first attempt to deeply investigate two alternative constructions of the empirical distributions of the traffic features (namely, one based on the flow destination address and the other one on its volume) and to study their impact on the system performance.

3   Theoretical Background

This section provides an overview of the theoretical background, focusing on sketches and Shannon entropy, so as to allow the reader to easily understand the rest of the paper.

3.1   Reversible Sketches

A sketch is a probabilistic data structure (a two-dimensional array) that can be used to summarise a data stream, by exploiting the properties of the hash functions [12]. Sketches differ in how they update hash buckets and use hashed data to derive estimates.

In more detail, a sketch is a two-dimensional $D \times W$ array $T_{D \times W}$, where each row d $(d = 0, \cdots, D - 1)$ is associated to a given hash function h_d. These functions give an output in the interval $(0, \cdots, W - 1)$ and these outputs are associated to the columns of the array. As an example, the element $T[d][w]$ is associated to the output value w of the hash function h_d.

When a new item arrives, the following update procedure is carried out for all the different hash functions:

$$T[d][h_d(i_t)] \leftarrow T[d][h_d(i_t)] + c_t \tag{1}$$

where i_t denotes the key (e.g., in our case the list of destination IP addresses) and c_t the corresponding weight (e.g., the number of bytes or flows received by that IP address).

Given the use of the hash functions, such data structures are not reversible, which makes impossible to identify the IP addresses responsible of an anomaly, after the detection. To overcome such a limitation, in our system we have used an improved version of the sketch, that is the reversible sketch [13].

3.2   Entropy Definition

In this section we recall the Shannon entropy definition; taking into account the nature of traffic data under test, we will focus on discrete distributions with a finite number L of elements.

Roughly speaking, the entropy of a random variable (RV) X (or its distribution), often called Shannon entropy [14], is a measure of the uncertainty (or variability) associated with the RV.

In more detail, let $P = \{p_1, p_2, \ldots, p_L\}$ be the probability distribution of the discrete RV X.

Then its Shannon entropy is defined as follows:

$$H(X) = \mathbb{E}\left[-\log_2 P(X)\right] = -\sum_{l=1}^{L} p_l \log_2 p_l \tag{2}$$

where $\mathbb{E}$ denotes the expectation operator, and is measured in bits (or shannon). Note that a change in the base of the logarithm just corresponds to a multiplication by a constant and a change in the unit of measure (nat for the natural logarithm and hartley (or ban) for the base 10 logarithm). In particular, when the natural algorithm is considered, (2) coincides with the well-known Boltzman-Gibbs entropy in statistical mechanics. According to the definition (2), Shannon entropy can be interpreted as the expectation of a particular function, known in the literature as self–information, which weights each p_l according to its logarithm.

It is well-known that $0 \leq H(X) \leq \log_2 L$, where the infimum corresponds to the degenerate distribution (i.e., $p_l = \delta_{k-l}$ for some integer k with $1 \leq k \leq L$) and the supremum is attained in case of uniform distribution (i.e., $p_l = 1/L \; \forall l$).

4 System Architecture

In this section we describe the architecture of the proposed system, pointing out the functionalities of each system block.

First of all the input data are processed by a module responsible of reading the network traffic (e.g., pcap [15] or NetFlow traces [16]) and of parsing them (e.g., by using the Flow-Tools [17], in case of NetFlow data). In more detail this first module will output a distinct file for each considered time-bin (let us assume we have N distinct time-bins), each file containing a list of keys i_t observed in the time-bin and the associated weights c_t.

After the data have been correctly formatted, they are passed as input to the module responsible for the construction of the reversible sketch tables [13]. In our system, such sketch tables will contain a histogram of size L in each bucket.

Indeed, given that we need to compute the entropy associated to a given traffic aggregate, maintaining a simple counter in each bucket of the sketch is not enough. Hence, instead of having a "standard" two-dimensional array, in our system we have implemented a three-dimensional data structure $T[d][w][l]$, in which the third dimension is used to store histograms.

Formally, for each new data, the update procedure of the sketch is either described by

$$\text{Method 1:}\quad T[d][h_d^1(i_t)][h_d^2(i_t)] \leftarrow T[d][h_d^1(i_t)][h_d^2(i_t)] + c_t \tag{3}$$

or by

$$\text{Method 2:}\quad T[d][h_d^1(i_t)][h_d^2(c_t)] \leftarrow T[d][h_d^1(i_t)][h_d^2(c_t)] + 1 \tag{4}$$

depending on the chosen definition of the empirical distribution of the traffic features, either based on the flow destination address i_t (Method 1) or on the feature c_t volume (Method 2).

Note that in our implementation, both the hashing schemes H^1 and H^2 include D distinct four-universal hash functions [18], which give output in the interval $[0; W-1]$ and $[0; L-1]$, respectively. This results in sketches $\in \mathbb{N}_{D \times W \times L}$, where D, W, and L can be varied (in the experimental tests they have been set to $D = 16$, $W = 512$, as justified in [19], and $L = 96$).

At this point, given that we have N distinct time-bins, we have obtained N distinct sketches $T_{D \times W \times L}^t$, where $t \in [1, N]$ identifies the time-bin under analysis.

It is important to highlight that, apart from the effectiveness of performing a random aggregation, with respect to "classical" aggregation techniques, the use of sketches also has another advantage. Indeed, as described in [10], entropy does not always allow to discriminate two (also very different) histograms (as an example, think of two scrambled versions of the same histogram). Hence, an attacker could realize a "mimicry" attack, in which the resulting histogram, yet very different from the reference one, leads to the same (or very similar) entropy value. In our case, given that the hashing scheme used to construct the sketch introduces some randomness (and it is in general unknown), such an attack is unfeasible.

Once the sketches have been constructed, they are passed in input to the block that is responsible for the actual anomaly detection phase, where two distinct sketches are considered: the reference sketch T^{ref}, which is the last observed non anomalous sketch, and the current sketch T^t.

At this point, for each bucket of the current sketch $T^t[d][w][\cdot]$ the system computes the entropy associated to the stored histogram and then the difference between such a value and the entropy associated to the same bucket in the reference sketch. Then, such a value is compared with a threshold to decide if there is an anomaly or not.

For each time-bin, the output of this phase is a binary matrix ($A \in \mathbb{N}_{D \times W}$) that contains a "1" if the corresponding sketch bucket is considered anomalous at that time-bin , "0" otherwise.

Note that, given the nature of the sketches, each traffic flow is part of several random aggregates (namely D aggregates), corresponding to the D different hash functions. This means that, in practice, any flow will be checked D times to verify if it presents any anomaly (this is done because an anomalous flow could be masked in a given traffic aggregate, while being detectable in another one).

Due to this fact, a voting algorithm is applied to the matrix A. The algorithm simply verifies if at least H rows of A contain at least one bucket set to "1" (H is a tunable parameter and it has been set $H = \frac{D}{2} + 1$). If so, the system reveals an anomaly, otherwise the matrix A is discarded and the reference sketch is updated. In case an anomaly is revealed, the responsible IP addresses are identified (by using the reversible sketch functionalities).

5   MAWILab Dataset

Our system has been extensively tested and evaluated using the traffic traces of the project MAWILab [20]. Taking into account the MAWI labels, we consider as "false positives" the flows that are not labeled as "anomalous" or "suspicious" in the MAWI archive, but that are anomalous according to the tested IDS. Instead, regarding the "false negative" definition, as discussed in [19], it depends on the actual interpretation of the MAWILab labels, and can be defined in several ways:

- "all": the number of unrevealed flows labeled as "anomalous";
- "fn 2/3/4 detector": the number of unrevealed flows labeled as "anomalous" and detected at least by two/three/four of the four detectors used in MAWI classification;
- "fn attack": the number of unrevealed flows labeled as "anomalous" belonging to the "attack" category (known attacks);
- "fn attack special": the number of unrevealed flows labeled as "anomalous" belonging to the "attack" category or the "special" category (attacks involving well-known ports);
- "fn unknown": the number of unrevealed flows labeled as "anomalous" belonging to the "unknown" category (unknown anomalous activities);

– "fn unknown 4 detector": the number of unrevealed flows labeled as "anomalous" belonging to the "unknown" category and detected by all the four detectors used in MAWI classification.

Given these definitions, in the following we discuss the results achieved by our system when taking into consideration, as traffic descriptors, either the number of flows with the same destination IP address or the quantity of traffic received by each IP address expressed in bytes.

6   Performance Evaluation

Before actually discussing the system performance, we have carried out several experimental tests to evaluate the effects of using different traffic descriptors and different definitions of histograms.

Figures 1 and 2 represent the scatter plots of entropy computed over the same traffic aggregates, when taking into consideration flows and bytes for Method 1 and 2, respectively. In more detail, we consider one row of the sketch and each point represents one bucket for the chosen time-bin: indeed, its coordinates are given by the values of the entropy associated to the histogram of the first descriptor (x axis) and second one (y axis).

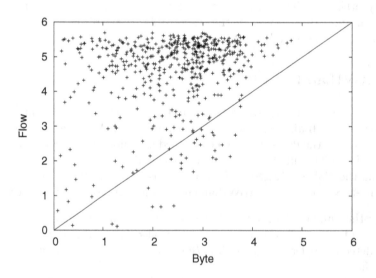

Fig. 1. Scatter plot: Method 1 – Byte vs. Flow

The basic idea is that two variables that are strongly related to each other should present a regular behavior in the scatter plot. In particular, if the scatter plot presents a linear pattern, the use of the variables should take to the same system performance.

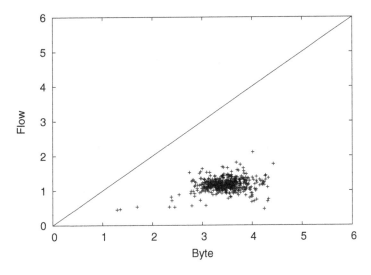

Fig. 2. Scatter plot: Method 2 – Byte vs. Flow

As it can be clearly seen from the figures, the scatter plots show that the couples of variables do not have a clear relationship. This result indicates, as will be confirmed by performance evaluation tests, that the different traffic descriptors can lead to quite different results. Moreover, the different "shape" of the two scatter plots also indicates that the use of different histogram definitions should take to different performance.

The subsequent figures show the actual performance, in terms of detection probability P_D and false alarm probability P_{FA}, of our system.

First of all, in Fig. 3 we present the performance achieved applying Method 1 to flows, when varying the interpretation of the MAWILab labels, as discussed in Sect. 5. The plots clearly show that the system performance are strongly influenced by the different interpretation of the labels, going from unacceptable (e.g., "fn attack") to very good (e.g., "fn unknown 4 detector"). This variability, already known in the literature [19], can be justified by the very little number of flows belonging to some categories (e.g., "fn attack") compared with the total number of anomalies. In any case, the lack of effectiveness in detecting known attacks is not a major concern for the applicability of the anomaly detection systems. Indeed, anomaly-based IDSs are typically used in conjunction with misuse-based systems, which are effective in revealing known attacks, but are unable to find the unknown ones (for which the signatures are not present yet!). The latter are instead well detected by the proposed algorithms.

Then, Fig. 4 presents the results obtained when applying Method 2 to flows. Apart from the previous considerations about the relationship between the performance and the MAWILab labels, which are still valid, it is very interesting to observe that the use of Method 2 (corresponding to the histogram based on

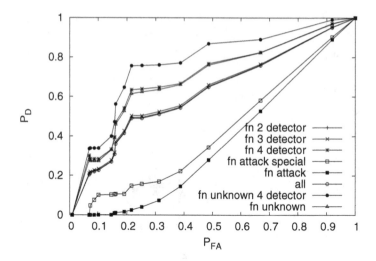

Fig. 3. ROC curve: Method 1 (Flow)

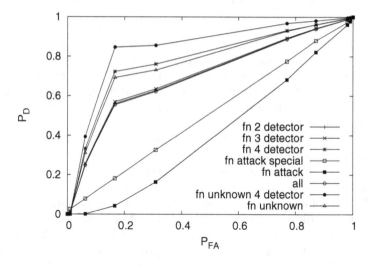

Fig. 4. ROC curve: Method 2 (Flow)

the volume feature) takes to better performance (indeed fixing $P_{FA} = 20\%$, P_D increases from about 78% to about 85%).

An analogous analysis is reported in Figs. 5 and 6, where we show the performance achieved over bytes when applying Method 1 and 2, respectively. In this case the differences in the performance offered by the two methods are even more evident: indeed, while Method 1 leads to unacceptable performance, Method 2 is able to offer almost optimal results.

All the results are summarized in Table 1, where, for sake of completeness, we report the values of the AuC (Area-under-the-Curve), for all the discussed

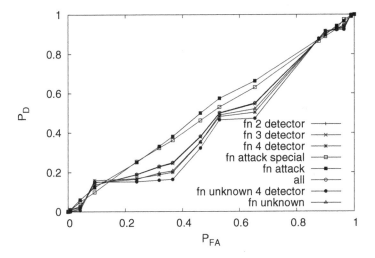

Fig. 5. ROC curve: Method 1 (Byte)

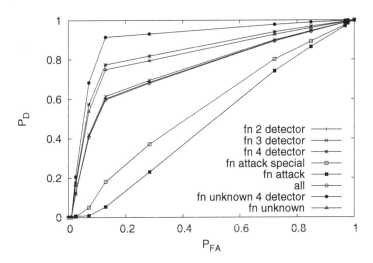

Fig. 6. ROC curve: Method 2 (Byte)

Table 1. AuC values

Method	AuC							
–	All	fn 2	fn 3	fn 4	Attack	Attack sp.	Unknown	Unknown 4
Flow – Method 1	0.61949	0.621851	0.627007	0.699259	0.362535	0.418059	0.693165	0.768803
Flow – Method 2	0.697516	0.700294	0.705378	0.781268	0.404348	0.510531	0.765937	0.842355
Byte – Method 1	0.455644	0.455011	0.453766	0.436463	0.515499	0.496697	0.440622	0.421189
Byte – Method 2	0.749288	0.751888	0.758025	0.835591	0.4803	0.557955	0.819242	0.90413

methods and for all of the different MAWILab labels. From the table we can easily conclude that the best performance (over the "fn unknown 4" label), are obtained by Method 2 using byte as traffic descriptor.

7   Conclusions

In this paper we have presented an information-theoretic approach to anomaly detection, considering two different traffic descriptors and two alternative constructions of the corresponding empirical distributions, one based on the traffic destination address and the other one on its volume. In more detail, after random aggregations via sketches, the Shannon entropy is employed as a "compact characterization" of the distribution for the current time-bin and then a threshold-based voting mechanism is used to detect anomalies, i.e. attacks. Indeed, it is well-known that random aggregation and entropy are powerful tools in the framework of intrusion detection, but little attention has been paid to possible ways in which the underlying empirical distribution (for which the entropy is calculated) can be constructed.

The extensive evaluation phase, carried over the MAWILab traffic traces, has demonstrated that the use of different traffic descriptors and histogram definitions may have a significant impact on the system performance. In more detail, the experimental results suggest that the system offers almost optimal performance, when using histograms based on the feature volume applied to bytes, with $P_D = 93\%$ and $P_{FA} = 17\%$.

Acknowledgment. This work was partially supported by Multitech SeCurity system for intercOnnected space control groUnd staTions (SCOUT), a research project supported by the FP7 programme of the European Community.

References

1. Callegari, C., Gazzarrini, L., Giordano, S., Pagano, M., Pepe, T.: When randomness improves the anomaly detection performance. In: Proceedings of 3rd International Symposium on Applied Sciences in Biomedical and Communication Technologies (ISABEL) (2010)
2. Wagner, A., Plattner, B.: Entropy based worm and anomaly detection in fast IP networks. In: 14th IEEE International Workshops on Enabling Technologies: Infrastructure for Collaborative Enterprise (WETICE 2005), pp. 172–177, June 2005
3. Callegari, C., Giordano, S., Pagano, M.: On the use of compression algorithms for network anomaly detection. In: 2009 IEEE International Conference on Communications, pp. 1–5, June 2009
4. Navaz, A.S.S., Sangeetha, V., Prabhadevi, C.: Entropy based anomaly detection system to prevent ddos attacks in cloud. CoRR abs/1308.6745 (2013)
5. Ghaffari, F., Abadi, M.: Droidmalhunter: a novel entropy-based anomaly detection system to detect malicious android applications. In: 2015 5th International Conference on Computer and Knowledge Engineering (ICCKE), pp. 301–306, October 2015

6. Marchetti, M., Stabili, D., Guido, A., Colajanni, M.: Evaluation of anomaly detection for in-vehicle networks through information-theoretic algorithms. In: IEEE 2nd International Forum on Research and Technologies for Society and Industry (RTSI 2016) (2016)

7. Lakhina, A.: Diagnosing network-wide traffic anomalies. In: ACM SIGCOMM, pp. 219–230 (2004)

8. Lee, W., Xiang, D.: Information-theoretic measures for anomaly detection. In: Proceedings of the 2001 IEEE Symposium on Security and Privacy, SP 2001. IEEE Computer Society, Washington, DC (2001)

9. Callegari, C., Giordano, S., Pagano, M.: Entropy-based network anomaly detection. In: IEEE International Conference on Computing, Networking and Communication (ICNC) (2017)

10. Zhang, L., Veitch, D.: Learning entropy. In: Domingo-Pascual, J., Manzoni, P., Palazzo, S., Pont, A., Scoglio, C. (eds.) NETWORKING 2011. LNCS, vol. 6640, pp. 15–27. Springer, Heidelberg (2011). doi:10.1007/978-3-642-20757-0_2

11. Subhabrata, B.K., Krishnamurthy, E., Sen, S., Zhang, Y., Chen, Y.: Sketch-based change detection: methods, evaluation, and applications. In: Internet Measurement Conference, pp. 234–247(2003)

12. Cormode, G., Muthukrishnan, S.: An improved data stream summary: the count-min sketch and its applications. J. Algorithms **55**(1), 58–75 (2005)

13. Schweller, R., Gupta, A., Parsons, E., Chen, Y.: Reversible sketches for efficient and accurate change detection over network data streams. In: Proceedings of the 4th ACM SIGCOMM Conference on Internet Measurement, IMC 2004, pp. 207–212. ACM, New York (2004)

14. Shannon, C.E., Weaver, W.: The Mathematical Theory of Communication. University of Illinois Press, Illinois (1949)

15. pcap (format). http://imdc.datcat.org/format/1-002W-D=pcap. Accessed 11 Jan 2017

16. Claise, B.: Cisco Systems NetFlow services export version 9. RFC 3954 (Informational), October 2004

17. Flow-Tools Home Page. http://www.ietf.org/rfc/rfc3954.txt

18. Thorup, M., Zhang, Y.: Tabulation based 4-universal hashing with applications to second moment estimation. In: SODA 2004: Proceedings of the Fifteenth Annual ACM-SIAM Symposium on Discrete Algorithms, Philadelphia, PA, USA, pp. 615–624. Society for Industrial and Applied Mathematics (2004)

19. Callegari, C., Casella, A., Giordano, S., Pagano, M., Pepe, T.: Sketch-based multidimensional IDS: a new approach for network anomaly detection. In: IEEE Conference on Communications and Network Security, CNS 2013, National Harbor, MD, USA, 14–16 October 2013, pp. 350–358 (2013)

20. MAWILab http://www.fukuda-lab.org/mawilab/. Accessed Apr 2016

Information Technology for Botnets Detection Based on Their Behaviour in the Corporate Area Network

Sergii Lysenko[✉], Oleg Savenko, Kira Bobrovnikova, Andrii Kryshchuk,
and Bohdan Savenko

Department of Computer Engineering and System Programming,
Khmelnitsky National University, Instytutska, 11, Khmelnitsky, Ukraine
{sirogyk,savenko_oleg_st,savenko_bohdan}@ukr.net,
bobrovnikova.kira@gmail.com, rtandrey@rambler.ru
http://ki.khnu.km.ua

Abstract. A new information technology for botnets detection based on the analysis of the botnets' behaviour in the corporate area network is proposed. Botnets detection is performing combining two ways: using network-level and host-level analysis. One approach makes it possible to analyze the behaviour of the software in the host, which may indicate the possible presence of bot directly in the host and identify malicious software, and another one involves monitoring and analyzing the DNS-traffic, which allows making conclusion about network hosts' infections with bot of the botnet. Based on this information technology an effective botnets detection tool BotGRABBER was constructed. It is able to detect bots, that use such evasion techniques as cycling of IP mapping, "domain flux", "fast flux", DNS-tunneling. Usage of the developed system makes it possible to detect infected hosts by bots of the botnets with high efficiency.

Keywords: Bot · Botnet · Botnet detection · DNS-traffic · Botnet behaviour · Botnet's evasion techniques · Malicious software · Artificial immune systems · Detectors · Positive selection algorithm · Approximate string matching algorithm

1 Introduction

Nowadays the botnets are the reason of such cybercrimes as DDoS attacks, banking fraud and cyber-espionage. Such criminal activities cause significant damage for economy. Botmasters employ various techniques to develop, control, maintain and conceal its complex C&C infrastructures. The main problems of the botnets detection is that they use P2P techniques [1], evasion techniques such as cycling of IP mapping, "domain flux", "fast flux", DNS-tunneling which increase the resilience against take-down actions. In order to manage and control the infected hosts the vast majority of botnets use DNS. In addition, some botnets use the encryption techniques for the communication payload in order to prevent its detection [2].

© Springer International Publishing AG 2017
P. Gaj et al. (Eds.): CN 2017, CCIS 718, pp. 166–181, 2017.
DOI: 10.1007/978-3-319-59767-6_14

2   Related Works

Research community pays close attention to the problem of the botnets detection and there are a great number of approaches devoted to its solving.

One of the way to explore the botnets' behaviour is the usage of the honeynets. The paper [3] is devoted to an analysis of a one-year-long period of operation of a honeynet composed of 6 Dionaea honeypots emulating Windows services. The analysis focused on the frequency of attacks according to the location of individual honeypots (sensors) as well as to the geographical location of attackers. From the statistical processing of the results, it was demonstrated that the most frequently attacking malware was well-known Conficker worm. Moreover, attacking OS were studied with the conclusion that Windows is the most frequent OS. Regarding the geographical location of the attackers, several non-western countries and autonomous systems were indicated as being the most frequent origin of the attacks.

The paper [4] presents recent results in honeynet made of Dionaea, Kippo (emulating Linux services) and Glastopf (emulating website services) honeypots. The most important result consists in the fact that the differentiation among honeypots according to their IP address is relatively rough (usually two categories, i.e. academic and commercial networks, are usually distinguished, but the type of services in commercial sites is taken into account, too).

Mentioned approaches are focused on the botnets behaviour investigation, but are not able to block them botnets' malicious activities.

Approach [5] is based on two components: (1) passive monitoring of communication characteristics and (2) DNS registration behaviour analysis. DNS registration analysis allows detecting the preparatory actions of deployment of the C&C infrastructure and the bots. Therefore, it allows botnets early detection and consequently facilitates proactive botnets mitigation. Moreover, as many end users are unable to detect and clean infected machines, this approach tackles the botnet phenomenon without requiring any end user involvement, by incorporating ISPs (Internet Service Providers) and domain name registrars. In addition, this will enable the discovery of similar behaviour of different connected systems, which allows detection in cases where bots are registered under domains that are not willing to cooperate. The main disadvantage of such approach is that it is infeasible without the provider support.

In [6] a scalable approach, which is not dependent on the volume of DNS-traffic was proposed. It is able to detecting malicious behaviour within large volumes of DNS-traffic. It leverages a signal processing technique, power spectral density (PSD) analysis, to discover the major frequencies resulting from the periodic DNS queries of botnets. The PSD analysis allows detecting sophisticated botnets regardless of their evasive techniques, sporadic behaviour, and even normal users' traffic. The method allows dealing with large-scale DNS data by only utilizing the timing information of query generation regardless of the number of queries and domains. Finally, the approach discovers groups of hosts which show similar patterns of malicious behaviour.

Main disadvantages of such approach are: relying on the similar periodic query patterns, it is suited for large networks, is not able to detect botnets that use DNS-tunneling.

The domain names in the DNS-queries can be randomly or algorithmically generated and their alphanumeric distribution is significantly different from legitimate ones. In [7] a negative reputation system that considers the history of both suspicious group activities and suspicious failures in DNS-traffic to detect domain-flux botnets was developed. The main goal is to automatically assign a high negative reputation score to each host that is involved in these suspicious domain activities. To identify randomly or algorithmically generated domain names, three measures, namely the Jensen-Shannon divergence, Spearman's rank correlation coefficient, and Levenshtein distance was used. But the proposed approach is only intended to be used for detection domain-flux evasion technique.

The common problem of the mentioned and other known approaches is that they are not able to solve not zero-day problem concerning botnets detection.

3   Previous Work

In [8] a new DNS-based technique for botnets detection in the corporate area networks was presented. Technique includes two ways for botnet detection: the passive and active DNS monitoring. It made it possible to identify the botnets, which use such evasion techniques as cycling of IP mapping, "domain flux", "fast flux", DNS-tunneling [9–18]. Such system, named BotGRABBER, was based on a cluster analysis of the features obtained from the payload of DNS-messages, which indicated the usage of the evasion techniques by botnets. The result of clustering was a degree of membership of the feature vectors to one of four clusters, where the membership of feature vector to cluster indicated the queries executing using the evasion techniques [8].

For the purpose of the amendment of possible uncertainty of clustering results, technique involves the additional features obtained by active DNS probing.

The main disadvantages of the proposed approach is that the active DNS probing may be "visible" to botmaster. Also, the involvement of cluster analysis in many cases has limited opportunities to describe the behaviours of the bots which use evasion techniques. This causes impossibility of the botnets detection as behaviour could be referred to "malicious" and "normal" clusters at the same time. Also, the technique could detect hosts that are infected with botnets, but is not able to directly localize the malicious software in the infected host.

4   Information Technology for Botnets Detection Based on the Botnet Behaviour in the Corporate Area Network

In this article, a new information technology for botnets detection based on the analysis of the botnets' behaviour in the corporate area network is presented.

The technology is designed to eliminate the disadvantages of the approach described above and to extend the capabilities of the botnets detection system BotGRABBER.

For this purpose, we propose perform botnets detection by combining two ways: using network-level and host-level analysis. They have to improve and refine the results of each other. On the one hand, the approach makes it possible to analyse the behaviour of the software in the host, which may indicate the possible presence of bot directly in the host and identify malicious software. On the other hand, it involves monitoring and analysing the DNS-traffic, which allows making conclusion about network hosts' infections with bot of the botnet. Thus, new information technology for botnets detection is able to increase the botnets' detection efficiency in the corporate area networks.

Two stages operate in parallel in such way:

1. If the suspicious software is detected in the host of the corporate area network, then it is blocked. After that, the approach provides the querying of the results of the network-level analysis about the suspicious DNS-requests from potentially infected host and from other hosts of the network for further analysis.
2. If suspicious DNS-requests are detected by the means of the network-level analysis, the list of the infected hosts in the network is produced. Then, with the knowledge of the suspicious behaviour of the software in the infected host we can localize bot and then block it.

Developed information technology includes two techniques for botnets detection. Let us consider each of them.

4.1   Botnets Detection Technique Based on the Bots Behaviour Analysis in the Host

In order to detect the suspicious software with the features of the botnets' bots, a new technique based on its behaviour in the host was developed. It includes such steps:

1. construction the set of bots' behaviours at the different stages of its life-cycle as the patterns (presented as the bit strings);
2. system calls monitoring for the host's software;
3. presentation of the monitored software behaviour as a bit string;
4. comparison the constructed bit string concerning to software's behaviour with the set of bit strings concerning to the botnets' behaviours' patterns at different stages of bots life-cycle;
5. block software functioning estimated as malicious;
6. querying about the results obtained by the means of the network-level analysis about the suspicious DNS-requests from potentially infected host and from other hosts of the network for further analysis.

Behavioural Model of the Bot. In order to detect the bots of the botnets in the hosts we have to explore its features and to built its behavioural model. Developed behavioural model takes into account the features of bot and formalizes its functioning process in the host during the life-cycle, which includes five stages: (1) infection; (2) initial registration or connection to C&C-server; (3) performance of the malicious activity; (4) maintenance; (5) termination of bot's functioning. The scheme of the bot's life-cycle is presented on the Fig. 1.

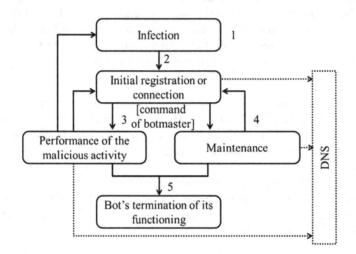

Fig. 1. The scheme of the bots' life-cycle

Thus, behavioural model of the bot we can present as a tuple:

$$M = \langle B, \Psi, L, F, Y, P, Z \rangle, \tag{1}$$

where $B = \{b_j^p\}_{j=1}^{N_B}$ - the set of the network protocols, used for botnet's control, N_B - number of the network protocols, $p \in P, P = \{1..65535\}$ - the set of the ports, used for botnet's control; $\Psi = \{\psi_j\}_{j=1}^{4}$ - the set of the botnets detection evasion techniques based on DNS; $L = \{l_j\}_{j=1}^{5}$ - the set of the life-cycle stages of the bot (l_1 - infection; l_2 - initial registration or connection; l_3 - malicious activity performance; l_4 - maintenance; l_5 - bot's termination of its functioning); $Z = \{z_j\}_{j=1}^{N_Z}$ - the set of the bots, which belong to botnet, N_Z - number of the bots; $F = \{f_j\}_{j=1}^{N_F}$ - the set of the bots' functions, corresponding to the life-cycle stage of the bot, N_F - number of the bots' functions; f_1 - function of the host infection $l_1 \Longrightarrow Y \xrightarrow{f_1} \{h_{inf} \mid h_{inf} \in H\}$, where Y - the set of the destruction actions of the bot, H - the set of the hosts of the network, h_{inf} - host, infected by the bot; f_2 - the function of the joining of the infected host to the botnet $l_2 \Longrightarrow Y \times \{h_{inf} \mid h_{inf} \in H\} \xrightarrow{f_2} Z'$; f_3 - function of the updated version the bot of the botnet $l_3 \Longrightarrow z \times Y \xrightarrow{f_3} z'$; f_4 - function of the command execution for destructive activity $l_4 \Longrightarrow Z \times \{p \mid p \in P\} \xrightarrow{f_4} Y$, where P - the set of the botmaster' commands which can be performed by bots of the botnet; f_5 - function of the termination of the bots functioning $l_5 \Longrightarrow Z \setminus \{z \mid z \in Z\} \xrightarrow{f_5} Z'$.

With the knowledge of the behaviour and main features of the botnets' bots, we can construct the set of bots' behaviours at the different stages of its life-cycle $DB = \{\Phi_{l_j}\}, j = \overline{1,5}$. Let us present the known behaviour pattern and a monitored software's behaviour as bit strings $\Phi = y_1 y_2 ... y_n$ and $T = t_1 t_2 ... t_m$ respectively, where $y_i \in Y, t_i \in Y$. Let us denote the boolean function of the string matching $\Omega(\Phi, T)$ between the known behaviour pattern and behaviour of the observed software which indicates matching or mismatching.

Comparison of the constructed bit string concerning to software's behaviour with the set of bit strings concerning to the botnets' behaviours' patterns. On the next stage we monitor software's behaviour in the host and present it as the bit string and compare it with known bots' ones (patterns).

All groups of behaviours are presented as patterns (namely as bit strings). Thus, the task of identification of the suspicious behaviour of some software is to find the string match of the observed behaviour with the string presented in our base of the botnets' behaviours. In order to solve this problem the approximate string matching algorithm was used. It deals with the k differences problem solving. If we are given two strings, the sequence $T = t_1 t_2 ... t_m$ and the pattern $\Phi = y_1 y_2 ... y_n$ in some alphabet Σ, and an integer k, the algorithm enables finding all substrings Φ' of T with the edit distance at most k from Φ. The edit distance intends the minimum number operations for editing (the differences) which are required for converting Φ' to Φ. An editing operation means either an insertion, a deletion or a change of a character. The k mismatches problem is a special case with the change as the only editing operation [19]. The preprocessing of the pattern needs time $O(mn)$. Table 1 demonstrates the experimental results of the approximate string matching for different values of length R and parameter k. Thus, when we have $k = 0$, that means an exact match, there is no solution. However, when the value $k = 2$, $k = 3$, the number of solutions is sufficiently low. Increasing of k leads to growing of solutions number, while the search time for matches is also increasing. The experiments proved that the task of the suspicious behaviour detection is possible while parameter $k = 4$.

Presenting $\Omega(\Phi_{l_i}, T)$ as the fact of identification of the suspicious behaviour of some software which is similar to some stage of bots life cycle l_i, the procedure of the monitoring can be presented as an Algorithm 1.

4.2 DNS-Based Technique for Botnets Detection Using the Artificial Immune System

For the purpose of the botnets detection in the network a new DNS-based technique, which is based on the usage of the artificial immune systems (AIS), was developed. It is aimed for anomaly detection in the DNS-traffic. The technique involves such steps:

if $((\Omega(\Phi_{l_1}, T) \text{ and } \Omega(\Phi_{l_2}, T)) \text{ or } \Omega(\Phi_{l_2}, T) \text{ is true})$ **then**
    mark software as suspicious, querying the results of the network −
    level analysis about anomaly in the DNS − traffic
    and possible infection of the host;
    if (*the results of the network − level analysis indicates, that*
    the host is infected) **then**
      | *block_the_software*
    end
end
if $(\Omega(\Phi_{l_3}, T) \text{ or } \Omega(\Phi_{l_4}, T) \text{ is true})$ **then**
  | *block_the_software*
end

Algorithm 1. Functioning of the botnets detection technique based on the bots behaviour analysis in the host at the stage of monitoring

Table 1. The experimental results of the approximate string matching for different values of length R and parameter k

	Alphabet Σ	Length of sequence R	k-difference parameter	Number of found strings
Φ_1	430	35	0	0
	430	35	2	0
	430	35	3	0
	430	35	4	1
	430	35	5	2
Φ_2	430	78	0	0
	430	78	2	0
	430	78	3	1
	430	78	4	3
	430	78	5	8
Φ_3	430	93	0	0
	430	93	2	1
	430	93	3	4
	430	93	4	17
	430	93	5	24

1. construction of the set of patterns for bots' behaviours in the DNS-traffic (presented as the bit strings);
2. gathering incoming DNS-traffic of the network;
3. analysis the fields of TTL (time to live period) of the incoming DNS-message about certain domain name;
4. building a feature vector on the base of the extracted features from the incoming DNS-messages about certain domain name;

5. anomaly detection in the DNS-traffic based on the usage of the artificial immune systems;
6. having the list of the infected hosts in the network, perform host-level analysis of the software in the infected host, localize and then block it.

Let us consider the steps of the technique in details.

Construction of the set of patterns for bots' behaviours in the DNS-traffic. In order to detect DNS-tunneling, such features, obtained from the payload of DNS-messages, are analysed [15–18]:

1. l_N - a length of the requested domain name, $l_N \in [75, 255]$;
2. n_U - a number of the unique characters in the domain name, $n_U \in (27, 37]$;
3. e_N - entropy of the domain name, $e_N \geq f_{Eb32}$, where f_{Ebn} - a dependence function of the DNS-message field entropy of its length, n - base of coding [16];
4. f_{UR} - usage uncommon record types of the DNS, that are not commonly used by a typical client (e.g. TXT, are most often used for tunneling (excluding mail servers), KEY or NULL);
5. e_R - an entropy of the DNS-records, which are contained in the DNS-messages (CNAME, TXT, NS, MX, KEY, NULL etc.), $e_R \geq f_{Eb64}, e_R \geq f_{Eb256}$;
6. l_P - a maximum size of the DNS-messages, $l_P > 300$.

In order to detect evasion techniques such as fast-flux service network, "domain flux" and cycling of IP mappings, the following features are used [9–14]:

1. n_{IP} - a number of IP addresses concerned with the domain name, $n_{IP} \in (5, \infty)$;
2. s_{IP} - an average distance between the IP addresses concerned with the domain name, $s_{IP} \in (65535, \infty)$;
3. n_A - a number of A-records corresponding to the domain name in the incoming DNS-message, $n_A \in (5, \infty)$;
4. s_A - an average distance between the IP addresses in the set of A-records corresponding to the domain name in the incoming DNS-message, $s_A \in (65535, \infty)$;
5. n_{UA} - a number of unique IP addresses in the sets of A-records corresponding to the domain name in the DNS-messages, $n_{UA} \in (8, \infty)$;
6. s_{UA} - an average distance between unique IP addresses in the sets of A-records corresponding to the domain name in the DNS-messages, $s_{UA} \in (65535, \infty)$;
7. n_D - a number of domain names that share IP address corresponding to the domain name, $n_D \in [8, \infty]$;
8. $t_{mod}, t_{med}, t_{aver}$ - a value of the TTL-period, mode $t_{mod} \in [0, 900]$, median $t_{med} \in [0, 900]$, average value $t_{aver} \in [0, 900]$;
9. f_S - a sign of the success of the DNS-request.

Also, the technique deals with two features: the group flush of the local DNS-cache f_f [20] and synchronous DNS-requests by hosts of the network f_q [8, 20].

Thus, we are able to present the pattern, which describe the bots' behaviour, as the set of the bit strings:

$$
p_i = \left\langle \begin{array}{c} l_{N_1}...l_{N_a} \; n_{U_{a+1}}...n_{U_{a+b-1}} \; e_{N_{a+b}}...e_{N_c} \; t_{mod_{c+1}}...t_{mod_{c+d-1}} \\ t_{med_{c+d}}...t_{med_e} \; t_{aver_{e+1}}...t_{aver_{e+f-1}} \; n_{A_{e+f}}...n_{A_g} \; n_{IP_{g+1}}...n_{IP_{g+h-1}} \\ s_{IP_{g+h}}...s_{IP_i} \; s_{A_{i+1}}...s_{A_{i+j-1}} \; n_{UA_{i+j}}...n_{UA_k} \; s_{UA_{k+1}}...s_{UA_{k+l-1}} \\ n_{D_{k+l}}...n_{D_m} \; f_{UR_{m+1}}...f_{UR_{m+n-1}} \; e_{R_{m+n}}...e_{R_o} \; l_{P_{o+1}}...l_{P_{o+l-1}} \\ f_{S_{o+p}}...f_{S_q} \; f_{q_{q+1}}...f_{q_{q+r-1}} \; f_{f_{q+r}}...f_{f_s} \end{array} \right\rangle ,
$$

$$(2)$$

where $a..s$ - the index numbers of the bits in the coded sequence of the pattern.

Gathering incoming DNS-traffic of the network. The incoming DNS-traffic is gathered by the set of network sniffers connected to the switch with port mirroring.

NOTE. In order to reject the legitimate DNS-requests and to reveal known malicious domain names technique operates with the "white" and "black" lists of the domain names.

Analysis the fields of TTL of the incoming DNS-messages about certain domain name. Based on the values of the TTL fields such incoming DNS-messages are processed: each first captured DNS-message about certain domain name within the TTL-period DNS; each repeated DNS-message received by host within the TTL-period, when the source of message is non-local DNS-server, and TTL-period referred to this message differs from the rest of TTL-period within which this message have received.

In the monitor stage of technique the feature vectors on the base of the extracted features from the incoming DNS-messages about certain domain name are to be built. It has the same form as pattern (2).

Anomaly detection in the DNS-traffic. In the network-level analysis, the botnets detection is carried out with the usage of the artificial immune systems apparatus [21]. It operates with the concept of "self-nonself" and enables the invaders patterns detection (abnormal DNS-traffic, which indicates the presence of botnets in the corporate area network).

In order to perform the "self-nonself" distinction, the concept of positive selection was employed [22]. The anomaly detection involves the following stages: building invader pattern, generation of detectors, monitoring and anomaly recognition.

The invaders patterns correspond to the behaviour of botnets in the DNS-traffic (abnormal DNS-traffic) and are based on the knowledge about the features of the botnets' DNS-traffic.

The value of each feature in pattern were normalized according to the scatter in the data. Normalized data was divided into intervals $d = (max - min)/(2^m - 2)$, where m defines the number of binary numbers used to encode. Each pattern was encoded in binary form according to the interval n, to which the features value belongs to $n \in [1; 2^m - 2)$. If the value of the feature is out of the interval $[min; max]$, it was encoded as m by "0" and "1" respectively.

Using the positive selection algorithm a set of detectors D, which have high affinity with the invader patterns were generated.

The aim of the recognition stage is to identify high affinity of the detector with the input data. It indicates the anomaly detection, hence the presence of botnets in the network.

In order to determine the value of the affinity the r-contiguous bits matching rule was used. Two strings of length l are said to match under this rule if they exactly match in at least r contiguous bit positions [23]. The string matching with detector regarded as the detection of the anomaly behaviour in the DNS-traffic.

NOTE. Feature vectors are normalized and coded the same way as patterns.

In order to reach high probability of the anomaly detection for the generated set of patterns N_s, the optimal values of the detectors number N_r and parameter r were explored. Table 2 and Fig. 2 show that a significantly higher probability of anomaly detection P_f can be reached with a value of $r = 32$, but in this case it is necessary to generate a large number of detectors.

So, the usage of the AIS apparatus enables the anomaly detection of the DNS-traffic with high efficiency.

In total the scheme of the information technology functioning is presented in Fig. 3.

Table 2. Probability of the anomaly detection with different values of the parameters for the positive selection algorithm

m	l	$N_s * 1000$	$N_r * 100$	Parameter, r	Probability of anomaly detection, P_f, %
2	64	10	10	8	0,88
2	64			16	14,15
2	64			32	33,51
2	64		50	8	0,98
2	64			16	29,87
2	64			32	85,77
2	64		100	8	1,02
2	64			16	54,25
2	64			32	99,3
2	64	100	100	8	2,01
2	64			16	15,3
2	64			32	25,6
2	64		500	8	2,55
2	64			16	23,36
2	64			32	72,88
2	64		1000	8	2,9
2	64			16	75,36
2	64			32	99,8

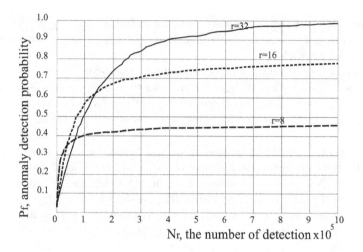

Fig. 2. Dependence of the anomaly detection probability P_f on the value r and the number of detectors N_r

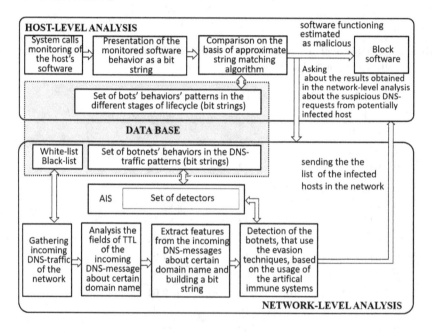

Fig. 3. The scheme of the information technology functioning

5   Experiments

In order to determine the efficiency of the proposed information technology for botnets detection based on the botnet behaviour in the corporate area network (improved BotGRABBER system), a series of experiments were held. As a base for the investigational study, the campus network of Khmelnitsky National University (Ukraine) was deployed.

A malicious software that emulated bots' behaviour in the DNS-traffic, was developed. Created software had the features of the botnets' bots with centralized architecture. Emulated bots' behaviour involved four types of scenarios which performed evasions techniques actions. Each experiment performed one scenario.

In order to simulate cycling of IP mappings for malicious domains the set of domain names were registered and were used in the capacity of C&C-server of the botnets. The domain name was periodically associated with IP address from the set of distributed IP addresses. Thus, botnets C&C-server had the opportunity to change the location periodically. In order to connect to the C&C-server the bots initiated DNS-requests about C&C-server domain name after each DNS TTL-period expiration or after the hosts' restart.

In order to simulate the "domain flux" evasion technique the list of domain names was generated. Such generation was performed by using the domain generation algorithms (DGA). Then the part of generated domain names were randomly selected and registered. One of the registered domain name was periodically associated with one IP address of the host that had the ability to perform the botnets C&C-servers' activities. Thus, the C&C-server was able to migrate periodically to the new domain names from the list and to change the locations periodically.

In order to connect with the C&C-server, the bots used DGA to generate the list of domain names and then sent the DNS-requests about those domain names, until they received a successful DNS-response with the IP address from C&C-server.

Also in order to imitate "domain flux" evasion technique, the domain names were generated using the certain words combinations of dictionary.

In order to simulate the fast-flux service network one host was used as the botnet C&C-server. Also, the set of domain names were registered, which were used as domain names of the C&C-server. For the C&C-server evasion, the network of proxy servers were used. C&C-server was hidden by the frontend "fast-flux" proxy network nodes, which served the "flux-agents" and were able to redirect bots' equests and data to and from backend C&C-server. "Flux-agents" used distributed IP addresses. Within the specified DNS TTL-period, the domain name was associated with a new subset of IP addresses from the set of "flux-agents" IP addresses. Using the round robin DNS algorithm, this subset was cyclically changed. Thus, when bot tried to connect to the C&C-server, it was actually connected to a "flux-agents".

In order to simulate the DNS-tunneling evasion technique the "fake" DNS-server was used. It was registered as an authoritative name server for a particular DNS zone. Using local or other DNS-server bots were able to perform DNS-requests about domain names in this zone. In the fields of the DNS-requests bots encrypted the information for C&C-server. According to the domain name resolution process, bots' DNS-requests were redirected to "fake" authoritative name server. In turn, bots received the commands from the "fake" authoritative name server. Such commands were coded in the fields of the DNS-answers. In order to transmit the encrypted information different encryption techniques and different types of DNS-records were used.

Each bot performed 600 DNS-requests during its functioning.

Generated software had the ability to perform different kinds of scenarios of the botnets functioning corresponding to its life-cycle stages. Thus, each experiment involved various ways of host infection (downloading from website, downloading from the mailbox, copying from flash drive). The initial connection to C&C-servers used different system ports. Also, each bot performed different malicious activities. All of them supported the maintenance function. After 24 h of its functioning bot deleted itself.

Having a great amount of ways to construct malicious activity, 36 experiments were held (9 experiments for 4 types of evasion techniques).

NOTE. All experiments were held within the campus network and were not able to cause any damage to hosts and network stable work.

The central aim of the experiments was to find out the pros and cons of the developed technology. In order to investigate the efficiency of proposed IT we held four type of experiments. We were interested in results produced by old BotGRABBER [8]; by botnets detection technique based on the bots behaviour analysis in the host (host-level analysis), described in the Sect. 4.1; by DNS-based technique for botnets detection using the artificial immune system (network-level analysis), described in the Sect. 4.2; by new BotGRABBER which combined both host- and network-level analysis.

Table 3 demonstrates four types of results of experiments mentioned above. As we can see, old BotGRABBER system in general demonstrated good and even better results in comparison with the botnets detection technique based on the host-level analysis. However, old BotGRABBER demonstrated worse results for "domain flux" botnets' detection. If we take a look at host-level botnets detec-

Table 3. Experimental results of the botnets detection: true positives (TP) and false positives (FP)

Name of evasion technique	Detection by old BotGRABBER [8]		Detection by new BotGRABBER (only host-level analysis)		Detection by new BotGRABBER (only network-level analysis)		Detection by new BotGRABBER (host- & network-level analysis	
	TP, %	FP, %	TP, %	FP, %	TP, %	FP, %	TP, %	FP, %
Cycling of IP mapping	99	2	91	3	99	1	99	1
"domain flux"	89	6	77	6	90	6	96	6
"fast flux"	97	4	84	5	95	3	99	4
DNS-tunneling	94	4	80	5	90	4	96	4
Total	94,75	4	83	4,75	93,5	3	97,5	3,75

tion results, used separate, we can see that they are quite low. Finally, usage of the BotGRABBER system, which includes both host- and network-level analysis demonstrated greater detection results in comparison with old one. Thus, the results of information technology for botnets detection based on the botnet behaviour in the corporate area network (implemented as a BotGRABBER system) demonstrated high efficiency (up to 97%). In addition there was an interesting fact concerning to developed information technology. As it includes two parallel techniques, the question was: what kind of level alarmed about botnet presence in the network during the experiments first? After 20 experiments, we found out that host-level analysis alarmed first 7 times, and network-level analysis - 13 times.

It worth mentioning, that "old" and "new" BotGRABBER systems showed similar false positives values and it was about 3–5%.

6 Conclusions

In this article, the new information technology for botnets detection based on the analysis of the botnets' behaviour in the corporate area network is presented. Botnets detection is performing combining two ways: using network-level and host-level analysis. One approach makes it possible to analyze the behaviour of the software in the host, which may indicate the possible presence of bot directly in the host and identify malicious software, and another one involves monitoring and analyzing the DNS-traffic, which allows making conclusion about network hosts' infections with bot of the botnet.

Based on this information technology an effective botnets detection tool BotGRABBER was improved. It is able to detect bots that use such evasion techniques as cycling of IP mapping, "domain flux", "fast flux", DNS-tunneling.

Usage of the developed system makes it possible to detect infected hosts by bots of the botnets and to localize the malicious software in the infected host with high efficiency up to 97% and demonstrates low false positives at about 3–5%. The feature of the proposed approach is that the botnet detection is "invisible" to botmaster.

7 Future Work

Developed information technology has many advantages, but some its results ought to be improved. Thus, the future work should involve new approaches such as artificial intelligence which could eliminate present problems.

Acknowledgments. This research was supported by a TEMPUS SEREIN project (Project reference number 543968-TEMPUS-1-2013-1-EE-TEMPUS-JPCP). Additionally, we thank the Khmelnytsky National University for providing access to the DNS-traffic during the early phases of this work.

References

1. Komar, M., Kochan, V., Sachenko, A., Ababii, V.: Improving of the security of intrusion detection system. In: 2016 International Conference on Development and Application Systems (DAS), pp. 315–319 (2016)
2. Harsha, T., Asha, S., Soniya, B.: Feature selection for effective botnet detection based on periodicity of traffic. In: Ray, I., Gaur, M.S., Conti, M., Sanghi, D., Kamakoti, V. (eds.) ICISS 2016. LNCS, vol. 10063, pp. 471–478. Springer, Cham (2016). doi:10.1007/978-3-319-49806-5_26
3. Sochor, T., Zuzcak, M.: Attractiveness study of honeypots and honeynets in internet threat detection. In: Gaj, P., Kwiecień, A., Stera, P. (eds.) CN 2015. CCIS, vol. 522, pp. 69–81. Springer, Cham (2015). doi:10.1007/978-3-319-19419-6_7
4. Sochor, T., Zuzcak, M., Bujok, P.: Analysis of attackers against windows emulating honeypots in various types of networks and regions. In: International Conference on Ubiquitous and Future Networks, pp. 863–868 (2016)
5. Dietz, C., Sperotto, A., Dreo, G., Pras, A.: How to achieve early botnet detection at the provider level? In: Badonnel, R., Koch, R., Pras, A., Drašar, M., Stiller, B. (eds.) AIMS 2016. LNCS, vol. 9701, pp. 142–146. Springer, Cham (2016). doi:10.1007/978-3-319-39814-3_15
6. Kwon, J., Lee, J., Lee, H., Perrig, A.: PsyBoG: a scalable botnet detection method for large-scale DNS traffic. In: Computer Networks, vol. 97, pp. 48–73 (2016)
7. Sharifnya, R., Abadi, M.: DFBotKiller: domain-flux botnet detection based on the history of group activities and failures in DNS traffic. Digit. Invest. **12**, 15–26 (2015)
8. Pomorova, O., Savenko, O., Lysenko, S., Kryshchuk, A., Bobrovnikova, K.: Anti-evasion technique for the botnets detection based on the passive DNS monitoring and active DNS probing. In: Gaj, P., Kwiecień, A., Stera, P. (eds.) CN 2016. CCIS, vol. 608, pp. 83–95. Springer, Cham (2016). doi:10.1007/978-3-319-39207-3_8
9. Schiller, C., Binkley, R., Botnets, J.: The Killer Web Application, p. 464. Syngress Publishing, Burlington (2007)
10. Yadav, S., Reddy, A.L.N.: Winning with DNS failures: strategies for faster botnet detection. In: Proceedings of the 7th International ICST Conference on Security and Privacy in Communication Networks, pp. 446–459 (2011)
11. Salusky, W., Danford, R.: Know your enemy: fast-flux service networks. The Honeynet Project (2007). http://www.honeynet.org/book/export/html/130
12. Nazario, J., Holz, T.: As the net churns: fast-flux botnet observations. In: Conference on Malicious and Unwanted Software (Malware 2008), pp. 24–31 (2008)
13. DAMBALLA: Botnet Communication Topologies. Understanding the intricacies of botnet command-and-control. https://www.damballa.com/downloads/r_pubs/WP_Botnet_Communications_Primer.pdf
14. Bilge, L., Kirda, E., Kruegel, C., Balduzzi, M.: EXPOSURE: finding malicious domains using passive DNS analysis. In: NDSS, pp. 1–17 (2011)
15. Farnham, G., Atlasis, A.: Detecting DNS tunneling. SANS Institute InfoSec Reading Room, pp. 1–32 (2013)
16. Dietrich, C.J., Rossow, C., Freiling, F.C., Bos, H., van Steen, M., Pohlmann, N.: On Botnets that use DNS for command and control. In: Proceedings of European Conference on Computer Network Defense, pp. 9–16 (2011)
17. Guy, J.: A study of DNS, 30 January 2009. http://armatum.com/blog/2009/a-study-of-dns/

18. Guy, J.: DNS part ii: visualization, 13 February 2009. http://armatum.com/blog/2009/dns-part-ii/

19. Tarhio, J., Ukkonen, E.: Approximate BoyerMoore string matching. SIAM J. Comput. **22**(2), 243–260 (1993)

20. Pomorova, O., Savenko, O., Lysenko, S., Kryshchuk, A., Bobrovnikova, K.: A technique for the Botnet detection based on DNS-traffic analysis. In: Gaj, P., Kwiecień, A., Stera, P. (eds.) CN 2015. CCIS, vol. 522, pp. 127–138. Springer, Cham (2015). doi:10.1007/978-3-319-19419-6_12

21. Dipankar, D.: Artificial immune systems. In: Encyclopedia of Sciences and Religions, pp. 136–139 (2013)

22. Zhang, F., Qi, D.: A positive selection algorithm for classification. J. Comput. Inf. Syst. 207–215 (2012)

23. Goswami, M., Bhattacharjee, A.: Detector generation algorithm for self-nonself detection in artificial immune system. In: 2014 International Conference for Technology on Convergence of Technology (I2CT), pp. 1–6 (2014)

Utilization of Redundant Communication Network Throughput for Non-critical Data Exchange in Networked Control Systems

Andrzej Kwiecień[1], Michał Maćkowski[1], Jacek Stój[1(✉)], Dariusz Rzońca[2], and Marcin Sidzina[3]

[1] Silesian University of Technology, Gliwice, Poland
{andrzej.kwiecien,michal.mackowski,jacek.stoj}@polsl.pl
[2] Rzeszow University of Technology, Rzeszów, Poland
drzonca@prz-rzeszow.pl
[3] The University of Bielsko-Biala, Bielsko-Biała, Poland
marcin.sidzina@gmail.com

Abstract. Redundancy is the main method for achieving high reliability level in networked control systems NCSs. It is often applied in network interfaces in order to maintain operability of communication subsystems even when faults occur. In most cases, redundant communication buses realize exactly the same functionality as the initial non-redundant bus. That makes the system more reliable but the additional throughput of the redundant buses is not exploited even if the system condition would allow it. The idea of taking advantage of that throughput had made the authors to work on multi-network interface node which on the other hand maintain the high reliability of the communication subsystem, and on the other, makes it possible to manage the additional communication resources more efficiently, and therefore increases some parameters of the communication network.

Keywords: Communication · Efficiency · Network · Networked Control Systems · NCS · Real-time · Redundancy · Temporal characteristic · Throughput

1 Introduction

Computer systems implemented in industry operate during the actual time that the industrial process occurs. Proper operation of these systems depends not only on the right result of computation but also on the time when the result is generated. Only with these two conditions satisfied, the system is applicable. There are two the most important factors that determine the usability of a given system: the temporal characteristics mentioned above and the reliability. The first is obtained by the usage of real-time components. As we consider Networked Control Systems NCSs, it applies also to communication protocols that have to be deterministic in the time domain [1,2]. The second feature is usually achieved by using redundancy of the most critical elements of the system [3].

© Springer International Publishing AG 2017
P. Gaj et al. (Eds.): CN 2017, CCIS 718, pp. 182–194, 2017.
DOI: 10.1007/978-3-319-59767-6_15

Redundancy may be applied to any element of a given NCS – programmable electronic devices PES, such as programmable logic controllers PLC or communication interfaces i.e. buses and communication modules. Multiplication of system elements is done to maintain operability of the system in spite of failures of the redundant elements. When the reliability of communication systems is concerned, additional buses are introduced to the NCS. Their role is often to duplicate the operation of the initial non-redundant communication bus [4]. In other words, using redundant devices and redundant network connections there is an additional throughput introduced to the communication subsystem. However, it is used only to multiply the exact set of communication tasks and data exchanges that would be defined in the same system without redundancy. For example, in a system with double communication bus redundancy, the same communication process is usually executed on both buses. In that solution, it is feasible to maintain the requested data flow in spite of a fault of one of the buses.

The parallel transmission of the same information via two or more independent communication links may guarantee required reliability of the system, but it is not the best solution from the point of view of communication efficiency. In fault-free conditions, the additional throughput delivered by the redundant bus could be used for implementing of some extra functionality or improving the communication system efficiency. This observation made the authors to develop a multi-network interface node MuNetIN capable of maintaining high level of communication bus reliability together with higher network efficiency, while still maintaining requirements of a real-time system.

2 Related Works

Industrial real-time system is one where the correctness of the system depends not only on the logical results, but also on the time at which the results are produced. In hard real-time systems, results must be produced within an intended timing constraint, otherwise, the results will lose their usability [4–7]. Reliability and availability of the communication in this case are the essential issues [8]. It is worth to mention that application of redundancy leads to additional financial costs, but it also decrease the temporal characteristics of the real-time system [9].

To allow automation networks to support simultaneously different protocols, IEC working group produced the IEC 62439 [10] standard suite that defines redundancy methods applicable to industrial networks with different topologies and recovery times. One of solutions is dual bus with only one bus active at a time. In the case of communication failure, the systems switches to the redundant network. It is used in 'Media Redundancy Protocol' MRP of IEC 62439-2 [10,11]. In contrast to this hot-standby switchover redundancy, the 'High availability Seamless Redundancy' HSR and 'Parallel Redundancy Protocol' PRP of IEC 62439-3 [12,13] are active redundancy approaches that work without reconfiguration timeouts when a single failure in one of its two redundant network structures occurs [14,15]. PRP is the redundancy in the network solution in which nodes use both networks simultaneously. This offers zero disruption time, making PRP suited for all hard real-time applications [16].

In case of HSR redundancy there are some trials to increase the traffic performance by eliminating duplicated frames in the network [17, 18]. However, the presented solution not only reduces the number of messages that are being sent, but also provides means for sending some more user data than it is possible in a system without media redundancy.

There are also some works where authors present and analyze the dynamic message transmission scheduling which is based on message priority and not on the basis of the condition and link bandwidth [19, 20]. However, the solution presented in this paper for communication network that supports management of multi-interfaced communication links, especially in industrial networks is a different approach and is also worth considering. It is an improved solution comparing the previously realized works presented in [21, 22].

3   Motivation - Application Example

During some practical applications, the authors encountered a problem with frequent data acquisition in a networked control system. Among other devices, there was a separately excited generator monitored by a PLC together with a SCADA station. After unsuccessful excitation of the generator, the maintenance team of the generator had the need to analyze the changes of the analogue inputs (AI) of the PLC, so that they would be able to figure out what was the source of the faulty operation of the generator. The states of the AIs were expected to be acquired and archived on the SCADA station with as great frequency as possible for later analysis. The possible AIs sampling frequency was determined first of all by the automata cycle of the PLC which was about 10ms. That gave 100 samples of every analogue input every second. However, sending data with such a frequency to the SCADA station by the present communication link was not possible. Therefore, there was an idea of recording the state of the AIs from the last second in the internal memory of the PLC in order to send the data to the SCADA station less frequently grouped in greater data sets (every set with a series of AIs states from one second). Unfortunately, taking into consideration the number of AIs, which was couple of dozens, the throughput of the communication link was not great enough. The solution would be to utilize the redundant communication bus between the PLC and SCADA station more efficiently. That would allow acquisition of the AIs with greater frequency whenever the communication subsystem operates properly, i.e. none of the redundant components were faulty. What is important, faults occurring during generator excitation where unrelated to faults in the communication subsystem. This experience was a motivation to do some research on this problem. The results are described in the following sections.

4   Multi-network Interface Node

As already mentioned, the origins of multi-network interface node MuNetIN are systems with communication bus redundancy, where the network controllers are connected to two or more independent buses realizing quite the same functionality. The MuNetIN node is equipped with three networks, but the management of those connections is different than it would be in a redundant system. The communication process on redundant system is usually managed independently and the set of communication tasks is the same on every redundant bus. Meanwhile, in the presented solution there is one manager common for all buses and the communication process on every bus may vary.

4.1 General Idea

Three MuNetIN buses have the same physical layer with Master-Slave protocol. The MuNetIN node is the Master node. The Slave subscribers may be connected to one, two or all three buses. The slaves addressing is unique for every slave and common on every bus. For example, when slave S1 is connected to buses A and B with an addr (the same for both buses) then the address addr is also considered occupied on the third C bus (Fig. 1).

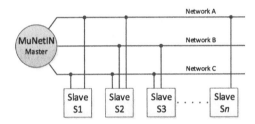

Fig. 1. Communication buses in multiple network interface node MuNetIN

In systems with redundancy, communication with every Slave would be executed to use every available network connection. For the above example, communication with slave S1 would be done on both A and B buses consuming twice the throughput necessary to send all user data. The MuNetIN on the other hand, would realize the data exchange only on all buses the given subscriber is connected to, but only in parts so the payload is distributed. That should increase the efficiency of the communication system.

Greater communication efficiency allows implementation of additional features to the system, such as additional data exchanges that are not critical to the operation of the whole industrial computer system, but useful from the point of view of maintenance or supervising activities. That requires to divide of the network data exchanges into two groups – one with exchanges critical to the system operation and the other with non-critical exchanges. The exchanges from the latter group would be executed only in fault-free circumstances, taking into consideration the current temporal parameters of the communication system, so that the real-time constraints are satisfied. The membership in these groups is based on data exchanges priorities defined by the system developer.

4.2   Triple Network Interface Communication System

In Master-Slave networks, the communication is based on request-response queries. The Master station, according to previously prepared list of queries, sends requests to the Slaves. The queries may be of several types with general functionality of two kinds – read or write queries.

In typical applications, all queries are exchanged on one communication bus and the Master station only determines which query should be send at a given time. The MuNetIN node has triple communication interfaces TNICS, so it also has to decide which interface the query should be sent through. Therefore, it requires one common data exchange scheduler which would control the communication process. In the MuNetIN node, a single Master station is implemented for all buses with the controls algorithms extended for the usage of multiple network connections. Moreover, the MuNetIN node has to include precise information about the availability of every Slave subscriber on every network.

Having information about Slaves availability, it is possible to transmit the queries on multiple communication interfaces. To realize the master functionality, a program module was implemented to service to multiple interfaces – Queries Distributor QD (see Fig. 2). The QD module checks the busy status of the network interfaces. When an interface is not occupied, then the QD asks the Queries Scheduler QS for a query to be realized on that interface. The QS may choose either periodic or aperiodic query to be realized. It may be either critical or not critical (optional), too. Then the distributor passes over the received query definition to a Query Handler QH. The QD at this point sets the bus status to busy and the QH sends it through the network. The busy status is cleared by the QH after realization of the query.

4.3   Query Scheduler

The queries in Master-Slave communication networks have two transmission rules – periodic and aperiodic. For every periodical query it is defined how often

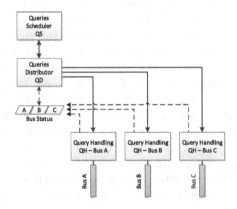

Fig. 2. Query distribution

it should be realized (e.g. every 100 ms). According to that parameter, the master creates a schedule of the queries transmission. Realization of all the queries from the schedule (some of the queries possibly more than once) is called the network communication cycle NC.

In Master-Slave networks, the aperiodic queries are usually sent immediately (out of the order of the scheduled periodical queries) or after realization of all periodic queries in the network cycle. When there is a necessity of sending more than one aperiodic query, the queries are sequenced in FIFO order. The maximum number of aperiodic queries in one NC is limited hence the real-time constraints defined for the communication network are satisfied.

The Master station that operates on one communication interface, may realize the queries based on a static schedule created once on the system startup. Having three independent interfaces it is no longer possible. The scheduling have to be performed dynamically taking into consideration the current status of all interfaces.

In the MuNetIN node, the query scheduler QS, on the Queries Distributor QD requests, selects from the list an appropriate query in two steps. Firstly, it excludes from the list all the queries that cannot be realized on the bus given by the QD. Secondly, it chooses from the result list the query with the shortest deadline which is calculated basing on the defined periodicity and the subtraction of the current time and the timestamp of the last successful realization of the given exchange. The aperiodic queries are passed over to the QD for realization when it would not jeopardize the periodic exchanges but no later than after realization of all queries from the schedule.

5   Formal Description of Distribution Algorithm

5.1   Periodic Tasks

Let BD (bus/device) denote a matrix which defines connections between buses and devices. Assuming there is n devices and m buses we are considering the matrix

$$BD = \begin{bmatrix} bd_{1,1} & \cdots & bd_{1,n} \\ \vdots & \ddots & \vdots \\ bd_{m,1} & \cdots & bd_{m,n} \end{bmatrix},$$

where

$$bd_{i,j} = \begin{cases} 1 & \text{if device } j \text{ is connected to bus } i, \\ 0 & \text{otherwise.} \end{cases}$$

Let matrix DT (device/task) similarly define devices served in particular communication tasks. Assuming there is n devices and r tasks, the matrix DT is as follows:

$$DT = \begin{bmatrix} dt_{1,1} & \cdots & dt_{1,r} \\ \vdots & \ddots & \vdots \\ dt_{n,1} & \cdots & dt_{n,r} \end{bmatrix},$$

where
$$dt_{i,j} = \begin{cases} 1 & \text{if device } i \text{ is involved in task } j, \\ 0 & \text{otherwise.} \end{cases}$$

We are considering communication tasks related to exactly one device, thus every column of matrix DT contains exactly one 1, i.e.:

$$\begin{cases} \forall_{j=1...r} \exists_i dt_{i,j} = 1 \\ \forall_{j=1...r} dt_{x,j} = dt_{y,j} = 1 \Rightarrow x = y \end{cases}$$

Matrix BT (bus/task) is a matrix product $BT = BD \cdot DT$ defining tasks which may be served on particular buses.

$$BD \cdot DT = BT = \begin{bmatrix} bt_{1,1} & \cdots & bt_{1,r} \\ \vdots & \ddots & \vdots \\ bt_{m,1} & \cdots & bt_{m,r} \end{bmatrix},$$

where
$$bt_{i,j} = \begin{cases} 1 & \text{if task } j \text{ may be served on bus } i, \\ 0 & \text{otherwise.} \end{cases}$$

Additionally, each from r tasks has preset periodicity, i.e. maximal time between serving successive task. Let P be a vector defining periodicity of the tasks. $P = [p_1 \cdots p_r]$, where: p_i denotes periodicity of task i.

Communication tasks are served by system described in Sect. 4.3, where Queries Distributor chooses successive task to serve. Distributor is called periodically, for every bus, as soon as the bus becomes idle. Selection of the next task is based on periodicity in order to choose the task which may be handled on a given bus, and whose time period left to obey the declared periodicity of that task is minimal. Therefore, additional information about time left to deadlines is necessary. Let T be a vector defining time left for serving each task. $T = [t_1 \cdots t_r]$, where t_i denotes maximal time, in which task i may be served to obey periodicity defined in P.

Matrices BD, DT, BT, P are constant, T is variable changed by Distributor. Initially $T = P$. According to passing time T decreases. Choosing task j to serve involves changing t_j to initial value $t_j = p_j$, defined by periodicity of task j.

Naturally, considering the system with defined time for serving single communication task t_{single}, identical for each task, it is possible to simplify the model to handle discrete time, where time flow is modeled by decreasing each element of T by t_{single}, after sending tasks in each bus.

Choosing the task to serve on bus i, Distributor is looking for a minimum value in these elements of T, where corresponding elements in vector $BT_{i,1..r}$ are 1, i.e. among the tasks which may be served on a given bus. Additionally, tasks, which are just being served on another bus, should be omitted. In order to decide if the query related to the task i has been just sent or not, such information will be stored in vector $Q = [q_1 \cdots q_r]$, where:

$$q_i = \begin{cases} 1 & \text{if task } i \text{ is just served,} \\ 0 & \text{otherwise.} \end{cases}$$

Hence, Distributor for given i should choose such j, that:

$$t_j = \min_{j \in \{1..r\}} \{t_j : bt_{i,j} = 1 \wedge q_j = 0\}.$$

5.2 Aperiodic Transactions

Aperiodic query may be sent if such transaction will not jeopardize timely realization of periodic tasks. The main problem is to define such condition. Let $p_{\text{MAX}} = \max_P p_i$ be maximal periodicity defined in P. We are considering a system with the following constraints:

- n devices, m buses, and r periodic tasks,
- a pool of aperiodic queries, without time constraints,
- constant time for serving single communication task t_{single}, identical for each query.

In such conditions communication will be performed in time slots, defined by t_{single}. Let S define number of available slots during p_{MAX} time, for all buses:

$$S = \frac{p_{\text{MAX}} * m}{t_{\text{single}}}.$$

Each task i during p_{MAX} time is handled approximately $\frac{p_{\text{MAX}}}{p_i}$ times. Thus, to handle all periodic tasks in given periodicity constraints, during p_{MAX} time on average $\sum_{i=1}^{r} \frac{p_{\text{MAX}}}{p_i}$ slots are needed.

Let cf be a coefficient, calculated in every cycle, similarly as above, but basing on t_i instead of p_i. Such coefficient might be understood as a number of slots needed for all periodic tasks, if periodicity of each task was equal to its actual time to deadline. According to passing time, such coefficient for each task increases until the task is served, then it returns to its initial value.

$$cf = \sum_{i=1}^{r} \begin{cases} \frac{p_{\text{MAX}}}{t_i} & \text{if } t_i \text{ is not zero,} \\ p_{\text{MAX}} & \text{otherwise.} \end{cases}$$

In the cycles where $cf < S - m$, i.e. dedicating m slots for aperiodic queries (each query being served in one slot) does not expose deadlines for periodic transactions to risk, successive aperiodic queries from the pool will be handled on each bus.

6 Algorithm Realization Example

Let us assume that connections between devices and buses are defined in the following matrix:

$$BD = \begin{bmatrix} 1 & 1 & 0 & 1 \\ 0 & 1 & 1 & 1 \\ 1 & 1 & 1 & 0 \end{bmatrix}$$

According to Sect. 5, each column of the matrix defines connection of a particular device to buses. Slave 1 is connected to buses A and C, slave 2 to all three buses, slave 3 to B and C, whereas slave 4 is connected to buses A and B.

DT matrix defining devices involved in particular tasks, is as follows:

$$DT = \begin{bmatrix} 1 & 1 & 0 & 0 & 0 & 0 & 0 & 0 \\ 0 & 0 & 0 & 1 & 0 & 1 & 1 & 0 \\ 0 & 0 & 1 & 0 & 0 & 0 & 0 & 0 \\ 0 & 0 & 0 & 0 & 1 & 0 & 0 & 1 \end{bmatrix}$$

Each column defines the device serving given task. The first task is served by slave 1, the second also by slave 1, the third by slave 3, the fourth by slave 2, etc.

BT matrix is calculated as matrix product of BD and DT.

$$BT = \begin{bmatrix} 1 & 1 & 0 & 1 & 1 & 1 & 1 & 1 \\ 0 & 0 & 1 & 1 & 1 & 1 & 1 & 1 \\ 1 & 1 & 1 & 1 & 0 & 1 & 1 & 0 \end{bmatrix}$$

Each column defines buses, where given communication task may be served. The first and second tasks may be served on buses A or C, the third on B or C, the fourth on all buses, etc.

Periodicity vector for the tasks is as follows:

$$P = \begin{bmatrix} 50 & 50 & 50 & 100 & 100 & 150 & 150 & 300 \end{bmatrix}$$

Tasks 1–3 have to be served at least every 50 time units, 4 and 5 at least every 100 time units, tasks 6 and 7 at least every 150 time units, and 8 at least every 300 time units.

Time for serving a single communication task equals $t_{single} = 10$. The number of slots calculated according to Sect. 5.2 equals $S = 90$. Resulting list of queries is shown in Table 1. For brevity, each aperiodic query has been denoted as Q+, however such value should be understood as one of the queries from the pool of aperiodic queries defined in Sect. 5.2, but not necessarily the same query each time.

The initial value of time to deadline for each task equals its periodicity (column 1 in Table 1). Coefficient cf calculated for these values equals 29.0. It is shown in the cf row in Table 1 and in Fig. 3 in the first graph from the bottom. The value of 27 is less than $S - m = 87$ (the $cf = 87$ is marked with dotted line in the bottom graph), so in the first cycle aperiodic queries Q+ are handled on each bus. The moments when queries Q+ may be sent are marked by gray rectangles with dots in the graph. No periodic queries have been sent, thus time to deadline for each task decreases by 10 (t_{single}) each, as shown in column 2 of Table 1 and in graphs Q1–Q8 depicting the change of deadlines of each query in time.

Table 1. List of queries

Time to deadline		1	2	3	4	5	6	7	8	9	10	11	12	13	14	15
	Q1	50	40	30	20	10	50	40	30	20	10	50	50	40	30	20
	Q2	50	40	30	20	10	50	40	30	20	10	50	40	30	20	10
	Q3	50	40	30	20	10	50	40	30	20	10	50	40	30	20	10
	Q4	100	90	80	70	60	50	40	30	20	10	0	100	90	80	70
	Q5	100	90	80	70	60	50	40	30	20	10	0	100	90	80	70
	Q6	150	140	130	120	110	100	90	80	70	60	50	40	30	20	10
	Q7	150	140	130	120	110	100	90	80	70	60	50	40	30	20	10
	Q8	300	290	280	270	260	250	240	230	220	210	200	190	180	170	160
cf		29.0	34.5	43.2	59.7	106.6	37.2	45.4	58.8	84.9	161.4	631.5	43.6	55.8	79.3	145.4
Query	Bus A	Q+	Q+	Q+	Q+	Q1	Q+	Q+	Q+	Q+	Q1	Q4	Q+	Q+	Q+	Q2
	Bus B	Q+	Q+	Q+	Q+	Q3	Q+	Q+	Q+	Q+	Q3	Q5	Q+	Q+	Q+	Q3
	Bus C	Q+	Q+	Q+	Q+	Q2	Q+	Q+	Q+	Q+	Q2	Q1	Q+	Q+	Q+	Q6

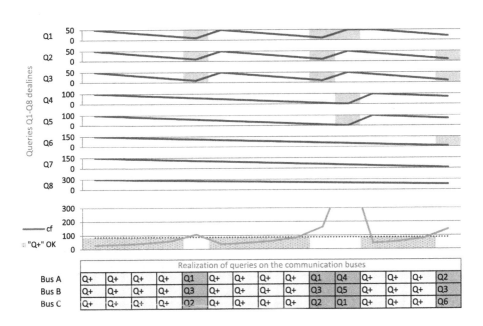

Fig. 3. Deadlines and queries realization

In the second cycle cf equals 34.5 which is still less than $S - m$. Therefore, aperiodic exchanges are sent again. Similarly, in the cycles 3rd and 4th. In the 5th cycle (column 5) the value of the cf is greater than $S - m$ and aperiodic exchanges with minimal time to deadline are chosen, namely Q1, Q3, and Q2 respectively for buses A, B, and C. The fact of sending the queries on buses is depicted in the bottom part of Fig. 3 (by table with grayed cells). Time to deadline for these queries is reset to the initial value (50), time to deadline for other queries Q4–Q8 decreases by 10 (column 6). In the 6th cycle cf equals 37.2, so again aperiodic queries are handled, and time to deadline for each task decreases by 10. Periodic queries are sent again in the 10th and 11th cycle as the cf is higher than $S-m$. After that again aperiodic queries are sent 3 more times.

For brevity, only 15 cycles have been shown, however numerous cycles and different examples have been analyzed during the algorithm development.

7   Conclusions and Future Work

The main idea presented in the paper is the concept of an algorithm for queries distribution to be implemented in multi network interface node MuNetIN which is dedicated to network control systems NCS. The goal is to take advantage of redundant communication interfaces with real-time communication protocols. The algorithm should allow both, increased reliability and greater communication network efficiency.

The performed simulation tests show that the algorithm is able to schedule periodic exchanges so that they are realized with at least the periodicity defined for each query. Moreover, it is possible to make room for realization of aperiodic exchanges. Nevertheless, the presented solution needs further development. At this point, the periodic exchanges and aperiodic exchanges are always "grouped" and being realized in "bursts" (e.g. couple of Q+ exchanges on all buses). Besides, in the presented example the exchange Q1 was always realized on the bus A. It would be better to realize it also on every other bus, to which the slave associated with that exchange is connected to. That would allow some diagnostics of the state of the network connections.

What is the strong point of the algorithm, the choice of queries for realization in the proposed algorithm is dynamic. As a result, the Master node is capable of adjusting to change network topology. Importantly, the cause of the change could be planned (moving a system node from one network to another) or unplanned (failures affecting the accessibility of the system nodes). Any changes in the topology will be automatically serviced by the query distributor so that the communication link is maintained using the available resources in most efficient way.

References

1. Gaj, P., Malinowski, A., Sauter, T., Valenzano, A.: Guest editorial: distributed data processing in industrial applications. IEEE Trans. Ind. Inform. **11**(3), 737–740 (2015)
2. Gaj, P., Jasperneite, J., Felser, M.: Computer communication within industrial distributed environment - survey. IEEE Trans. Ind. Inform. **9**(1), 182–189 (2013)
3. Birkholz, H., Sieverdingbeck, I.: Link-failure assessment in redundant ICS networks supported by the interconnected-asset ontology. In: 2014 IEEE International Workshop Technical Committee on Communications Quality and Reliability (CQR), pp. 1–6, May 2014
4. Kopetz, H.: Real-Time Systems: Design Principles for Distributed Embedded Applications. Springer, Heidelberg (2011)

5. Sauter, T.: The three generations of field-level networks-evolution and compatibility issues. IEEE Trans. Ind. Electron. **57**(11), 3585–3595 (2010)
6. Flak, J., Gaj, P., Tokarz, K., Wideł, S., Ziebinski, A.: Remote monitoring of geological activity of inclined regions - the concept. In: Kwiecień, A., Gaj, P., Stera, P. (eds.) Computer Networks. Communications in Computer and Information Science, vol. 39, pp. 292–301. Springer, Heidelberg (2009)
7. Popovic, M., Mohiuddin, M., Tomozei, D.C., Boudec, J.Y.L.: iPRP: parallel redundancy protocol for IP networks. In: 2015 IEEE World Conference on Factory Communication Systems (WFCS), pp. 1–4, May 2015
8. Yu, X., Jiang, J.: Hybrid fault-tolerant flight control system design against partial actuator failures. IEEE Trans. Control Syst. Technol. **20**(4), 871–886 (2012)
9. Kwiecień, A., Stój, J.: The cost of redundancy in distributed real-time systems in steady state. In: Kwiecień, A., Gaj, P., Stera, P. (eds.) Computer Networks. Communications in Computer and Information Science, vol. 79, pp. 106–120. Springer, Heidelberg (2010)
10. International Electrotechnical Commission: IEC 62439-2: Industrial communication networks: High availability automation networks - part 2: Media redundancy protocol
11. Zuloaga, A., Astarloa, A., Jiménez, J., Lázaro, J., Araujo, J.A.: Cost-effective redundancy for ethernet train communications using HSR. In: 2014 IEEE 23rd International Symposium on Industrial Electronics (ISIE), pp. 1117–1122, June 2014
12. International Electrotechnical Commission: IEC 62439-3 (2012): "industrial communication networks: High availability automation networks" - part 3: Parallel redundancy protocol (PRP) and high availability seamless redundancy (HSR)
13. Giorgetti, A., Cugini, F., Paolucci, F., Valcarenghi, L., Pistone, A., Castoldi, P.: Performance analysis of media redundancy protocol (MRP). IEEE Trans. Ind. Inform. **9**(1), 218–227 (2013)
14. Kirrmann, H., Weber, K., Kleineberg, O., Weibel, H.: HSR: zero recovery time and low-cost redundancy for industrial ethernet (high availability seamless redundancy, IEC 62439-3). In: 2009 IEEE Conference on Emerging Technologies Factory Automation, pp. 1–4, September 2009
15. Rentschler, M., Heine, H.: The parallel redundancy protocol for industrial IP networks. In: 2013 IEEE International Conference on Industrial Technology (ICIT), pp. 1404–1409, February 2013
16. Hoga, C.: Seamless communication redundancy of IEC 62439. In: 2011 International Conference on Advanced Power System Automation and Protection, vol. 1, pp. 489–494, October 2011
17. Nsaif, S.A., Rhee, J.M.: Improvement of high-availability seamless redundancy (HSR) traffic performance. In: 2012 14th International Conference on Advanced Communication Technology (ICACT), pp. 814–819, February 2012
18. Nsaif, S.A., Kim, S., Rhee, J.M.: Quick removing (QR) approach using cut-through switching mode. In: 2016 18th International Conference on Advanced Communication Technology (ICACT), pp. 170–147, January 2016
19. Saadi, I.M.A.: Dynamic message transmission scheduling using can protocol. Int. J. Sci. Technol. Res. **2**(9), 158–162 (2013)
20. Wey, C.L., Hsu, C.H., Chang, K.C., Jui, P.C.: Enhancement of controller area network (CAN) bus arbitration mechanism. In: 2013 International Conference on Connected Vehicles and Expo (ICCVE), pp. 898–902, December 2013

21. Kwiecień, A., Kwiecień, B., Maćkowski, M.: Algorithms for transmission failure detection in a communication system with two buses. In: Gaj, P., Kwiecień, A., Stera, P. (eds.) Computer Networks. Communications in Computer and Information Science, vol. 608. Springer International Publishing, Cham (2016)
22. Kwiecień, A., Kwiecień, B., Maćkowski, M.: A failure influence on parameters of real-time system with two buses. Computer Networks. Communications in Computer and Information Science, vol. 608. Springer International Publishing, Cham (2016)

Software Defined Home Network
for Distribution of the SVC Video
Based on the DASH Principles

Slawomir Przylucki[1]($\boxtimes$), Artur Sierszen[2], and Dariusz Czerwinski[1]

[1] Lublin University of Technology, 38A Nadbystrzycka Street, 20-618 Lublin, Poland
spg@spg51.net, d.czerwinski@pollub.pl
[2] Lodz University of Technology, 116 Zeromskiego Street, 90-924 Lodz, Poland
artur.sierszen@p.lodz.pl
http://www.pollub.pl
http://www.p.lodz.pl

Abstract. With the spread of Dynamic Adaptive Streaming over HTTP (DASH) systems, appeared the problem related to the resources management in end-users' networks. The article proposes a new solution for the distribution of Scalable Video Coding (SVC) video streams in home networks. It relies on a Software Defined Network (SDN) controller, which based on the parameters of transmitted SVC video and the usage status of network resources allows for dynamic traffic shaping. At the same time, it was assumed that the proposed solution is compatible with Server and Network Assisted DASH (SAND) standard and its implementation is possible without the use of additional, dedicated hardware. The article presents the analysis of influence of proposed SDN on the improvement of the SVC video quality received by DASH clients in a typical home network. Test results confirm the superiority of the proposed system over the conventional solutions.

Keywords: Home network · Software Defined Network (SDN) · Scalable Video Coding (SVC) · Dynamic Adaptive Streaming over HTTP (DASH) · Server and Network Assisted DASH (SAND) · Docker containers

1 Introduction

Nowadays, the importance of multimedia transmissions for the entire market of Internet services, is no longer subject to discussion. The undoubted success of services based on the transmission of video is not only due to the rapid development of network infrastructure, both within the Internet backbone and access networks. An important role in their wide acceptance also played the development of techniques for efficient video encoding, new methods for video streaming and scalable content distribution. In the article, special attention is paid to two techniques: HTTP Adaptive Streaming (HAS) [1] and Scalable Video Coding (SVC) [2].

© Springer International Publishing AG 2017
P. Gaj et al. (Eds.): CN 2017, CCIS 718, pp. 195–206, 2017.
DOI: 10.1007/978-3-319-59767-6_16

Despite the obvious advantages of the HAS approach [3], there are still issues that need attention in the implementation of video streaming services. The most important challenges can certainly include issues related to ensuring the fairness allocation of available bandwidth to individual video streams [4].

Most of the current proposals to guarantee fair allocation of resources were focused on systems implemented in distribution networks or networks of local Internet providers [5]. Much less attention was paid to home networks or networks in general, which are attached to the end-user of DASH services. Such an approach does not seem to be justified in the context of the rapid increase in the number of individual users of video streaming services. Unfortunately, in home networks, dedicated hardware and complex algorithms for monitoring and traffic shaping, could not be easily applied [6]. A promising basis for such solutions is the idea of Software Defined Network (SDN) [7]. So far, two basic restrictions significantly affect the popularity of the solutions of this type. The first was the limited availability of network devices supporting the protocol OpenFlow. The second reason was related to the complex process of installation and configuration of SDN controllers. Currently, the first cause becoming less important due to the availability of efficient, home networking equipment. In turn, the second limitation can be effectively eliminated using the lightweight virtualization. In this context, particularly promising seems to be the idea of microservices and especially the containerization techniques based on Docker containers [8].

The remainder of the paper is organised as follows. Related works are presented in Sect. 2. Section 3 contains the description of the proposed SDHN solution for video stream distribution in a home network. The test scenario and performance evaluation are presented in Sect. 4. Finally, a brief conclusion is made in Sect. 5.

2   Background and Related Works

The fundamental principle of the HAS systems is based on the distribution of an audio-video content that is stored on HTTP servers in the form of multiple representations. These representations are associated with different video encoding parameters and thus require a different amount of available bandwidth along the transmission path. Depending on the utilization of network resources between the server and client and also considering the parameters of the receiving devices, it becomes possible to choose the best representation and thus to adapt to changing network conditions. The effectiveness of these solutions has led standardization organizations to try to develop a uniform standard for systems HAS. As a result, in 2012, ISO/IEC published the standard: Dynamic Adaptive Streaming over HTTP (DASH) [1]. In the case of DASH, the progressive download of a video representation was replaced by the so-called segment approach. This solution is based on segmentation of each representation into chunks containing the video fragments of the fixed length. A description of segmented data structure, stored on the server DASH, is contained in a dedicated Media Presentation Description files (MPD). DASH clients, based on the information contained in these files, can switch between different representations (chunks) during the video playback. The DASH adaptation algorithm, described in the standard, unfortunately does not contain mechanisms related to the reducing the impact of the mutual influence of the individual video streams. Each DASH client realizes the process of adaptation in isolation from the others, competing with the other receivers. In other words, DASH client are trying, in very selfish way, to secure the best quality of the received video. This leads to decreasing the values of Quality of Experience (QoE). This degradation is observed at all receivers in a given network segment [9]. Recently these issues were reflected in the new standard developed by ISO/IEC, which was called Server and network assisted DASH (SAND) [10].

The introduction of multi-layer coding simplifies the preparation, storage and video distribution in DASH systems [11]. The idea of SVC is to use the hierarchical structure of the video stream, which allows receivers to select only the portion of the video data while maintaining accuracy and continuity of the decoding process. This feature has been obtained by defining within the structure of a video stream, the following types of layers: the base layer and the enhanced layers [12]. The base layer is always necessary for decoding of a video. It provides video playback with minimal values of the parameters (resolution, image smoothness, image quality). The enhanced layers allow for completion of data from the base layer and thereby to obtain better video quality. An important advantage of the SVC-DASH systems is that the video content is encoded only once. SVC-DASH clients realize the process of adaptation by selection of the amount of decoded layers from a single representation of video content. This process is often called as switching the Operation Point (OP) [13,14].

Recent years have brought increased interest in the possibility of using the SDN principles to the process of adaptation in DASH systems [6,15,16]. In this context, the most promising areas of the research are the methods for traffic shaping in general and ensuring a fair distribution of SVC video streams as a special case [17,18]. Simultaneously, thanks to advances in the field of virtualization, it became possible to implement these solutions on the equipment available for most users of home computer networks [19].

3 Proposed Solution

The proposed solution is designed for typical home networks. Thus, following assumptions have been made:

- Multiple DASH clients are connected to an external network (typically, the ISP network) through the gateway. The link between the home gateway and ISP is the bottleneck.
- The proposed SDHN should be compatible with the principles of SAND. The SDN controller should work with any network devices supporting the protocol OpenFlow 1.3 and above.
- The SDHN controller should be installed on any hardware platform, supporting Docker environment, version 1.9 and above.

From the perspective of the DASH clients, the SDHN is a kind of service responsible for ensuring the best possible parameters of video stream. Thus, the functionality of the proposed SDNH can be called as QoS as a Service (QaaS). The structure of the SDNH controller consists of four functional components:

- Network resources module (NRM).
- QoS controller (QoSC).
- Traffic shaping module (TSM).

First two modules are associated with a few fundamental functionalities. Thus, their structure can be further divided into blocks and databases. The NRM consists of two functional blocks, respectively: monitoring block and policy block. The QoSC contains fairness distribution block (FDB) and two databases, respectively: DASH clients database and video stream database. The TSM is responsible for generating the final rules for each identified video flow. The parameters required by these rules are provided by QoSC. The TSM module is typical component of SDN controller and by using the OpenFlow, it controls the tasks performed by switch (e.g. OpenFlow vSwitch). The structure of SDHN controller and the internal relationship between its components are presented in Fig. 1. The elements which use (are compatible with) the SAND principles are indicated by black backgrounds.

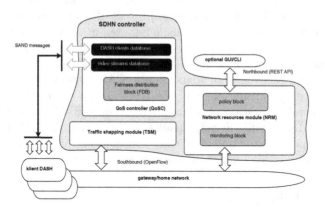

Fig. 1. The architecture of the proposed SDHN solution.

3.1   Network Resources Module

The main task associated with this module is gathering information about all the traffic within a home network. The source of data is gateway and mutual communication is performed based on the OpenFlow protocol. The monitoring block, which is a part of this module, is also responsible for classification of video flow. This task included the identification of new flow and termination of given video transmission. Considering that target area of application is a home

network, the identification process base on a destination IP address. The second
block under this module is the policy block. In general case, it includes any policy
associated with DASH service. A common use of this module is to implement
the predefined Service Level Agreement (SLA), resulting from a specific network
operator tariffs. The software implementation of the policy block allows to define
restrictions on whole DASH service or on the individual DASH clients.

3.2   QoS Controller

The QoS controller is the element that has benefited greatly from the SAND
architecture. Two of its elements directly use the Status and Metrics messages
as defined in SAND specification [10]. These messages are exchanged between
DASH clients and QoSC. In the case of proposed solution, the representation of
SAND DANEs are DASH clients database and video streams database. The first
database consists of parameters of all registered clients (set of operating points
such as desired bandwidth and quality) but also their QoS metrics (average
throughput, buffer level, initial playout delay). The second database comprise
of data originated from the MDFs associated to requested video flow (stream
structure, chunk descriptions, OPs parameters). The third component of the
QoSC, fairness distribution block (FDB) uses these databases to calculate fair
provisioning of available bandwidth.

Fairness Distribution Block. The algorithm implemented in the FDB is based
on the utility function (UF) described in our earlier research [14]. Generally, this
utility function is non-linear function which represents the relationship between
video stream bitrate and relative utilization (RU) indicator. The RU indicator
represents the ratio of the number of the OPs that DASH client can (or wishes to)
received to all available OPs for given SVC stream. One of the most important
advantages of the method of creation of the UF is fact that it requires only the
data contained in the MDF file.

Next task of the QoSC is to define the set of stream bitrates that ensures
fairness distribution of available bandwidth across all registered DASH clients.
Other words, the QoSC should find the optimal value of Y_{optimal} satisfying the
Formula (1).

$$Y_{\text{optimal}} = \max[\min_i(UF_i(x_i))] \qquad \Sigma(x_i) \leq Blimit \qquad (1)$$

where UF_i be a utility function, x_i the bitrate for i-th video stream and $Blimit$
represents the total amount of available bandwidth for all streams.

To solve this optimization problem, the integer-programming algorithm was
used [20]. We discussed this algorithm in our earlier article [14]. The results
are the optimum bitrate values. However, these values do not necessarily correspond to the OPs bitrates defined within SVC stream. For this reason, when it
is necessary, calculated values of bitrate must be changed to the nearest, lower
value of the bitrate associated with the particular OP. Finally, the calculated

values are sent to the DASH clients as SAND PER Messages (message: SharedResourceAssignment) [10]. Summarizing, the QoSC block performs sequentially the following three tasks:

- reads the data contained in DASH clients and video stream databases, creates the utility function,
- finds solution to optimization problem (Formula (1)),
- downgrades the optimal bitrates to the bitrate (lower) of the closest OP.

4   Test Scenario, Evaluation and Results

The proposed SDHN solution has been tested against the advantages it brings in comparison to the standard DASH solutions. Thus, two software solutions were configured. The first one was consistent with DASH principles. This means that DASH clients were responsible for the process of dynamic adaptation according to DASH standard and no traffic shaping algorithms were implemented inside the home network. This solution, later in the article, is referred to as legacy DASH. The second one is the implementation of the proposed SDHN and it introduced two additional elements, the SDN controller (QoSC and TSM modules) and databases (NRM module), respectively. Both new components were implemented as Docker containers. This solution is referred as SDHN DASH.

Considering specific features of home networks in context of the video streaming services, we decided to conduct the two, separate tests. The first test (test A) assumed that a couple of DASH clients were switching on and off (playback/pause) during the whole period of test. The second test (test B) was a representation of continuous video playback of all clients, during the entire test. These tests correspond to typical scenarios in the home networks with multiple video receivers.

4.1   The Testbed Configuration

In order to assess the effectiveness of the SDHN solution as well as to compare it with legacy DASH, we developed a evaluation testbed as shown in Fig. 2.

All the network components were connected to the TP-LINK TL-WR1043ND home switch. That switch served as a gateway. The software platform for the operation of the switch was OpenWRT with OpenFlow 1.3 [21]. The bandwidth on the link between gateway and HTTP server which store SVC videos, was limited to 18 Mb/s by means of TC (Linux traffic control tool) rules. DASH Server was implemented based on DASH-JS library [22]. The implementations of SDHN controller was divided into two Docker containers. All of them were installed automatically, from local repository configured on the x86 64-bit, 8 GB RAM PC. All necessary resources (images, dockerfiles, docker-compose files) can be stored on any local or remote repository (e.g. DockerHub) and installation on Docker-ready systems requires no user intervention. The testbed includes a few DASH client. The base for them was the scootplayer [23], the an experimental MPEG-DASH request

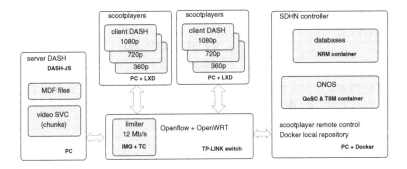

Fig. 2. The structure of the testbed.

engine with support for accurate logging. It was modified to support the SAND messages listed in Sect. 3. The DASH clients were implemented as separate, virtual instances of LXD (Linux Container Deamon) containers. All tests (the clients activities) were orchestrated by remote control scripts.

The "Elephants Dreams" video sequence (1080 p, 24 f/s) was used for all tests. The video was encoded using the reference software JSVM (Joint Scalable Video Model) what resulted in two SVC video files with resolutions 1080 p and 720 p, respectively. In the encoding process the Profile High was used. The parameters of the individual layers within the SVC stream were as follows:

- three temporal layers (T0 - 6 f/s, T1 12 f/s and T2 24 f/s),
- four SNR(MGS) layers (initial QP: SNR0 - 42, SNR1 - 42, SNR2 - 32 and SNR3 - 32).

4.2   Evaluation Metrics

To evaluate SDHN solution, we choose three metrics. Their aim were to reflect the feature of our video distribution system when multiple SVC-DASH video flows share the same bottleneck link.

Fairness index. Publications of recent years bring many proposal of metrics for fairness assessment [17,24]. We decided to base on the wide-used Jains Index [25] with modification which reflects the structure of SVC video stream. The fairness index $FairI$ according to Jains approach is expressed by the Formula (2).

$$FairI_a = \frac{\sum_{i=1}^{n} a_i{}^2}{n \cdot \sum_{i=1}^{n} a_i{}^2} \tag{2}$$

Here, n is the number of competing users. The parameter a signifies the fairness metrics which depends upon the application. In the case of SVC streams, the metric can be associated with fraction of demand. The DASH adaptation process causes the change of OP (change of demand). Also, clients have unequal demands. Thus it is reasonable to measure fairness by closeness of the allocations to respective demands. As the result, we propose the following definition of the fairness metric 3.

$$a_i = \begin{cases} \frac{OP_i}{OPmax_i} & \text{if } OP_i < OPmax_i \\ 1 & \text{otherwise} \end{cases} \tag{3}$$

Here, $OPmax_i$ is the number of available (accepted) OPs inside the SVC stream. In other words, it is the maximum demand of ith DASH client. OP_i is the number of received OP by i^{th} DASH client.

Fluctuation index. Based on our experience [14], the bitrate fluctuations (switches between video representation) have great influence on the perceived quality (QoE) of video on clients' side. The system for distribution of video in the DASH service should prevent or at least minimize the occurrence of bitrate-switches. To asses this issue, the Fluctuation Index $FlucI$ has been used. Because the bitrate-switches (the OPs switches) may occur only between chunks thus the index is defined by the Formula 4.

$$FlucI = \sum_{i=1}^{n} \frac{F_n}{C_n} \tag{4}$$

Here, F_n is the number of switches for SVC video stream n and C_n is a total number of received chunks.

Link Utilization Index. It is obvious that the limited capacity of outgoing link should be utilized as good as possible. In order to asses this issue, we used well-known idea of measure of utilization on a network layer. Thus, the Link Utilization Index LUI is defined by the Formula 5.

$$LUI = \frac{\sum_{i=1}^{n} B_n}{Blimit} \tag{5}$$

Here, $Blimit$ is the bandwidth of the bottleneck and B_n is the summary bitrate of all SVC video streams.

4.3   Results of SDHN Evaluation

The evaluation carried out included five DASH clients. Two of them were configured for the maximum 1080 p video resolution and others, for the video of maximum 720 p resolution. Both tests (A and B) were repeated fifty times. In the case of Test B (SDNH DASH), five schedules of clients' activities were used (each repeated ten times). Every single test lasted 150 s (the fifty chunks, each 5 s long). The example of such schedule and corresponding bitrates of video streams are presented in Fig. 3.

The schedule on the Fig. 3 defines two strong congestions. First one (between 30th and 90th second) grows slowly while the next one (between 100th and 135th second) is of the burst type. The behaviour of legacy DASH and SDHN DASH solution is presented in Fig. 4.

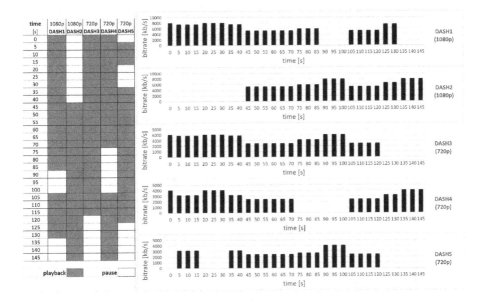

Fig. 3. The example of the schedule of the test B (left) and corresponding bitrates of the SVC video streams received by five DASH clients (right).

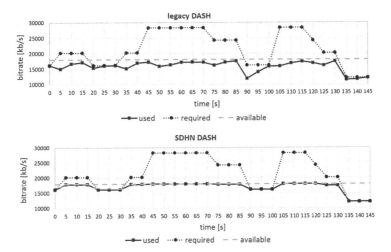

Fig. 4. The utilized bandwidth, the total demand on bandwidth and the bottleneck capacity for two video distribution systems: legacy DASH (upper) and proposed SDHN DASH (lower).

In the most of tested cases, the legacy DASH was sensitive to congestions. It is worth noting that the highest levels of overload (resulting from exaggerated trials of adaptation) were seen when congestions had been disappearing (between 85th and 100th second). As it was expected, the SDHN DASH system was able to calculate near-optimal bitrate values what in combination with playout buffer, allowed for a significant reduction of fluctuations (number of switching). An additional effect is much better link utilization. The values of the Fluctuation Index and Link Utilization Index, for two compared video distribution system and for all 50 test, are presented in the Table 1.

Table 1. The mean values of the *FlucI* and *LUI* indexes.

Scenario – system	Indexes	
	FlucI	LUI
Test A – legacy DASH	2.39	0.93
Test A – SDHN DASH	0.16	1.0
Test B – legacy DASH	2.88	0.88
Test B – SDHN DASH	0.89	0.94

In all the tests conducted, the proposed SDHN solution demonstrated superiority over the standard DASH system. Particularly, in the case of continuous playback, SDHN offers almost perfect solution. Of course, the 100% utilisation of the link capacity is not always possible due to the parameters associated with chunks (parameters of the video encoding).

The fairness defined according to the formulas 2 and 3 proved to be better in the case of proposed solution. The graph of the values of the *FairI* index value during test B (the test according to schedule presented in Fig. 3) is shown in Fig. 5.

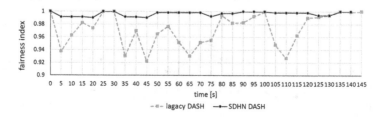

Fig. 5. The fairness Index for the test B.

For the standard DASH distribution system (legacy DASH), the appearance of congestion results in significant fairness degradation. In case of relatively longer congestions, the fluctuations of *FaitI* is also observed. These phenomena are much weaker for proposed SDHN system.

5   Conclusion and Future Work

The combination of the SVC video coding and DASH principles is undoubtedly a step towards greater flexibility of video streaming services. However, this approach requires the new adaptation methods which are able to provide the high level of fairness during the process of video distribution. This issue has been analysed from the providers perspective. The article presented the different point of view. We proposed the video distribution system based on DASH and employing SDN which is dedicated to the typical home network. Through a combination of previously developed utility function for SVC-DASH and SAND architecture, we elaborated the software solution which can be implemented in the relatively simple SDN controller. Therefore, it was possible to orchestrate and implement the proposed solution by means of Docker containers. Home network users can automatically install our SDHN solution if only have Openflow compliant switch. No additional, specialized hardware is required beyond the typical PC. In order to make the proposed solution even more practical, we are working on a version optimized for single-chip computers and selected, commercial gateways.

The tests carried out have shown the superiority of proposed SDHN solution in terms of the guarantee fairness distribution of SVC video streams and the use of available bandwidth on a link to an external network. At the same time, the SDHN allowed for substantial reduction of bitrate switching, which is important to maintain high, perceived quality of received video.

References

1. ISO/IEC 23009-1:2014: Information technology - Dynamic adaptive streaming over HTTP (DASH) - Part 1: Media presentation description and segment formats (2014)
2. Reichel, J., Schwarz, H., Wien, M., Vieron, J.: Joint Scalable Video Model 9 of ISO/IEC 14496-10: 2005/AMD3 Scalable Video Coding. Joint Video Team (JVT), Docs. JVT-X202 (2007)
3. Stockhammer, T.: Dynamic adaptive streaming over HTTP standards and design principles. In: Second Annual ACM Conference on Multimedia Systems, San Jose, pp. 133–144. ACM (2011)
4. Akhshabi, S., Anantakrishnan, L., Dovrolis, C., Begen, A.C.: What happens when HTTP adaptive streaming players compete for bandwidth. In: Proceedings of ACM Workshop on Network and Operating System Support for Digital Audio and Video, Toronto, pp. 9–14 (2012)
5. Timmerer, C., Griwodz, C., Begen, A.C., Stockhammer, T., Girod, B.: Guest editorial adaptive media streaming. IEEE J. Sel. Areas Commun. **32**(4), 681–683 (2014)
6. Abuteir, R.M., Fladenmuller, A., Fourmaux, O.: SDN based architecture to improve video streaming in home networks. In: Proceedings of IEEE 30th International Conference on Advanced Information Networking and Applications, Crans-Montana, pp. 220–226 (2016)
7. ONF: Open Networking Foundation Software-Defined Networking (SDN) Definition. https://www.opennetworking.org/sdn-resources/sdn-definition
8. Docker Inc. and Docker Project. https://www.docker.com/
9. Akhshabi, S., Begen, A., Dovrolis, C.: An experimental evaluation of rate-adaptation algorithms in adaptive streaming over HTTP. In: Second Annual ACM Conference on Multimedia Systems, San Jose, pp. 157–168. ACM (2011)

10. ISO/IEC FDIS 23009-5:2016: Information Technology - Dynamic adaptive streaming over HTTP (DASH) - Part 5: Server and network assisted DASH (SAND) (2016)
11. Kalva, H., Adzic, V., Furht, B.: Comparing MPEG AVC and SVC for adaptive HTTP streaming. In: IEEE International Conference on Consumer Electronics, Las Vegas, pp. 158–159 (2012)
12. Unanue, I., Urteaga, I., Husemann, R., Del Ser, J., Roesler, V., Rodriguez, A., Sanchez, P.: A tutorial on H264/SVC scalable video conding and its tradeoff between quality, coding efficiency and performance. In: Lorente, J.D.S. (ed.) Recent Advances on Video Coding. InTech, Rijeka (2011). Chap. 1
13. Schwarz, H., Marpe, D., Wiegand, T.: Overview of the scalable video coding extension of the H.264/AVC standard. IEEE Trans. Circ. Syst. Video Technol. **17**(9), 1103–1120 (2007)
14. Przylucki, S., Czerwinski, D., Sierszen, A.: QoE-oriented fairness control for DASH systems based on the hierarchical structure of SVC streams. In: Gaj, P., Kwiecień, A., Stera, P. (eds.) Computer Networks. Communications in Computer and Information Science, vol. 608, pp. 180–191. Springer, Cham (2016)
15. Mustafa, B., Nadeem, T.: Dynamic traffic shaping technique for HTTP adaptive video streaming using software defined networks. In: Proceedings of 12th Annual IEEE International Conference on Sensing, Communication, and Networking, Seattle, pp. 178–180 (2015)
16. Cetinkaya, C., Ozveren, Y., Sayit, M.: An SDN-assisted system design for improving performance of SVC-DASH. In: Federated Conference on Computer Science and Information Systems, Lodz, pp. 819–826 (2015)
17. Mu, M., Broadbent, M., Farshad, A., Hart, N., Hutchison, D., Ni, Q., Race, N.: A scalable user fairness model for adaptive video streaming over SDN-assisted future networks. IEEE J. Sel. Areas Commun. **34**(8), 2168–2184 (2016)
18. Tang, S., Hua, B., Liu, S.: A heuristic algorithm for optimal discrete bandwidth allocation in SDN networks. In: Proceedings of IEEE Global Communications Conference, San Diego, pp. 1–7 (2015)
19. Xingtao, L., Yantao, G., Wei, W., Sanyou, Z., Jiliang, L.: Network virtualization by using software-defined networking controller based Docker. In: Proceedings of IEEE Information Technology, Networking, Electronic and Automation Control Conference, Chongqing, pp. 1112–1115 (2016)
20. Dakin, R.J.: A tree-search algorithm for mixed integer programming problems. Comput. J. **8**(3), 250–255 (1965)
21. GitHub Inc. https://github.com/CPqD/ofsoftswitch13/wiki
22. ITEC – Dynamic Adaptive Streaming over HTTP. http://www-itec.uni-klu.ac.at/dash/
23. Scootplayer. http://scootplayer.readthedocs.io/en/latest
24. Liu, L., Zhou, C., Zhang, X., Guo, Z.: A fairness-aware smooth rate adaptation approach for dynamic HTTP streaming. In: IEEE International Conference on Image Processing (ICIP), Quebec, pp. 4501–4505 (2015)
25. Jain, R.K., Chiu, D.-M.W., Hawe, W.R.: A Quantitative Measure of Fairness and Discrimination for Resource Allocation in Shared Computer System. In: DEC Research Report TR-301 (1984)

Teleinformatics and
Telecommunications

Minimum Transmission Range Estimation for Vehicular Ad Hoc Networks in Signalised Arterials

Bartłomiej Płaczek$^{(\boxtimes)}$ and Marcin Bernas

Institute of Computer Science, University of Silesia,
Będzińska 39, 41-200 Sosnowiec, Poland
`placzek.bartlomiej@gmail.com`, `marcin.bernas@gmail.com`

Abstract. In this paper the connectivity is analysed of vehicular ad hoc networks (VANETs) in arterial roads with signalised intersections. A method is proposed to derive an estimate of minimum transmission range (MTR), i.e., a minimum value of the vehicles transmission range that allows all vehicles in a given road section to be connected via multihop communication. The introduced method takes into account the impact of traffic signals coordination on formation of vehicle platoons. It is applicable for both the under-saturated and the over-saturated traffic conditions. The main contribution in this paper is formulation of a relationship, which allows the upper bound of MTR to be estimated for VANETs in unidirectional and bidirectional signalized arterial roads. The conducted experiments confirm that the derived estimate fits well with results of simulation experiments.

Keywords: Vehicular ad hoc networks · Connectivity · Minimum transmission range · Traffic signal control

1 Introduction

Vehicular ad hoc network (VANET) is considered as an important component of the next generation automotive systems, which enables inter-vehicle communication for a number of safety, traffic management, and comfort-related applications [1–4]. Using the communication between vehicles, VANET could provide a safer and a more convenient driving experience. From the perspective of the VANET applications development, network connectivity is very important since it might be difficult to provide reliable services in the case of disconnections. Due to the dynamically changing topology of VANET, the connectivity has direct influence on channel contention and vehicle communications [5].

Transmission range is a critical parameter, which significantly affects VANET connectivity [6]. On the one hand, the connectivity of VANET is expected to be improved as the transmission range increases [7]. On the other hand, a longer transmission range in a dense wireless network can lead to a huge amount of interferences between neighbour nodes and a high network overhead. Thus, the

© Springer International Publishing AG 2017
P. Gaj et al. (Eds.): CN 2017, CCIS 718, pp. 209–220, 2017.
DOI: 10.1007/978-3-319-59767-6_17

network will suffer from inefficiency due to packet collisions and losses [8]. To alleviate this issue, the transmission range in VANET should be locally adjusted to the current traffic situation. So far, several transmission range adjustment techniques for VANET have been investigated [8,9]. Such techniques require accurate estimation of a transmission range value, which guarantees connectivity of all vehicles in a considered area.

In this paper, the minimum transmission range (MTR) is analysed for VANET in arterial roads with signalised intersections. MTR corresponds to the minimum common value of the vehicles transmission range, which enables all vehicles to be connected via multihop communication in a single network cluster. A method is proposed to derive an estimate of the upper bound MTR. The analytical estimate of the upper bound MTR is verified in a simulation environment for unidirectional and bidirectional signalized arterial road.

The paper is organized as follows. A short review of the related literature is provided in Sect. 2. The method of MTR estimation is introduced in Sect. 3. Simulation experiments and their results are described in Sect. 4. Finally, conclusions are given in Sect. 5.

2   Related Works and Contribution

A number of efforts have been already accomplished to investigate the MTR for VANET in highways. In [10] a cellular automata model of the uninterrupted highway traffic was considered and an analytical lower-bound for the MTR was derived. It was shown that a non-homogeneous distribution of vehicles, due to traffic jams, results in the need for a higher transmission range in comparison to that required for homogeneous vehicles distribution. Based on the above-mentioned results, a traffic density estimate was used in [11] to develop an algorithm, which adjusts the transmission range of vehicles dynamically according to the local traffic conditions on a highway.

The multihop connectivity of one-dimensional VANET, with given either vehicle locations or traffic densities, was discussed in [12]. A model was proposed, which can be used as a guideline for determining the required transmission range, in order to achieve very high network connectivity. However, that approach is limited to simple traffic scenarios without traffic lights.

Inter-platoon connectivity for a bidirectional highway was analysed in [7]. The authors have assumed that all vehicles drive in platoon-based patterns, which facilitate better traffic performance as well as information services. Impact was investigated of the transmission range on expected time of safety message delivery among platoons. The simulation experiments have shown that the transmission range is a critical parameter, which significantly affects the inter-platoon communication in VANET.

The highway scenario was also considered in [9] to introduce a transmission range control algorithm for safety applications. According to that algorithm, the transmission range was adapted by taking into account vehicle density, VANET load and packet generation rate of each vehicle. Another scheme, which enables dynamic transmission range adjustment, based on both local vehicle density and distance to destination, was presented in [8]. That study was devoted to the broadcast-type communication between vehicles equipped with directional antennas.

A connectivity robustness model was proposed in [13] to assess vehicle-to-vehicle communication in connected vehicle environments. That model accounts for the transmission range and multiple other factors affecting the connectivity in VANETS. The results in [13] confirm that overall robustness of connectivity increases for higher values of transmission range.

In [14] a cellular automata approach was applied to construct an urban traffic mobility model. Based on the developed model, characteristics of global traffic patterns in urban areas were analysed. It was shown that even though the inter-vehicle spacing of both highway and urban traffic can be approximated by the exponential distribution, the connectivity pattern of a vehicle is very different in these two scenarios. The simulation results presented in [14] indicate that the parameters of traffic control at signalised intersections such as cycle duration, green split, and coordination of traffic lights have a significant effect on inter-vehicle spacing distribution and the required transmission range.

A simulation environment was also used in [6] to investigate the MTR for a road network with signalised intersections. Based on the simulation experiments, it was shown that the signalised intersections cause significant fluctuations of inter-vehicle distance and thus the signal-controlled traffic requires higher communication ranges to ensure connectivity. The higher communication range values were needed to connect the vehicles stopped at traffic lights with those moving freely.

According to the authors' knowledge, the problem of analytical MTR determination for arterial streets with signalised intersections has not been studied so far in the VANET-related literature. The main contribution in this paper is a relationship formulation, which allows the upper bound of MTR to be estimated for VANETs in signalised arterial roads.

3 A Method for Minimum Transmission Range Estimation

In arterial streets with traffic signals, vehicles are grouped into platoons at exits of the signalised intersections. The vehicle platoons move from one signalized intersection to the next. A well-coordinated traffic signal system moves platoons of vehicles so that they arrive during the green phase of the downstream intersection. In order to evaluate the minimum transmission range, which allows maintaining connectivity of all vehicles in a signalised arterial road, it is necessary to consider the distances between the vehicle platoons.

In Fig. 1 the progression of platoons along a signalised artery is illustrated by means of time-space diagrams. Vehicle trajectories are plotted on the time-space diagrams as the sloped lines. Stopped vehicles are shown as horizontal lines. The horizontal bars show indications of the traffic signals at intersections. An example of the time-space diagram for unidirectional artery is presented in Fig. 1(a). In this example the vehicles move northbound (from bottom to top of the diagram). A time-space diagram of bidirectional artery is presented in Fig. 1(b), where the dashed lines correspond to the vehicles that moves southbound (top to bottom).

It should be noted that the vehicle trajectories in time-space diagrams (Fig. 1) are idealized for presentation purposes. The slope of the trajectories in these diagrams corresponds to space-mean speed of the vehicles. Vertical distance between the trajectories represents average space headway between vehicles. Moreover, for sake of clarity, the examples in Fig. 1 do not include vehicles travelling outside platoons. The simplified presentation of time-space diagram, which is based on the average traffic stream parameters, does not involve loss of generality. The simulation model, which was used for evaluation of the proposed method (Sect. 4), involves random variations of vehicle speed, presence of vehicles outside platoons as well as platoon dispersion.

For the unidirectional artery (Fig. 1(a)), the longest transmission range is necessary to connect the vehicle platoons, when one platoon approaches the end of the considered road section (signal 3) and the second platoon enters this road section (approaches signal 1) Thus, the MTR can be evaluated according to the following formula:

$$r = d - w, \tag{1}$$

where d denotes length of the considered road segment and w is length of the vehicle platoon.

In case of the bidirectional arterial road (Fig. 1(b)), the longest communication range is necessary to connect the two vehicle platoons that move in opposite directions and approach ends of the analysed road section. It means that lengths of the two platoons (w_1 and w_2) have to be taken into account when evaluating the MTR. Thus, Eq. 1 is used for this case with $w = w_1 + w_2$.

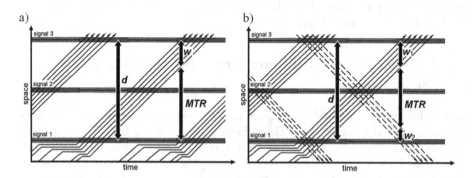

Fig. 1. Minimum transmission range in signalized arterial: (a) unidirectional traffic, (b) bidirectional traffic.

Length of a vehicle platoon (w) can be evaluated as follows:

$$w = (n - 1) \cdot h, \tag{2}$$

where n is number of vehicles in the platoon and h is average space headway between vehicles. The average space headway is equal to the inverse of traffic density ($h = 1/k$). When taking into account the fundamental equation of traffic flow [15], the average space headway can be expressed as:

$$h = 1/k = v/s, \tag{3}$$

where k is traffic density, v is space-mean speed of the vehicles, and s is satura-tion flow rate, i.e. the maximum number of vehicles that can pass through the intersection during a time unit, when the traffic signal is green. The saturation flow rate is used in (3) because the platoon is created if a group of vehicles leave an intersection when the traffic signal turns green. It should be also noted that in Eq. (1) the length of a single vehicle is ignored ($w = 0$ if $n = 1$) for sake of simplicity. Substituting (3) into (2) yields,

$$w = \max(0, n - 1) \cdot v/s \tag{4}$$

The platoon is a group of vehicles leaving an intersection when red light ends. Therefore, the number of vehicles in the platoon (n) can be estimated as a number of vehicles that have been stopped at an intersection due to the red signal. Assuming that R is duration of the effective red signal and q is intensity of incoming traffic, the number of vehicles stopped during time period can be calculated by employing an analytical approach with fluid approximation, as depicted in Fig. 2.

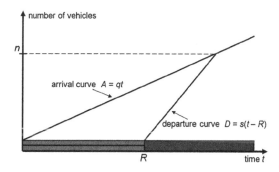

Fig. 2. Number of vehicles stopped at an intersection due to red signal (n). (Color figure online)

Based on the arrival curve and the departure curve presented in Fig. 2, it can be found that the total number of stopped vehicles, which corresponds to the number of vehicles in the platoon, can be estimated as follows:

$$n = \frac{R \cdot q}{1 - q/s} \tag{5}$$

The above considerations are related to the MTR estimation for under-saturated traffic conditions, whereby vehicle queues are only created during the red signal and are dissolved during green signal. In oversaturated traffic conditions, when the intensity of traffic flow (q) exceeds the intersection capacity (c), vehicle queues fill approaches of the intersection and spill over from one intersection to another. In such case, the queues persist and cannot be cleared totally during green signal.

According to the proposed approach, the MTR for oversaturated traffic is estimated based on the length of the widest gap between vehicles, which can be created when vehicles leave queue during green signal. The expected number of vehicles that pass an intersection during effective green time (G) is $G \cdot s$. Average road length occupied by one vehicle in the queue equals $1/k_{jam}$, where k_{jam} denotes maximum vehicle density in traffic jams. Thus, the MTR for oversaturated traffic conditions is estimated as $G \cdot s/k_{jam}$.

Finally, based on the above considerations, the upper bound of MTR for VANET in a signalised arterial road of length d is given by

$$r(q) = \begin{cases} d - \left(\dfrac{R \cdot q}{1 - q/s} - 1 \right) \cdot v/s, & q \le c, \\ G \cdot s/k_{jam}, & q > c, \end{cases} \tag{6}$$

where c denotes capacity of the signalized intersection and the remaining symbols were defined earlier in this section. The capacity is calculated by using the following formula:

$$c = s \cdot \frac{G}{R + G}. \tag{7}$$

4   Evaluation in Simulation Environment

The analytical relationships that were derived in the previous section are verified through simulations of vehicle traffic for both unidirectional and bidirectional arterial road.

4.1   Simulation Setup

In this study the stochastic cellular automata (CA) model of urban traffic was used for the traffic simulation [16]. Each traffic lane is represented by a one-dimensional cellular automaton. The traffic lanes are divided into cells of equal lengths (7.5 m). An occupied cell in the cellular automaton corresponds to single vehicle. At each discrete time step (1 s) the state of CA is updated according to four steps (acceleration, deceleration, velocity randomization, and movement). These steps are necessary to reproduce the basic features of real urban traffic. Step 1 represents driver tendency to drive as fast as possible; step 2 involves reaction to other vehicles and to traffic lights; step 3 introduces non-deterministic acceleration due to random external factors and the overreaction of drivers while slowing down. Finally, in step 4 the vehicles are moved according to the new velocity calculated in steps 1–3. Steps 1–4 are applied for all vehicles.

According to the definition of step 3, the speed of all vehicles that have a velocity of at least 1, is reduced by one unit with a probability of p. Parameter p is referred to as the deceleration probability. It should be noted that without step 3, the model is deterministic [17]. Detailed definitions of steps 1–4 can be found in [14].

A schema of the simulated arterial road, with eight signalized intersections, is shown in Fig. 3. Total length of the arterial road is 7.5 km (1000 cells). The above-mentioned CA model of the arterial road was implemented in Matlab. The simulations were performed for two scenarios. Unidirectional traffic was considered in Scenario 1 and bidirectional traffic was analyzed in Scenario 2. Simulation data were collected in a section between 4th and 6th intersection (the shaded area in Fig. 3).

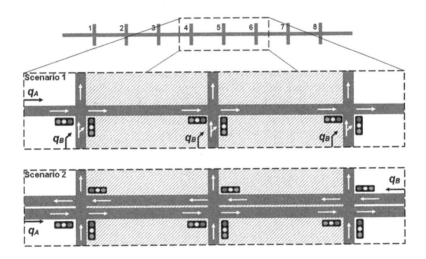

Fig. 3. Simulated arterial road.

Intensity of the traffic flow was determined for the network model by parameters q_A and q_B (in vehicle per second). At each time step vehicles were randomly generated with a probability equal to the intensity q_A or q_B in the entrances of the road model, as shown in Fig. 3. As a result, binomially distributed traffic flows were obtained that approximate a Poisson distribution. The simulation period was 10000 s for each considered combination of the traffic intensities. The traffic signal control was simulated assuming the green wave coordination strategy. Detailed parameters of the simulation model are presented in Table 1.

4.2 Experimental Results

Simulation experiments were performed to determine the MTR needed to connect all vehicles in the analysed road section. In general, the MTR is calculated

Table 1. Parameters of simulation model.

Parameter	Value
Total road length	1000 cells
Length of analyzed road section (d)	222 cells
Distance between intersections	111 cells
Effective green time (G)	38 s
Effective red time (R)	38 s
Mean free-flow speed (v)	2.8 cells per second
Maximum speed	3 cells per second
Deceleration probability	0.2
Saturation flow rate (s)	0.54 vehicles per second per lane
Capacity of intersection (c)	0.27 vehicles per second per lane
Maximum vehicle density (k_{jam})	1 vehicle per cell

by constructing the minimum spanning tree in a graph $K(N, E)$, which represents the VANET in the considered road section [10]. The vertices N in K correspond to vehicles and edges E represent communication links. An edge $e = (n_i, n_j)$ exists in K if the distance between vehicles i and j is shorter than or equal to $\min(r_i, r_j)$, where r_i, r_j denote transmission ranges of vehicles i and j, respectively. The MTR is the longest edge in the minimum spanning tree. In a single-lane VANET, the MTR is the widest gap between any two consecutive vehicles. The MTR was computed at each time-step of the simulation and stored along with the intensity of traffic flow.

Results of both the simulation experiments and the analytical MTR evaluation for unidirectional traffic (Scenario 1) are compared in Figs. 4, 5 and 6. The

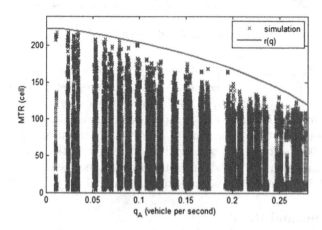

Fig. 4. MTR in unidirectional signalized artery (Scenario 1, $q_B = 0$ veh./s)

scatter plots present the MTR values that were determined for various traffic intensities during simulation. The curves in the plots correspond to the upper bound of MRT, which was estimated by using Eq. (6). These plots show clearly that the MTR values are concentrated below the derived upper bound for all analysed intensities of traffic flow.

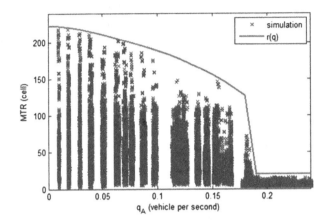

Fig. 5. MTR in unidirectional signalized artery (Scenario 1, $q_B = 0.5 q_A$).

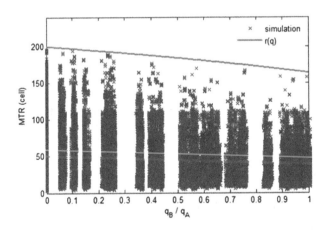

Fig. 6. MTR in unidirectional signalized artery (Scenario 1, $q_A = 0.1$ veh./s).

The results in Fig. 4 were obtained assuming that there are no vehicles turning at intersections from the side roads into the arterial road ($q_B = 0$ veh./s). In this case, without the traffic inflows from side roads, only free-flow traffic conditions are observed. Subsequently, the intensity of the traffic entering arterial from side roads was set to half of the traffic intensity on the arterial road, i.e., $q_B = 0.5 \cdot q_A$ (Fig. 5). For $q_A = 1.8$ veh./s such setting led to oversaturated traffic conditions since the total traffic intensity ($q_A + q_B = 1.5 \cdot q_A$) exceeds the capacity of intersections ($c = 0.27$ veh./s). This causes a significant decrease of the MTR, as shown in Fig. 5.

The performed MTR analysis has also taken into account different ratios of the side inflow intensity and the arterial traffic intensity (q_B/q_A). Figure 6 shows the MTR values that were obtained for $q_A = 0.1$ veh./s and q_B between 0 and 0.1 veh./s. It can be observed in these results that the MTR decreases when the side inflow intensity increases. The shorter MTR for higher q_B intensities are caused by the vehicles from side roads that join the platoons on the arterial road and extend their lengths.

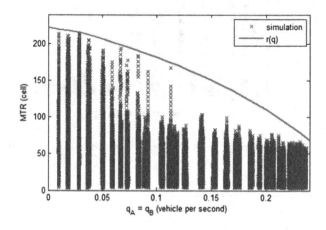

Fig. 7. MTR in bidirectional signalized artery (Scenario 2, $q_A = q_B$).

The MTR values determined for the bidirectional traffic (Scenario 2) are shown in Figs. 7 and 8. It should be noted that in this scenario q_A and q_B denote intensities of the traffic in arterial road for both directions. The traffic inflows from side roads are not present. The results in Fig. 7 were obtained for equal intensities of the traffic in both directions ($q_A = q_B$). The MTR values for different q_A and q_B intensities are presented in Fig. 8. In general, the results confirm that the proposed method enables correct upper bound evaluation for MTR also in case of the bidirectional arterial road.

When comparing the results obtained for Scenario 2 with those of Scenario 1 it can be observed that the MTR values decreases faster with the intensity for the bidirectional traffic. The reason is that the number of vehicles in the analysed section of the bidirectional road is higher than in the unidirectional road and the distances between platoons are shorter, as shown in Fig. 1b.

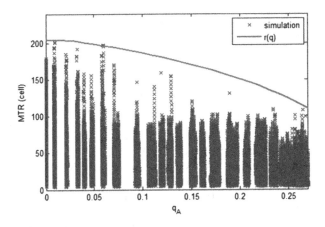

Fig. 8. MTR in bidirectional signalized artery (Scenario 2, $q_B = 0.1$ veh./s).

5   Conclusion

Simulation results show that the proposed method allows the upper-bound MTR to be accurately estimated for VANETs in unidirectional and bidirectional signalized arterials. The derived MTR estimate fits well with the results of simulation experiments for a wide range of traffic intensities. In case of under-saturated traffic, the MTR decreases when the traffic flow intensity increases. In comparison to the under-saturated traffic, a significantly lower MTR value is obtained for the over-saturated traffic conditions. For both the under-saturated and the over-saturated traffic conditions, the traffic signals timing has a strong impact on the estimated MTR. When the MTR is considered for under-saturated traffic, the distance between intersections is an additional important factor, which has to be taken into account. It was also observed that in case of bidirectional traffic the VANET connectivity is improved by lowering the MTR requirements.

In practice this method may be used by control nodes at intersections that are equipped with wireless communication and collect data reported by vehicles via VANET in order to optimize the traffic signals. Such control nodes could evaluate the MTR and broadcast this information to the vehicles. The proposed approach can be combined with advanced data collection strategies [18]. Further research will be necessary to integrate the method into a dynamic range control protocol for VANETs. Another interesting topic for the future research is to use the stochastic geometry model [19] for the MTR analysis.

References

1. Al-Sultan, S., Al-Doori, M.M., Al-Bayatti, A.H., Zedan, H.: A comprehensive survey on vehicular ad hoc network. J. Netw. Comput. Appl. **37**, 380–392 (2014)
2. Płaczek, B.: Selective data collection in vehicular networks for traffic control applications. Transp. Res. Part C: Emerg. Technol. **23**, 14–28 (2012)

3. Bernaś, M.: VANETs as a part of weather warning systems. In: Kwiecień, A., Gaj, P., Stera, P. (eds.) CN 2012. CCIS, vol. 291, pp. 459–466. Springer, Heidelberg (2012). doi:10.1007/978-3-642-31217-5_48

4. Bernaś, M.: WSN power conservation using mobile sink for road traffic monitoring. In: Kwiecień, A., Gaj, P., Stera, P. (eds.) CN 2013. CCIS, vol. 370, pp. 476–484. Springer, Heidelberg (2013). doi:10.1007/978-3-642-38865-1_48

5. Shao, C., Leng, S., Zhang, Y., Vinel, A., Jonsson, M.: Performance analysis of connectivity probability and connectivity-aware MAC protocol design for platoon-based VANETs. IEEE Trans. Veh. Technol. **64**(12), 5596–5609 (2015)

6. Kafsi, M., Papadimitratos, P., Dousse, O., Alpcan, T., Hubaux, J.P.: VANET connectivity analysis. In: IEEE Workshop on Automotive Networking and Applications (Autonet), no. LCA-CONF-2009-001 (2008)

7. Jia, D., Lu, K., Wang, J.: On the network connectivity of platoon-based vehicular cyber-physical systems. Transp. Res. Part C: Emerg. Technol. **40**, 215–230 (2014)

8. Soua, A., Ben-Ameur, W., Afifi, H.: Beamforming-based broadcast scheme for multihop wireless networks with transmission range adjustment. In: 10th Annual Conference on Wireless On-Demand Network Systems and Services (WONS), pp. 107–109 (2013)

9. Javed, M.A., Khan, J.Y.: Performance analysis of an adaptive rate-range control algorithm for VANET safety applications. In: 2014 International Conference on Computing, Networking and Communications (ICNC), pp. 418–423 (2014)

10. Artimy, M.M., Robertson, W., Phillips, W.J.: Minimum transmission range in vehicular ad hoc networks over uninterrupted highways. In: 2006 IEEE Intelligent Transportation Systems Conference, pp. 1400–1405 (2006)

11. Artimy, M.: Local density estimation and dynamic transmission-range assignment in vehicular ad hoc networks. IEEE Trans. Intell. Transp. Syst. **8**(3), 400–412 (2007)

12. Jin, W.L., Recker, W.W., Wang, X.B.: Instantaneous multihop connectivity of one-dimensional vehicular ad hoc networks with general distributions of communication nodes. Transp. Res. Part B: Methodol. **91**, 159–177 (2016)

13. Osman, O.A., Ishak, S.: A network level connectivity robustness measure for connected vehicle environments. Transp. Res. Part C: Emerg. Technol. **53**, 48–58 (2015)

14. Tonguz, O.K., Viriyasitavat, W., Bai, F.: Modeling urban traffic: a cellular automata approach. IEEE Commun. Mag. **47**(5), 142–150 (2009)

15. Dhingra, S.L., Gull, I.: Traffic flow theory historical research perspectives. Transportation Research E-Circular, E-C149 (2011)

16. Maerivoet, S., De Moor, B.: Cellular automata models of road traffic. Phys. Rep. **419**(1), 1–64 (2005)

17. Płaczek, B.: A traffic model based on fuzzy cellular automata. J. Cell. Automata **8**(3–4), 261–282 (2013)

18. Płaczek, B., Bernas, M.: Uncertainty-based information extraction in wireless sensor networks for control applications. Ad Hoc Netw. **14**, 106–117 (2014)

19. Steinmetz, E., Wildemeersch, M., Quek, T.Q., Wymeersch, H.: A stochastic geometry model for vehicular communication near intersections. In: Globecom Workshops (GC Wkshps), pp. 1–6. IEEE (2015)

The Possibilities and Limitations of the Application of the Convolution Algorithm for Modeling Network Systems

Adam Kaliszan[(✉)] and Maciej Stasiak

Chair of Communication and Computer Networks,
Faculty of Electronics and Telecommunications,
Poznan University of Technology, Poznań, Poland
{adam.kaliszan,maciej.stasiak}@put.poznan.pl
http://nss.et.put.poznan.pl/

Abstract. This article verifies the applicability of the convolution algorithm for modelling multi-service network systems with different call streams, including non-Poissonian streams. The study includes investigations of traffic streams that are state-dependent and independent. The service time is determined by the following distributions: Pareto, uniform, gamma, normal and exponential. The results of the simulation experiments are compared with the results of modelling by convolution algorithm. The results obtained in this study allow limits for the applicability of the convolution algorithm to be determined and identified.

1 Introduction

The works [1,2] propose a convolution algorithm to model full-availability resources that are systems with state-independent call admission process. In [3–7], a modification to convolution algorithms for modelling a number of selected systems with state-dependent call admission process, such as: the full-availability group with reservation [2], limited-availability group [4], threshold systems [4] and overflow systems [6] is presented. In all the above mentioned works the application of the convolution algorithm is limited to model systems that are offered mutually independent classes of traffic generated by a finite (Engset traffic, Pascal traffic) or infinite (Erlang traffic) number of traffic sources. A review of the literature also provides a number of methods in which methods other than convolution algorithms are used to model systems with finite and infinite number traffic sources and different new call admission mechanisms [8]. However, the issue of a possible application of convolution algorithms to model call streams other than Poisson call streams has not been addressed as yet.

Hanczewski et al. [7] proposes the model of a queueing system with cFIFO service discipline and considers an application of a convolution algorithm to model systems which are offered traffic classes with other than exponential distributions of time between subsequent events. The model for cFIFO queueing systems is an approximated model and hence on the basis of the results presented in [7] it is not possible to unequivocally evaluate the influence of non-Poissonian call streams on the accuracy of the convolution algorithm.

© Springer International Publishing AG 2017
P. Gaj et al. (Eds.): CN 2017, CCIS 718, pp. 221–235, 2017.
DOI: 10.1007/978-3-319-59767-6_18

Until now, no analysis of a possibility to model full-availability resources, the so-called full-availability group), with non-Poissonian classes of offered call streams has been carried out. This fact has induced the authors to start a further investigation into available possibilities to apply a convolution algorithm to analyse full-availability systems in which call streams are determined on the basis of one of the following distributions: uniform, normal, gamma, Pareto or exponential. The objective of the study was then to determine limitations in the application of these distributions in modelling telecommunications systems by convolution methods.

This article is structured as follows. Section 2 presents the traffic parameters that describe the full-availability group, while Sect. 3 provides a description of a call class with non-Poissonian streams. Then, Sect. 4 discusses the convolution algorithm for modelling a full-availability group with any call classes. Section 5 presents a simulation method for a determination of occupancy distributions in single-service systems with non-Poissonian traffic call streams. Section 6 compares the results of analytical calculations with the results of simulation experiments for different (diverse) distributions that describe call streams. Finally, Sect. 7 is a brief summary chapter that sums up the article.

2   Full-Availability Resources (The Full-Availability Group)

The basic model of resources with multi-service traffic is the so-called full-availability group. This group is offered m call classes from the set M that correspond to the services offered by a telecommunications system under investigation. In a full-availability group, a call of a given class is always admitted for service if only the group has unoccupied resources required for this call to be serviced. Each call class is characterised by a different number of resources d_i that are needed for a single call to be serviced, which is expressed in AU (Allocation Units). While constructing multi-service models for telecommunications systems it is assumed that the AU is the greatest common divisor (or its multiplicity) of the bit rates of all calls offered to the system [7]. After a determination of a given value that corresponds to the AU it is also possible to express the capacity of a full-availability group in AUs.

A full-availability group for a large number of Erlang traffic classes of Erlang type can be described by the model [9,10]. In the case of many other traffic classes (Engset, Erlang and Pascal), the so-called BPP traffic (after the names of call streams in Engset, Erlang and Pascal traffic, i.e. Bernoulli, Poisson and Pascal traffic streams, respectively), the model [11,12] or the model [13] can be used. In addition, the convolution algorithm [1,2] can be used to model a multi-service full-availability group with BPP traffic. The results of studies presented in [9–12,14–21] are based on an analysis of multi-dimensional Markov processes that occur in the full-availability system and therefore cannot be used to model non-Poissonian (i.e., those that are different from BPP) call streams. A possibility to apply convolution algorithms for modelling multi-service full-availability groups with non-Poissonian call streams have not been analysed as yet.

A full-availability group for one traffic class is described by the Erlang model (for Erlang traffic: Poisson stream, exponential service time) [22], Engset (for Engset traffic: Bernoulli call stream, exponential service time) [23] or Pascal (model) (for Pascal: Pascal call stream, exponential service time) [24].

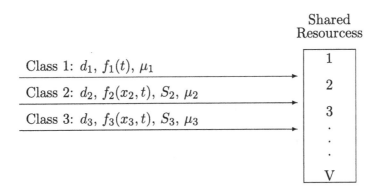

Fig. 1. A diagram of the full-availability group with multi-service traffic

A diagram of the full-availability group is shown in Fig. 1. In line with the adopted notation, $f_i(t)$ is the function of time distribution between subsequent calls of class i. The parameter x_i in function $f_i(x_i, t)$ indicates the dependence of a distribution on the number x_i of serviced calls of class i. For these distributions that are dependent on the number of serviced calls, the number of traffic sources S_i is a significant parameter. The parameter μ_i defines the average service intensity for calls of class i. In traffic engineering issues, particular traffic classes are characterised by the following parameters: the average traffic intensity A_i and the average number of AUs demanded by a single call of a given class, denoted in the article by the symbol d_i. A Poisson call stream of class i is characterised by the average call intensity λ_i, whereas Bernoulli and Pascal streams by the call intensity γ_i from one unoccupied source (sources in active state, i.e. in a state in which a subsequent call can be generated). The dependence between the call intensity λ_i and the average value of the intensity of offered Erlang traffic A_i can be expressed by the following formula:

$$A_i = \frac{\lambda_i}{\mu_i}. \tag{1}$$

In a Bernoulli stream, call arrival intensity decreases along with the number of serviced calls. The dependence between the number of traffic sources S_i, intensity γ_i on a source in active state and the average value of the intensity of offered traffic A_i can be written as follows [25]:

$$A_i = (S_i - A_i) \cdot \frac{\gamma_i}{\mu_i}. \tag{2}$$

In a Pascal stream, call arrival intensity increases along with the number of serviced calls. The dependence between the parameters S_i, γ_i and the average value of offered traffic A_i can be written with the following Formula [25]:

$$A_i = (S_i + A_i) \cdot \frac{\gamma_i}{\mu_i}. \tag{3}$$

3   Non-Poissonian Call Streams

Consider now those call classes that are derivatives of Poisson classes in terms of the independence of call intensity on the occupancy state in the system. In these classes, a call stream can have any distribution that defines time t_{int} between subsequent calls. The time t_{int} distribution can be described on the basis of two parameters: the expected value $E_i(t_{\mathrm{int}})$ and variance $D_i(t_{\mathrm{int}})$ of the time between subsequent calls. The expected value $E_i(t_{\mathrm{int}})$ is the inverse of the average intensity λ_i, which makes it possible to write Eq. (1) in the following form:

$$E_i(t_{\mathrm{int}}) = \frac{1}{A_i \mu_i}. \tag{4}$$

Let us define now the call classes that are derivatives of the Bernoulli class in terms of the dependence of the call stream on the occupancy state of the group. In these streams, along with the increase in the number of serviced calls, time t_{int} between consecutive calls also increases. Further on in the article, these classes will be called call classes with the state-dependent arrival process (with descending traffic pattern). Let t_{act} determines the time between a termination of call service and the arrival of a new call for a single traffic source, i.e. the time when the source is active. The expected value $E_i(t_{\mathrm{act}})$ determines thus the average value of active time of the source. The expected time $E_i(t_{\mathrm{act}})$ is the inverse of the average intensity of the source in active state γ_i. The time t_{act} can be described by any (randomly selected) distribution whose parameters depend on the expected value $E_i(t_{\mathrm{act}})$. As a result, the dependence between the expected value $E_i(t_{\mathrm{act}})$ and the average traffic intensity of offered traffic A_i can be expressed, on the basis of (2), as follows [25]:

$$E_i(t_{\mathrm{act}}) = \frac{S_i - A_i}{A_i \mu_i}. \tag{5}$$

It is also possible to define in a similar way classes that are state-dependent with ascending traffic pattern that are derivative classes from Pascal call classes. The dependence between the average traffic intensity and the expected value $E_i(t_{\mathrm{act}})$ can be written in this particular case, on the basis of (3), in the following way:

$$E_i(t_{\mathrm{act}}) = \frac{S_i + A_i}{A_i \mu_i}. \tag{6}$$

Let us consider now the way in which appropriate parameters for selected distributions used in the investigation into the convolution algorithm are determined. These distributions make a determination of the time between calls (state-independent call classes) or the time between a termination of a service and a subsequent call coming from the same source (state-dependent call classes) possible. The distribution parameters can be determined on the basis of the expected value $E_i(t_*)$ and variance $D_i(t_*)$, where $t_* = t_{int}$ for state-independent streams and $t_* = t_{act}$ for state-dependent streams.

In the uniform distribution the probability density function is written with the following formula:

$$f_i(t) = \frac{1}{t_{i,max} - t_{i,min}}, \tag{7}$$

where $t_{i,min}$ denotes the minimum value after which a call of class i arrives, whereas $t_{i,max}$ is the maximum value. The parameters $t_{i,min}$ and $t_{i,max}$ can be determined on the basis of the expected value and variance:

$$t_{i,min} = E_i(t_*) - \sqrt{3 \cdot D_i(t_*)}, \tag{8}$$

$$t_{i,max} = E_i(t_*) + \sqrt{3 \cdot D_i(t_*)}. \tag{9}$$

In the case of the normal distribution, the probability density function is determined directly on the basis of the expected value and variance for time t_*:

$$f_i(t) = \frac{1}{D_i(t_*)\sqrt{2\pi}} e^{-\frac{(t-E_i(t_*))^2}{2D_i(t_*)}} \tag{10}$$

In the gamma distribution the density function $f(x)$ is typically written in the following way:

$$f_i(t) = \frac{t^{\alpha_i-1} e^{-t/\beta_i}}{\beta_i^{\alpha_i} \Gamma(\alpha_i)}, \tag{11}$$

where α_i shape parameter (tail index), while β_i is the scale parameter. Those parameters (α_i and β_i) can be determined on the basis of the expected value and variance:

$$\alpha_i = \frac{[E_i(t_*)]^2}{D_i(t_*)}, \tag{12}$$

$$\beta_i = D_i(t_*)/E_i(t_*). \tag{13}$$

In the Pareto distribution the density function can be written as follows:

$$\underset{t>\beta_i}{\forall} \quad f_i(t) = \frac{\alpha_i \cdot \beta_i^{\alpha_i}}{t^{\alpha_i+1}}, \tag{14}$$

where α_i shape parameter (tail index), while β_i is the scale parameter. Due to a particular specificity of the distribution, the parameter α_i should be greater than 2. This value of the parameter α_i limits variance to finite value. The coefficients α_i and x_i can be determined on the basis of the following formulas:

$$\alpha_i = 1 + \sqrt{1 + \frac{D_i(t_*)}{[E_i(t_*)]^2}}, \tag{15}$$

$$\beta_i = \frac{E_i(t_*)\sqrt{1 + \frac{D_i(t_*)}{[E_i(t_*)]^2}}}{1 + \sqrt{1 + \frac{D_i(t_*)}{[E_i(t_*)]^2}}}. \tag{16}$$

4   Convolution Algorithm

In convolution algorithms the input data that describe the offered call class i include: the number of demanded resources d_i and occupancy distribution $[P]_V^{\{i\}}$. The single element $[P(n)]_V^{\{i\}}$ of the distribution determines the occupancy probability (of) n AUs by calls of class i in a system with the capacity V AUs that services calls of class i only. Figure 2 shows a diagram of a full-availability group for the convolution algorithm.

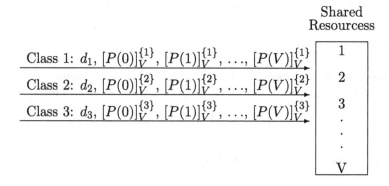

Fig. 2. A diagram of a full-availability group used for convolution modelling

In the next step of the convolution algorithm the distributions of single classes $[P]_V^{\{i\}}$ are aggregated to determine the occupancy $[P]_V^M$ for a multi-service system that services the set M of call classes. The distribution $[P]_V^M$ for the system to which all call classes are offered makes a determination of the QoS parameters, such as the blocking probability or loss probability, possible.

The convolution algorithm can be described by the three following steps:

1. Determination of the single occupancy distributions $[P]_V^{\{i\}}$ for each offered traffic class i,
2. Determination of the aggregated, normalised occupancy distribution $[P]_V^M$,
3. Determination of the blocking probability B_i.

In the first step, for each class i, the distribution $[P]_V^{\{i\}}$ of the occupancy probability for a given number of AUs by calls of class i in a system with the capacity V AUs is determined. The distribution $[P]_V^{\{i\}}$ can be determined on the basis of the probabilities $p_i(x)$ that determine the number of calls x serviced in the single-service system:

$$[P(n)]_V^{\{i\}} = \begin{cases} p_i(x) & \text{for } x = n/d_i \in N \\ 0 & \text{in all other cases.} \end{cases} \tag{17}$$

The probabilities $p_i(x)$ can be determined on the basis of

- single-service models (for BPP traffic),
- simulation experiment,
- measurements in a real system.

In the second step, with an application of the convolution operation, the non-normalised occupancy distribution of the system $[P]^M$ is determined.

$$[P]^M = [P]_V^{\{1\}} * [P]_V^{\{2\}} * \ldots * [P]_V^{\{m\}}. \tag{18}$$

The convolution operation $[P]^A * [P]^B$ can be written in the following way:

$$[P(n)]^{A \cup B} = \sum_{l=0}^{n} [P(l)]^A \cdot [P(n-l)]^B, \tag{19}$$

where A and B are the sets of single or previously aggregated call classes. Then, the distribution $[P]^M$ is normalised to the distribution $[P]_V^M$.

$$[P(n)]_V^M = G \cdot [P(n)]^M, \tag{20}$$

where G is the normalisation constant equal to:

$$G = 1 / \sum_{n=0}^{V} [P(n)]^M. \tag{21}$$

In the third step, appropriate system characteristics, e.g. the blocking probability B_i, can be determined:

$$B_i = \sum_{n=V-d_i+1}^{V} [P(n)]_V^M. \tag{22}$$

5   Occupancy Distributions for Non-Poissonian Call Streams

For the purpose of this article, the occupancy distributions $p_i(x)$ for single, non-Poissonian call streams have been determined in a simulation [26]. Appropriate time distributions between events used in the simulators used in the study were as follows: uniform distribution (7), normal distribution (10), gamma distribution (11) and Pareto (distribution) (14). The assumption was that in all considered systems the service time had exponential character, while the service intensity μ_i for each class i $(i \in M)$ was equal to 1 $[(\text{time unit})^{-1}]$.

$$\underset{i \in M}{\forall} \ \mu_i = 1. \tag{23}$$

Another assumption in the investigations discussed in this article was that for each considered class i $(i \in M)$ with non-Poisson distribution the ratio of square of expected value $[E_i(t_*)]^2$ to variance $D_i(t_*)$ was equal to 3.

$$\underset{i \in M}{\forall} \ \frac{[E_i(t_*)]^2}{D_i(t_*)} = 3. \tag{24}$$

On the basis of (4)–(6) for required value of traffic intensity of a given class, the expected value $E_i(t_*)$ can be determined. Then, taking into consideration the conditions (23) and (24) it is possible to determine all necessary parameters for new call arrival time distributions. Figure 3 shows the necessary parameters for the simulation of state-independent traffic streams in relation to the average intensity of traffic offered to the system. The graphs do not include the value $t_{\min}$ for the uniform distribution that, consistent with the adopted assumptions, is always equal to 0. An exemplary occupancy distribution in a system with the capacity 10 AUs and average intensity of offered traffic equal to 7 Erl. is shown in Fig. 4. Each of the curves presented in the diagram defines appropriate call streams in which the time between calls is described by the uniform, normal, gamma and Pareto distributions.

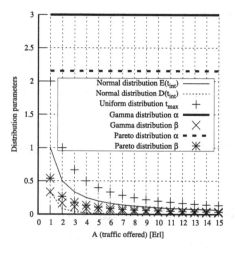

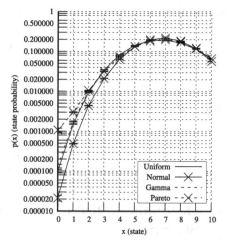

Fig. 3. The dependence of the parameters of call stream distributions under study on the average intensity of offered traffic

Fig. 4. Occupancy distribution in a single-service system with the capacity 10 AUs and average offered traffic intensity 7 Erl.

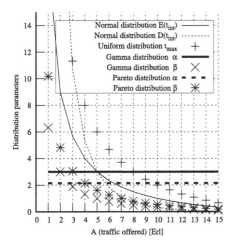

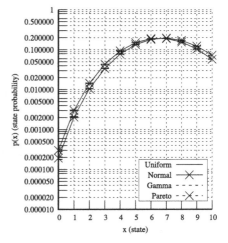

Fig. 5. The dependence of the parameters of the studied call stream distributions on the average intensity of offered traffic A_i

Fig. 6. The occupancy distribution in a single-service system with the capacity 10 AUs and average intensity of offered traffic 7 Erl., and the number of sources 20

Figures 5 and 7 show in turn the necessary parameters for a simulation of state-dependent streams. The assumption is that the number of traffic sources S_i is equal to 20.

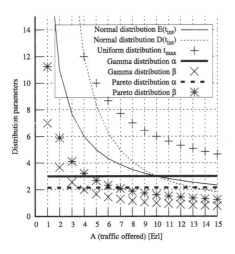

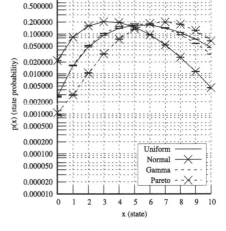

Fig. 7. The dependence of the parameters of the studied call stream distributions on the average intensity of offered traffic A_i

Fig. 8. The occupancy distribution in a single-service system with the capacity 10 AUs and average intensity of offered traffic 7 Erl., and the number of sources 20

The exemplary occupancy distributions for the systems with the capacity 10 AUs to which state-dependent streams generated by 20 sources are offered are shown in Figs. 6 and 8. The presented curves show the occupancy distribution for a system which is offered traffic with the average intensity of 7 Erl. Each of the curves presents appropriate call streams in which the activity time of a source is described by a uniform, normal, gamma and Pareto distributions.

6   A Comparison of the Results of the Simulation with the Convolution Algorithm

In order to determine the boundary possibilities of a possible application of convolution algorithms a comparison of the results obtained on the basis of analytical algorithms with the results of corresponding simulation experiments was made. The simulator applied for the purpose is described in [27]. For all

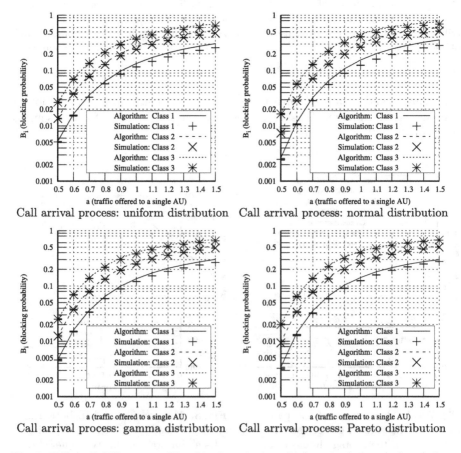

Fig. 9. Full-availability group $V = 20$, $d_1 = 1$, $d_2 = 2$, $d_3 = 3$, $A_1d_1 : A_2d_d : A_3d_d = 1 : 1 : 1$, state-independent call streams, exponential service time $\mu_1 = \mu_2 = \mu_3 = 1$

systems presented in this chapter the group capacity is equal to 20 AUs. The group was offered 3 call classes that demanded $d_1 = 1$, $d_2 = 2$, $d_3 = 3$ AUs, respectively. The assumption was that the proportions in the intensity of offered traffic $A_1d_1 : A_2d_d : A_3d_d$ were equal to 1 : 1 : 1. This means that calls of all classes occupy on average the same number of resources in the system with infinite capacity. The total traffic offered to the system was equal to aV:

$$\sum_{i=1}^{m} A_i d_i = aV, \tag{25}$$

where a is the average traffic offered per one AU.

Figure 9 for a selected number of time distributions t_{int} between subsequent calls shows the dependence between the blocking probability B_i for all call classes and offered traffic a that falls on a single AU. On the basis of the presented results

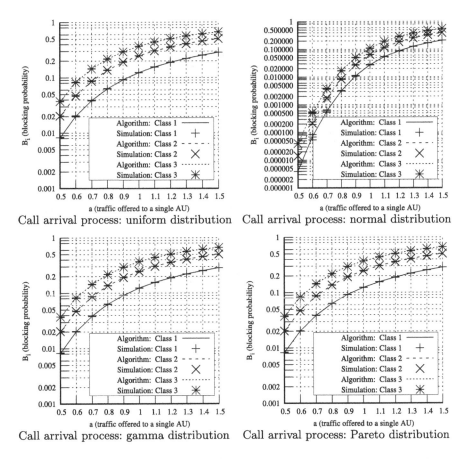

Fig. 10. Full-availability group $V = 20$, $d_1 = 1$, $d_2 = 2$, $d_3 = 3$, $A_1d_1 : A_2d_d : A_3d_d = 1 : 1 : 1$ descending state-dependent call streams, exponential service time $\mu_1 = 1$, $\mu_2 = 1$, $\mu_3 = 1$

it can be claimed that the convolution algorithm can be successfully applied to models to which non-Poissonian call streams are offered. Slight variances in the results obtained in the performed simulation experiment and the results of analytical calculations follow from the fact that the distributions under investigation have no sequence property. This means that the probability of a new call arrival in such streams depends on the moment of arrival of the preceding call and, in consequence, the convolution algorithm proves to be an accurate method only for traffic classes with Poisson call streams that have no sequence property.

Figure 10 shows the dependence between the blocking probability B_i and offered traffic a per a single AU for streams with state-dependent call arrival process (with descending traffic pattern). A comparison of Figs. 9 and 10 confirms that for call streams with state-dependent call arrival process in which the call arrival intensity decreases with the number of serviced calls, the accuracy of

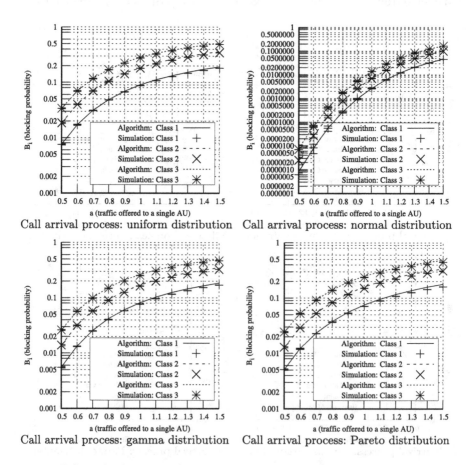

Call arrival process: uniform distribution   Call arrival process: normal distribution

Call arrival process: gamma distribution   Call arrival process: Pareto distribution

Fig. 11. Full-availability group $V = 20$, $d_1 = 1$, $d_2 = 2$, $d_3 = 3$, $A_1 d_1 : A_2 d_d : A_3 d_d = 1 : 1 : 1$ ascending state-dependent call streams, exponential service time $\mu_1 = 1$, $\mu_2 = 1$, $\mu_3 = 1$

the convolution algorithm increases. This results from the fact that the sources of calls are mutually independent, whereas the influence of sequence property appears "within the source" only. The total stream offered to the system is generated by a large number of call sources. Each source has its own memory and a rejection of a call from one source has no influence on the call generation process in the remaining sources.

Similar conclusions can be drawn for the streams with state-dependent call admission process in which the intensity of call inflow increases along with the number of serviced calls. Figure 11 shows the dependence between the blocking probability B_i and offered traffic a per a single AU.

The presented results confirm high accuracy of the convolution method for systems with state-dependant call streams.

7   Conclusions

The results of the study presented in the article show that the convolution algorithm can be successfully used to model systems with non-Poissonian call streams, including state-dependent and state-independent call streams. The approximation error is just slight and in extreme instances does not exceed 20%. This article presents the results for groups with low capacities ($V = 20$ AUs). The investigations carried out by the authors show that the approximation error decreases along with an increase in the capacity of the system. In their future studies the authors intend to determine the accuracy of the convolution algorithm for a case where the group is offered a mixture of call streams that differ in their time distributions between calls. The authors also intend to study the accuracy of the convolution algorithm for service times that are not exponential.

References

1. Iversen, V.: The exact evaluation of multi-service loss system with access control. In: Seventh Nordic Teletraffic Seminar, Lund, pp. 56–61, August 1987
2. Ross, K.: Multiservice Loss Models for Broadband Telecommunication Network. Springer, London (1995)
3. Głąbowski, M., Kaliszan, A., Stasiak, M.: Asymmetric convolution algorithm for blocking probability calculation in full-availability group with bandwidth reservation. IET Circ. Devices Syst. **2**(1), 87 (2008)
4. Głąbowski, M., Kaliszan, A., Stasiak, M.: Modeling product-form state-dependent systems with BPP traffic. Perform. Eval. **67**, 174–197 (2010)
5. Głąbowski, M., Kaliszan, A., Stasiak, M.: Two-dimensional convolution algorithm for modelling multiservice networks with overflow traffic. Math. Prob. Eng. Article ID 852082, 18 (2013)
6. Głąbowski, M., Kaliszan, A., Stasiak, M.: Generalized convolution algorithm for modelling state-dependent systems. IET Circ. Devices Syst. **8**(5), 378–386 (2014)
7. Hanczewski, S., Kaliszan, A., Stasiak, M.: Convolution model of a queueing system with the cFIFO service discipline. Mob. Inf. Syst. **2016**, 1–15 (2016)

8. Moscholios, I., Logothetis, M., Vardakas, J., Boucouvalas, A.: Performance metrics of a multirate resource sharing teletraffic model with finite sources under the threshold and bandwidth reservation policies. IET Netw. **4**(3), 195–208 (2015)
9. Kaufman, J.: Blocking in a shared resource environment. IEEE Trans. Commun. **29**(10), 1474–1481 (1981)
10. Roberts, J.: A service system with heterogeneous user requirements – application to multi-service telecommunications systems. In: Pujolle, G. (ed.) Proceedings of Performance of Data Communications Systems and their Applications, pp. 423–431. North Holland, Amsterdam (1981)
11. Głąbowski, M.: Recurrent method for blocking probability calculation in multiservice switching networks with BPP traffic. In: Thomas, N., Juiz, C. (eds.) EPEW 2008. LNCS, vol. 5261, pp. 152–167. Springer, Heidelberg (2008). doi:10.1007/978-3-540-87412-6_12
12. Głąbowski, M., Stasiak, M., Weissenberg, J.: Properties of recurrent equations for the full-availability group with BPP traffic. Math. Prob. Eng. (2012)
13. Delbrouck, L.: On the steady-state distribution in a service facility carrying mixtures of traffic with different peakedness factors and capacity requirements. IEEE Trans. Commun. **31**(11), 1209–1211 (1983)
14. Rácz, S., Gerö, B.P., Fodor, G.: Flow level performance analysis of a multi-service system supporting elastic and adaptive services. Perform. Eval. **49**(1–4), 451–469 (2002)
15. Vassilakis, V.G., Moscholios, I.D., Logothetis, M.D.: Call-level performance modelling of elastic and adaptive service-classes with finite population. IEICE Trans. Commun. **E91.B**(1), 151–163 (2008)
16. Moscholios, I., Vardakas, J., Logothetis, M., Boucouvalas, A.: Congestion probabilities in a batched poisson multirate loss model supporting elastic and adaptive traffic. Ann. Telecommun. - annales des télécommunications **68**(5), 327–344(2013)
17. Moscholios, I.D., Logothetis, M.D., Vardakas, J.S., Boucouvalas, A.C.: Congestion probabilities of elastic and adaptive calls in Elang-Engset multirate loss models under the threshold and bandwidth reservation policies. Comput. Netw. **92, Part 1**, 1–23 (2015)
18. Iversen, V.B.: Reversible Fair Scheduling: The Teletraffic Theory Revisited. Springer, Berlin (2007)
19. Sobieraj, M., Stasiak, M., Weissenberg, J., Zwierzykowski, P.: Analytical model of the single threshold mechanism with hysteresis for multi-service networks. IEICE Trans. Commun. **E95–B**(1), 120–132 (2012)
20. Głąbowski, M., Hanczewski, S., Stasiak, M., Weissenberg, J.: Modeling Erlang's ideal grading with multi-rate BPP traffic. Math. Prob. Eng. Article ID 456910, 35 (2012)
21. Głąbowski, M., Kaliszan, A., Stasiak, M.: Modelling overflow systems with distributed secondary resources. Comput. Netw. **108**, 171–183 (2016)
22. Erlang, A.: Solution of some problems in the theory of probabilities of significance in automatic telephone exchanges. Elektrotechnikeren (1917)
23. Engset, T.: On the calculation of switches in an automatic telephone system. Telektronikk 99–142 (1998)
24. Wallstrom, B.: A distribution model for telefone traffic with varying call intensity, including overflow traffic. Ericsson Technics (2) 183–202 (1964)
25. Iversen, V.: Teletraffic Engineering Handbook. ITU-D SG 2/16 and ITC Draft (2001)

26. Głąbowski, M., Kaliszan, A.: Simulator of full-availability group with bandwidth reservation and multi-rate bernoulli-poisson-pascal traffic streams. In: EUROCON 2007 - The International Conference on "Computer as a Tool", pp. 2271–2277. IEEE (2007)
27. Hanczewski, S., Kaliszan, A.: Simulation studies of queueing systems. In: 2016 10th International Symposium on Communication Systems, Networks and Digital Signal Processing (CSNDSP), pp. 1–6, IEEE, July 2016

An Efficient Method for Calculation of the Radiation from Copper Installations with Wideband Transmission Systems

Piotr Zawadzki[✉]

Institute of Electronics, Silesian University of Technology,
Akademicka 16, 44-100 Gliwice, Poland
Piotr.Zawadzki@polsl.pl

Abstract. The recent innovations in connecting hardware and smart encoding schemes make it possible the old copper installations are capable to support T1/E1 bit rates. It is obvious that old networks destined for POTS, will intensively dissipate energy transmitted by new wideband services and they should be regarded as a potential source of electromagnetic hazards. This note reports results of the study towards developing a computer-aided method for numerical modeling of an electromagnetic environment in a close vicinity of cables carrying signals of wideband transmission systems. The proposed method is formulated in the frequency domain and uses multiconductor transmission line theory for determination of field sources. Subsequently, fullwave approach is applied to find near electromagnetic field in the vicinity of the cable.

Keywords: Wideband transmission system · Electromagnetic hazards

1 Introduction

Bandwidth "hungry" computing applications, the increasing need for information sharing and the explosive growth in higher data transmission rates demand the greater speed and larger bandwidth from LANs (Local Area Networks) or local premises telephone installations. One method of improving the information infrastructure is to install an all-fibre network. However, the all-fibre premises distribution network has its price: implementing it requires the replacement of copper cables with fibre and the installation of more expensive active components, such as fibre optic interface devices. With recent innovations in the design of connecting hardware and smart encoding schemes UTP (Unshielded Twisted Pair) or FTP (Foiled Twisted Pair) cables are now capable of supporting high bit rates.

DSL (Digital Subscriber Line) refers to several types of advanced modems that enable fast access at speeds 300 times faster than most analog modems. Since DSL works on regular telephone lines they are considered a key means of opening the bottleneck in the "last mile" of the existing telephone infrastructure, as telephone companies seek cost-effective ways of providing much higher

© Springer International Publishing AG 2017
P. Gaj et al. (Eds.): CN 2017, CCIS 718, pp. 236–244, 2017.
DOI: 10.1007/978-3-319-59767-6_19

speed to their customers. It is obvious that old premise networks, destined for the POTS, will intensively dissipate energy for the new modulation schemes and they might become a source of not negligible electromagnetic radiation. The above observation leads to the conclusion that new services installed on existing networks should be treated not only as victims of the external noise, but also as potential sources of electromagnetic emissions. This, in turn, calls for the elaboration of the fast, reliable, computer aided method for prognosis of electromagnetic hazards arising in this new environment.

The proposed method is formulated in the frequency domain and uses multiconductor transmission line theory for determination of field sources. Subsequently, fullwave approach is applied to find near electromagnetic field in the vicinity of the cable.

2   Numerical Model

Numerical models for the radiation of the cable with the given parameters can be derived on a basis of computational intensive methods based on a direct Maxwell equations discretization. However everyday EMC engineering practice requires maybe less accurate but more efficient methods.

We use multiconductor transmission line theory to determine current distribution along the cable [1,2] and full-wave approach [3] to determine field of a specified current distribution.

2.1   Calculation of the Current Distribution

For a system of N conductors (plus reference conductor) telegraphers equation can be written in the matrix form

$$\frac{d\mathbf{V}(s)}{ds} + \mathbf{Z}'\mathbf{I}(s) = \mathbf{0} \tag{1a}$$

$$\frac{d\mathbf{I}(s)}{ds} + \mathbf{Y}'\mathbf{V}(s) = \mathbf{0} \tag{1b}$$

where $\mathbf{V}(s)$ and $\mathbf{I}(s)$ are the voltage and current vectors and $\mathbf{Z}'$ and $\mathbf{Y}'$ are the square $N \times N$ matrices describing per unit length line parameters. These two first order differential equations can be transformed into two second order equation for the current distribution

$$\frac{d^2\mathbf{I}(s)}{ds^2} - \mathbf{R}'\mathbf{I}(s) = \mathbf{0} \tag{2}$$

where $\mathbf{R}' = \mathbf{Y}'\mathbf{Z}'$. The solution of this equation has the form

$$\mathbf{I}(s) = \mathbf{Z_c}^{-1}\mathbf{S}^{-1}\left[\mathbf{E}^+(s)\mathbf{a} - \mathbf{E}^-(s)\mathbf{b}\right] \tag{3}$$

where

$$\mathbf{Z_c} = \mathbf{S}^{-1} \left[\frac{1}{\gamma_i} \right] \mathbf{S} \mathbf{Z}' \tag{4}$$

$$[\gamma_i] = \begin{bmatrix} \gamma_1 & 0 & \cdots & 0 \\ 0 & \gamma_2 & \cdots & 0 \\ \cdots\cdots\cdots\cdots\cdots \\ 0 & 0 & \cdots & \gamma_n \end{bmatrix} \tag{5}$$

$$\mathbf{E}^{\pm}(s) = \begin{bmatrix} e^{\mp\gamma_1 s} & 0 & \cdots & 0 \\ 0 & e^{\mp\gamma_2 s} & \cdots & 0 \\ \cdots\cdots\cdots\cdots\cdots\cdots\cdots \\ 0 & 0 & \cdots & e^{\mp\gamma_n s} \end{bmatrix} \tag{6}$$

and matrix $\mathbf{S}$ is a diagonalisation matrix of $\mathbf{R}'$, γ_i is a propagation constant of the i-th mode and vectors $\mathbf{a}$ and $\mathbf{b}$ have to be calculated from the boundary conditions.

2.2   Decomposition on Elementary Dipoles

Due to linearity of Maxwell equations linear extended source (i.e. cable) can be decomposed into collection of short sections and each section can be treated as elementary dipole. If the cable section length l holds the condition [3] $l < \lambda/10$ (λ is the wavelength) then one can neglect the spatial dependency of the current on a dipole. We have in this case the staircase approximation of the current distribution along the cable, and in the net effect the entire structure is replaced by a set of elementary dipoles.

The resulting field in point P can be determined as a sum of respectively retarded contributions. The rigorous approach to field calculation in the presence of a lossy ground requires computation of the so called Sommerfeld integrals [4]. The cost of computation of those integrals is prohibitively large and unacceptable in most practical implementations. Fortunately, the finite ground conductivity has in many cases only a small impact on obtained results and it is acceptable to use approximated method based on perfect soil assumption and method of electrical images.

2.3   Field of an Elementary Dipole

The elementary dipole is characterized by its directional vector $\mathbf{1}_l$ (see Fig. 1), length l that is small compared to all distances of interest, and current I_0. The solution of Maxwell equations in free space for this source configuration has the form [5,6]

$$\begin{aligned} E = -j\frac{\eta I_0 l}{4\pi k} &\left\{ \left[(kR)^2 - (1 + jkR) \right] \mathbf{1}_l \right. \\ &\left. - \left[(kR)^2 - 3(1 + jkR) \right] (\mathbf{1}_l \cdot \mathbf{1}_R) \mathbf{1}_R \right\} \frac{e^{-jkR}}{R^3} \end{aligned} \tag{7a}$$

$$\boldsymbol{H} = \frac{I_0 l}{4\pi} \left(\boldsymbol{1}_l \times \boldsymbol{1}_R\right) \left(1 + jkR\right) \frac{e^{-jkR}}{R^2}. \tag{7b}$$

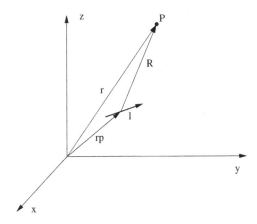

Fig. 1. Elementary dipole in a free space

3   Method Validation

Method validation can be performed by comparing its results with predictions of other methods that implement commonly accepted and verified models or by the comparison with independently conducted computations based on the same mathematical model of the reality. In the following sections we present method validation as a two stage process:

1. comparison with independent multiconductor transmission line analysis code developed by Excem [8]
2. comparison with MOMIC code implementing rigorous full-wave approach to scattering and radiation problems [7]

The mentioned above programs were used for verification of current and field distributions for some canonical examples. All field calculations are performed with perfect ground assumption. Keys for presented plots have the following meaning: line code denotes results obtained with the validated program, Excem code – results from [8], MOM code – results obtained by a rigorous approach to electromagnetic waves radiation based on solution of an integral equation for current distribution with method of moments [5,7].

3.1   Comparision with Excem Code

The distribution of sources uniquely determines the electromagnetic field. The effective, nonbalanced current flowing along the multiconductor cable is the field source in our case. The report [8] was chosen as a source of reliable data to validate current distribution along the cable. We considered four pair UTP cable with each pair loaded with $R_0 = 120\,\Omega$ and one pair forced with voltage $V_0 = 2\,\text{V}$. The cable has length $L = 10\,\text{m}$ and was positioned on height $h = 10\,\text{cm}$. The current distribution along cable for two frequencies is presented in Fig. 2.

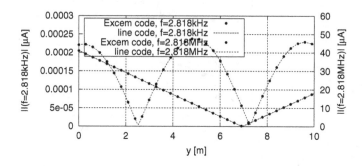

Fig. 2. Effective current distribution along multiconductor line.

3.2 Comparision with Method of Moments

Lets consider transmission line composed by a horizontal conductor over perfectly conducting ground (see Fig. 3) and grounded at both ends by loading impedances. This structure (together with its electrical image) was analysed by a rigorous approach and with proposed method. The observed quantities were: current distribution along the line for two fixed frequencies and electric field produced at point $P\,(L, L/2, 2h)$.

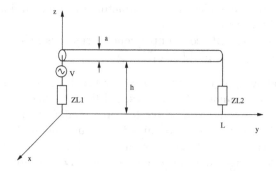

Fig. 3. Transmission line geometry, $L = 2\,\text{m}$, $h = 1\,\text{cm}$, $a = 1\,\text{mm}$, $Z_{L1} = Z_{L2} = 50\,\Omega$.

It is clearly seen from Fig. 4 that antenna mode gives no contribution to the current on the line, and in consequence, radiated fields predicted by the transmission line theory and rigorous approach are the same (see Fig. 5).

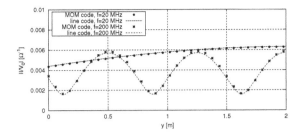

Fig. 4. Current distribution along horizontal conductor

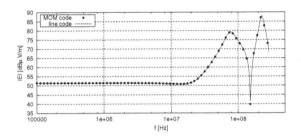

Fig. 5. Field at point $P\left(L, L/2, 2h\right)$.

4 Parametric Study

A motivation in developing a model is to have the capability to perform parametric studies, in which independent model parameters can be varied in order to gain a global understanding of the phenomenon. We used presented method to investigate the dependence of the field in the vicinity of the multiconductor transmission line on the following parameters:

– cable type,
– geometrical configuration of the line,
– line forcing.

4.1 Calculation Setup

The considered line is composed from sections of different length of UTP four pair cables S278 and S298. Each pair is terminated with the resistance $120\,\Omega$. Two cases of line forcing were considered: first pair forced by a sinusoidal voltage of amplitude 2 V, or all pairs forced with the voltage same as previously. This forcing conditions are identified on graphs by keys V1 and V4, respectively. Geometrical configurations further referenced as I-like and U-like are presented in Fig. 6. Configurations are parametrized by the cable section length L and the height h above ground, which is assumed to be ideal. Observation point was positioned at point P(1 m, $L/2$, 1 m). The calculations were performed for two heights and three lengths (Table 1).

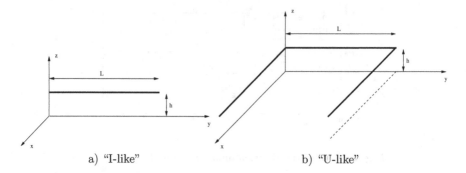

a) "I-like"   b) "U-like"

Fig. 6. Analysed geometries

Table 1. Parameters of calculations

Cable	Height	Geometry	Length	Forcing
S278	$h = 15\,\text{cm}$	I-like	$L = 5\,\text{m}$	V1
S298	$h = 1\,\text{m}$	U-like	$L = 10\,\text{m}$	V4
			$L = 15\,\text{m}$	

4.2 Discussion

The strength of the field at the resonant frequency strongly depends on cable type but resonant frequency itself does not (see Fig. 7).

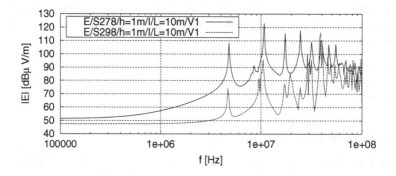

Fig. 7. Amplitudes at resonant frequencies.

The resonant frequencies are determined by geometrical configuration (see Fig. 8) of the system, and it is not possible to design modulation schemes to program them out i.e. make PSD function small at resonant frequencies. However, it is still possible to design new modulation schemes to avoid resonant frequencies of potential victims, for instance radio amateur receivers.

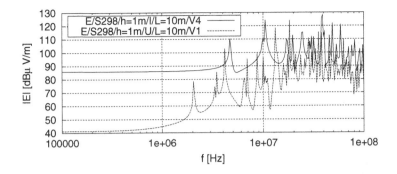

Fig. 8. Dependency of resonant frequencies on geometrical configuration.

Moreover, field amplitude at resonant frequency do not depend on geometrical cable configuration. The results for two geometries with the same total length $L_{tot} = 15$ m are presented in Fig. 9. This behavior can be explained by a special choice of the observation point. It is positioned close to the cable, and far ends of the cable do not give significant contributions to the field because those contributions vanish as $1/R$.

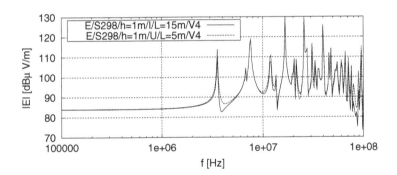

Fig. 9. Field dependency on geometrical configuration.

5   Conclusion

The proposed method is general and flexible. Virtually, it is capable to analyse any service supported over any cable, unless reliable cable model and PSD (Power Spectral Density) function of the transmitted signal are known. The method is also fast as the computation time grows with square of the number of pairs in a cable and linearly with the cable length. Moreover, method is able to analyse not only straight section of the cable, but also handles networks with cable bends.

Acknowledgements. Author acknowledges support by the Ministry of Science and Higher Education funding for statutory activities and Rector of Silesian University of Technology grant number 02/030/RGJ17/0025 in the area of research and development.

References

1. Paul, C.R.: Analysis of Multiconductor Transmission Lines. Wiley, New York (1994)
2. Tesche, F.M., Ianoz, M.V., Karlsson, T.: EMC Analysis Methods and Computational Models. Wiley, New York (1997)
3. Stutzman, W.L., Thiele, G.A.: Antenna Theory and Design. Wiley, New York (1981)
4. Somerfeld, A.: Uber die ausbreitung der wellen in der drahtlosen telegraphie. Annalen der Physics **28**, 665 (1909)
5. Zawadzki, P.: Elaboration of software for prediction of signals induced by a lightning discharge on telecommunication cables placed inside buildings. Technical report NT/LAB/QFE/143, CNET, Lannion, December 1996
6. Zawadzki, P.: Evaluation of the near field radiated by cables used for wideband transmission systems. Technical report NT/CNET/6338, CNET, Lannion, September 1999
7. Harrington, R.F.: Field Computation by Moment Methods. MacMillan, New York (1968)
8. Comaparaison de l'emission des cablages de telecommunication. Excem doc. no. 98042102A

A Videoconferencing System Based on WebRTC Technology

Robert Bestak[(⊠)] and Jiri Hlavacek

Czech Technical University in Prague, Technicka 2, Prague 6, Czech Republic
robert.bestak@fel.cvut.cz, jirka.hlavacek@gmail.com

Abstract. Last years, videoconferencing systems are rapidly evolving and they are becoming more and more popular as a real time communication tool among users. A technology such as VoIP can be used in conjunction with desktop videoconferencing systems to enable low-cost face-to-face business meetings without a necessity to travel, especially for internationally oriented companies that have offices all around the word. In this paper, we focus on and discuss a videoconferencing platform based on the WebRTC technology. We analyze an impact on the multiplexing server's CPU load and memory requirements for different number of communicating users while taken into account different HW/SW configurations of end-point devices.

Keywords: Real time communication · Videoconferencing systems · WebRTC · Measurement

1 Introduction

Historically, a real time communication, has only been available to large companies who could afforded to pay expensive licensing fees, and/or to buy specific proprietary depend HW. Since that, videoconferencing systems evolved to a low cost standard-based technology that is available to the general public. Besides the audio and video communications, current videoconferencing systems also support variety of other services, including instant messaging, documents sharing, screen sharing, video recording, etc. Technological developments in the 2010 s have further extended the capabilities of videoconferencing systems for use not only for in the static environment but in the mobile environment as well.

A videoconferencing technology typically employs both audio and video channels jointly interconnecting two (point-to-point) or several users (multi-point) located at different sites. The multi-point communication is possible via a unit called Multipoint Control Unit (MCU) that represents a bridge interconnecting the communicating users (similarly to the audio conference call). Either all participating users can call the MCU, or the MCU can call the involved users itself. A MCU can be characterized according to the number of simultaneous supporting calls, its ability to provide transposing of data rates and protocols, or the way it is implemented (stand-alone HW devices, vs. embedded unit in a dedicated videoconferencing system). Certain videoconferencing platforms employ

© Springer International Publishing AG 2017
P. Gaj et al. (Eds.): CN 2017, CCIS 718, pp. 245–255, 2017.
DOI: 10.1007/978-3-319-59767-6_20

a standards-based H.323 technique known as "decentralized multi-point" where each involved end-point device in a multi-point communication exchanges video and audio directly with other end-point devices without using a MCU.

Videoconferencing systems can be operated in two modes: (a) Voice-Activated Switch (VAS), or (b) Continuous Presence. In the VAS mode, the MCU switches end-point devices in such way that it can be seen by the other end-point devices by the levels of one's voice. For an example, if there are four users in a video-conference, the only one that will be displayed is the currently talking user. In the Continuous Presence mode, all users are displayed at the same time; the user streams are assembled together by MCU into a single video stream that is transmitted to all users.

Though, the videoconferencing systems gain increasing popularity, there are still several issues the systems have to face, among them: complexity (not all users are technically skill, users prefer a simple interface and usage), interoperability (not all systems can readily be interconnected - there are different standards, features and qualities), or bandwidth (it is not always possible to have a high quality connection to support a good quality conferencing communication).

In general, the videoconferencing systems can be classified into the following three categories:

- Dedicated systems: these systems have all required components packed into a single platform (console), which contains all necessary interfaces, codecs, and control unit.
- Desktop systems: they are designed as add-ons (HW boards or SW codecs) to common PCs and laptops, and they transform them into videoconferencing platforms. A range of different cameras and microphones, which contains the necessary codecs and interfaces, can be used.
- WebRTC platforms: they represent videoconferencing solutions that are available through standard web browsers. Typically, a browser uses the local camera and/or microphone to establish the connection.

Current approaches supporting real time communications are based either on utilizing a separate application or on employing a specific plug-in (e.g., Flash plug-in). By using a separate application means leaving the web browser and launching a dedicated application, i.e. the browser content and the real-time content are independent. The plug-in solution provides a tighter integration between the real-time and browser contents. However, plug-ins are proprietary solutions and they do not work in all environments. Contrary, the WebRTC platforms include all necessary audio and video components to support the real-time communications directly via web browsers.

The rest of paper is organized as follows. The next section briefly describes basic features of the WebRTC technology, including related works. A comparison of selected real-time communication platforms with WebRTC is provided in Sect. 3. Impact on the multiplexing server's CPU load and memory requirements for different number of communicating end-point devices are analyzed in Sect. 4. Finally, Sect. 5 concludes the paper.

2   Web Real-Time Communication

2.1   Basic Features

Web Real-Time Communication, or WebRTC, is a specification plug-in free, real time communication via a web browser. It provides communication protocols and defines application programming interfaces allowing two or more users to mutually communicate in real-time. The WebRTC technology is being standardized by World Wide Web Consortium (W3C) organization at the application level and by Internet Engineering Task Force (IETF) organization at the protocol level.

WebRTC is composed of three main elements:

- Web browser: To send and receive audio and video streams directly from a web browser, the web browser has to be enhanced with capabilities for controlling the local audio and video elements at the device at which the web browser is running (e.g., a PC or smartphone).
- Web application: A user is typically asked to download a Java script from a web server. This script runs locally at the user's device and interacts with the web server.
- Web server: The server provides the Java scripts for users and executes the necessary application logic.

Initially, the WebRTC technology was designed to provide a real time communication between two users. Later on, it was enhanced with the multi-connection support. There are 2 approaches how to support the multi-connections: (a) centralized mode, and (b) fully meshed mode. In the first case, a user agent is in charge of dialog and mixing of media streams. The assumption here is that only the host mixes streams and sends them to each end-point device; the host could be a dedicated server or a standalone device with sufficient computational power. To join the videoconference call, a new end-point device has to notify the host. In the centralized mode, the communication fully depends on the host; if a host leaves the conference, all connections are disconnected. In the second mode, fully meshed model, end-point devices are interconnected to each other and any end-point device can invite and leave the call without effecting the other end-points.

The WebRTC technology can be nowadays integrated within a SIP-based system and thus offering additionally functions [1,2]. A videoconferencing architecture can include a unit which allows legacy SIP user agents and SIP WebRTC clients to interoperate. A real-time chatting tool based on HTML5 and WebRTC can also be proposed; it includes a basic information management module and a communication module allowing text communication by means of a dedicated server and voice/video sessions through a point-to-point connection between web browsers.

2.2   Related Works

The WebRTC technology has received a lot of attention by researchers couple of last years.

In [3], a study of WebSockets and WebRTC usage from the point of mobile device power consumption in LTE networks is provided. Based on realized power consumption analysis, a couple of recommendations for standardization process are provided by authors.

Performance of WebRTC's adaptive video streaming capabilities in an IEEE 802.16e network environment is evaluated in [4].

The applicability of HTML5 and WebRTC technology for a P2P video streaming is investigated in [5]. The work describes a Bit Torrent P2P video streaming solution which is used to identify the performance bottlenecks in WebRTC-based P2P video streaming implementations.

In [6], the WebRTC-based communication performance is evaluated for case of star and full-mesh network topologies, with focus on the performance of the congestion control mechanism used with WebRTC.

A feasibility of live video streaming protocols via the WebRTC technology is investigated in [7]. The experiments illustrate a possibility to implement a pull-based P2P streaming protocol with WebRTC, at least, for small-scale P2P networks.

A performance of WebRTC on mobile devices, while taking into account different type of mobile devices, wireless network connectivity and web browser configurations, is evaluated in [8].

In [9], authors present a benchmarking tool, call WebRTCBench, which measures WebRTC peer connection establishment and communication performance. Authors discuss performance of the WebRTC technology across a range of implementations and devices.

WebRTC communications over the LTE mobile network simulated by NS-3 tool is investigated in [10]. Several multimedia WebRTC streams between the two end-point devices are analyzed and empirical CDFs of typical performance figures including throughput, jitter, and packet loss are derived under different LTE scenarios.

2.3   Security and Identification

As to the WebRTC security aspects, the WebRTC stack includes Datagram Transport Layer protocol that is designed to prevent eavesdropping and information tampering and to support real-time data encryption and association management service to Secure Real-time Transport Protocol.

An issue with a traditional desktop SW is whether a user can trust the application itself [11]. The installation of a new SW or a plug-in can potentially scrumptiously install a malware or other undesirable SW. Typically, users have no idea where the SW was created or from whom they are downloading the SW. Thus, a malicious third party has possibility in repackaging perfectly safe and trusted SW to include malware, and offering this package on free SW websites.

Advantageously, WebRTC is not a plug-in, nor is there any installation process for any of its components. The WebRTC platform is simply installed as part of WebRTC compatible browser (e.g., Chrome or Firefox). Therefore, there is no risk of installing malware or virus through the use of an appropriate

WebRTC application. Anyway, WebRTC applications should still be accessed via a HTTPS website signed by a valid authenticator.

Moreover, the WebRTC technology is done in such way that a user is explicitly asked a permission for the camera or microphone to be used. As the permission can be done on one-time or permanent access base, a WebRTC application cannot arbitrarily gain access or operate any device. Furthermore, if the microphone or camera is being used, the user interface is required to expressly show the user that the microphone or camera are being operated.

It is desirable that a user is able to verify the identity of their communicating partners, i.e. users naturally wants to be certain that they are communicating with users they believe that they are speaking to, and not to an imposter. The authentication of peers has to be performed independently from the signaling server as the signaling server itself cannot be trusted. To do this, web-based identity providers can be utilized, such as Facebook Connect, BrowserID, OAuth, etc. The role of these providers is to verify the user identity to other users, based on the authority of identity provider itself. This allows users to tie their authentication on other services to their main account on a trusted service. The implementations of each identity provider may differ due to independent development by different companies rather than being based on an open-source standard, but the underlying principle and functions remain the same.

3   Comparison of Videoconferencing Platforms

In this section, we provide a brief comparison of selected real-time communication platforms and we highlight differences between them and the WebRTC technology.

3.1   Skype

Comparing to WebRTC, Skype is an application. Skype is available for different platforms and provides video and audio services between two or more users. Additionally, users may exchange digital documents (images, text, video and any others), and text and video messages.

3.2   Hangouts

Hangouts is a communication platform, developed by Google, supporting instant messaging, video, audio, and SMS between two or more users. The service can be accessed online (Gmail or Google+ websites) using either PCs or mobile phones. Comparing to the WebRTC technology, Hangout uses a proprietary protocol, although with WebRTC elements.

3.3   WebEx

WebEx Meeting Center is a part of commercial solution developed by Cisco. It is a web conferencing platform delivered through the Cisco Collaboration Cloud.

The stream processing is done in the cloud and the multiplexing of media streams is not at the charge of end-point device. It uses a proprietary algorithm named Universal Communication Format that deals with a large range of media.

3.4   Jitsi

Comparing to previously mentioned solutions, Jitsi is a free and open source multi-platform supporting video, audio, and instant messaging. It is based on the OSGi framework (Collection of Java Library). Jitsi is not a web-oriented, a user is asked to download and install specific SW to run it. In terms of functions, libraries are not as rich as the WebRTC ones but the main abilities are effective. Similarly to WebEx, the media stream processing could be done at a server instead of at the end-point device.

The Jitsi videobridge, a videoconferencing solution supporting the WebRTC. It's a Selective Forwarding Unit, a centralized server, which is forwarding selected streams to other participating users in the videoconference call.

3.5   AnyMeeting

AnyMeeting is a web conferencing and webinar services allowing users to host and to attend web based conferences and meetings and to share their desktop screens with other users via web browser.

The above mentioned platforms are summarized and compared in Table 1. The comparison takes into account aspects such as the place of stream multiplexing, set of functions (text chat, screen sharing, etc.), web-browser compatibility, and multi-connection support.

Table 1. Comparison of videoconferencing platforms.

	Skype	Hangouts	WebEx	Jitsi	Jitsi video bridge	Any Meeting	WebRTC
Multiplexing	Client	Server	Server	Server	Client	Server	Client
Rich. functionalities	Middle	No	Yes	Yes	No	Midle	Yes
Requirements	App	Plugin	App	App	None	App	None
Multi-call setup	Limited	Yes	Yes	Yes	Yes	Limited	Yes

4   Measurements and Results

In this section, we investigate impact on the multiplexing server's CPU load and memory requirements for different numbers of used end-point devices; distributed processing issues are discussed in more details for example in [12,13]. We set up a testbed that considers 3 PCs with various HW and SW configurations as indicated in Table 2. The configuration of the server is indicated in the table as well. Each participant receives a return video with multiplexed streams of each participant in one video stream, i.e. the end-point device has to decode one flow instead of one flow per end-point device. The multiplexing is provided by Kurento media server [14,15].

Table 2. Parameters of PCs.

PC1	I5 2.5 GHz dual-core, Nvidia 610 M 1 GB, 4 GB RAM, Ubuntu 14.04 LTS (NO 720P)
PC2	I3 1.7 GHz dual-core, Nvidia 820 M 2 GB, 4 GB RAM, Win 10 (NO 120P)
PC3	I7 2.5 GHz dual-core, Nvidia GTX970 3 GB, 8 GB RAM, Win 7
Server	Intel(R) Xeon(R) CPU 5160 @ 3 GHz, 8 GB RAM, Ubuntu 16.04.1 LTS

In total, there have been tested 28 scenarios, which differs in the number of involved end-point devices, PCs (1, 2, and 3), and display resolution configuration (160×120, 320×240, 640×480, and 1280×720), see Table 3. Notice that the configuration id is constructed in such way that the first digit indicates PC (using Table 2) and the second digit is associated to the display resolution; for example the configuration 1222, refers to two interconnected end-point devices (PC1 and PC2) where the display resolution is set to 640×480. The resolution of the return video, for mono user configuration is used 800×600 and for the other cases 200×150 video per user is sent back to the end-point device. The PC1 and PC2 use all the available bandwidth, while the data rate of PC3 is limited to 600 kB/s.

All measurements are done by using Firefox browser, and the values are given as an arithmetic average of ten consequent measurements (which is a common way to evaluate a CPU load).

Figure 1 shows the server's CPU load for different testing scenarios (please notice that the server's CPU has 4vCPU, therefore the load can theoretically reach up to $4 \times 100\%$). As can be observed, the server's CPU load is the lowest for the PC3, where the bit rate is the lowest. As expected, the bit rate and therefore the video quality has a considerable impact on the server's CPU load. For increasing number of involved end-point devices, the server's CPU load increase as well (scenarios 102030–132333). From the point of display resolution, the server's CPU load strongly depends on the resolution and bit rate, often given by the HW/SW configuration of involved end-point devices. Figure 1 indicates that about one CPU per end-point device is needed.

The impact on server's RAM is illustrated in Fig. 2. As we can expect, the requirements on the RAM are directly proportional to the increasing number of communicating end-point devices, more or less no matter what is the HW/SW configuration of these end-point devices. Whereas, the display resolution has a negligible impact on the memory requirement.

Scenarios with one PC are special cases that illustrate the memory and CPU consumption when no multiplexing is performed, i.e. the scenarios represent the load needed for decoding and encoding of the video flow.

Table 3. Testing configurations.

Configuration	PC	Sent resolution	Received resolution by end-point	Bit rate [kB/s]
10	PC1	160 × 120	800 × 600	373
11	PC1	320 × 240	800 × 600	658
12	PC1	640 × 480	800 × 600	1470
13	PC1	1280 × 720	800 × 600	1470
20	PC2	160 × 120	800 × 600	514
21	PC2	320 × 240	800 × 600	514
22	PC2	640 × 480	800 × 600	1430
23	PC2	1280 × 720	800 × 600	2150
30	PC3	160 × 120	800 × 600	486
31	PC3	320 × 240	800 × 600	601
32	PC3	640 × 480	800 × 600	608
33	PC3	1280 × 720	800 × 600	590
1020	PC1+PC2	160 × 120	200 × 150	832
1121	PC1+PC2	320 × 240	200 × 150	1040
1222	PC1+PC2	640 × 480	200 × 150	2490
1323	PC1+PC2	1280 × 720	200 × 150	2050
1030	PC1+PC3	160 × 120	200 × 150	750
1131	PC1+PC3	320 × 240	200 × 150	1110
1232	PC1+PC3	640 × 480	200 × 150	1250
1333	PC1+PC3	1280 × 720	200 × 150	1220
2030	PC2+PC3	160 × 120	200 × 150	900
2131	PC2+PC3	320 × 240	200 × 150	1060
2232	PC2+PC3	640 × 480	200 × 150	1130
2333	PC2+PC3	1280 × 720	200 × 150	2620
102030	PC1+PC2+PC3	160 × 120	200 × 150	1220
112131	PC1+PC2+PC3	320 × 240	200 × 150	1870
122232	PC1+PC2+PC3	640 × 480	200 × 150	1900
132333	PC1+PC2+PC3	1280 × 720	200 × 150	2670

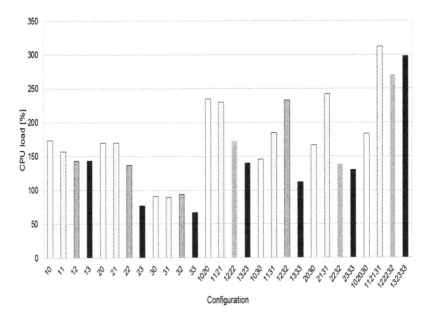

Fig. 1. Server's CPU load.

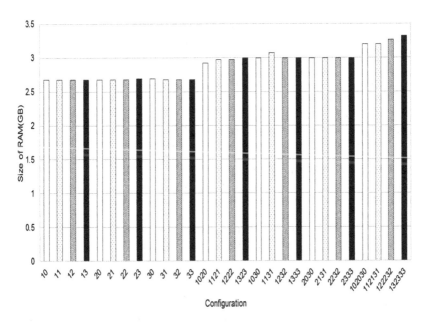

Fig. 2. Requirements on the size of server's RAM.

5   Conclusion

In this paper, we have discussed basic characteristic of the WebRTC technology, and compare the technology with some other exiting real-time videoconferencing systems. Additionally, we have evaluated impact on the multiplexing server's CPU load and requirements on RAM for different numbers of end-point devices running WebRTC sessions. In total, we have tested 28 different configurations. The obtained measurements illustrate a strong relation between the video resolution and bit rate of the involved end-point devices and the server's CPU load. The requirements on server's RAM are than directly proportional to the number of involved end-point devices, no matter what is the HW/SW configurations and considered display resolutions. Obtained results illustrate that the server has to be dimensioned in such way that about one CPU has to be considered per one end-point device.

The current scenarios only include PCs. In our next work, we plan to take into account a more heterogonous environment and to consider as end-points mobile devices as well.

Acknowledgments. This research work was supported by the Grant Agency of CES-NET grant no. 542/2014.

References

1. Segec, P., Paluch, P., Papan, J., Kubina, M.: The integration of WebRTC and SIP: way of enhancing real-time, interactive multimedia communication. In: 12th IEEE International Conference on Emerging eLearning Technologies and Applications (ICETA), pp. 437–442 (2014)
2. Amirante, A., Castaldi, T., Miniero, L., Romano, S.P.: On the seamless interaction between WebRTC browsers and SIP-based conferencing systems. IEEE Commun. Mag. **51**(4), 42–47 (2013)
3. Mandyam, G., Ehsan, N.: HTML5 connectivity methods and mobile power consumption. Technical report, Qualcomm (2012)
4. Fund, F., Wang, C., Liu, Y., Korakis, T., Zink, M., Panwar, S: Performance of dash and WebRTC video services for mobile users. In: 20th International Packet Video Workshop (PV), p. 18. (2013)
5. Nurminen, J., Meyn, A., Jalonen, E., Raivio, Y., Marrero, R.: P2P media streaming with HTML5 and WebRTC. In: IEEE Conference on Computer Communications Workshops, pp. 63–64 (2013)
6. Singh, V., Lozano, A., Ott, J.: Performance analysis of receive-side real-time congestion control for WebRTC. In: 20th International Packet Video Workshop (PV), p. 18 (2013)
7. Rhinow, F., Veloso, P., Puyelo, C., Barrett, S., Nuallain, E.: P2P live video streaming in WebRTC. In: World Congress on Computer Applications and Information Systems, p. 15 (2014)
8. Heikkinen, A., Koskela, T., Ylianttila, M.: Performance evaluation of distributed data delivery on mobile devices using WebRTC. In: IEEE International Wireless Communications and Mobile Computing Conference (2015)

9. Taheri, S., Beni, L.A., Veidenbaum, A.V., Nicolau, A., Cammarota, R., Qiu, J., Lu, Q., Haghighat, M.R.: WebRTCBench: a benchmark for performance assessment of WebRTC implementations. In: 13th IEEE Symposium on Embedded Systems For Real-time Multimedia (2015)
10. Carullo, G., Tambasco, M., Di Mauro, M., Longo, M.: A performance evaluation of WebRTC over LTE. In: 12th Annual Conference on Wireless On-Demand Network Systems and Services (2016)
11. Pałka, D., Zachara, M., Wójcik, K.: Evolutionary scanner of web application vulnerabilities. In: Gaj, P., Kwiecień, A., Stera, P. (eds.) CN 2016. CCIS, vol. 608, pp. 384–396. Springer, Cham (2016). doi:10.1007/978-3-319-39207-3_33
12. Piórkowski, A., Szemla, P.: Client-side processing environment based on component platforms and web browsers. In: Kwiecień, A., Gaj, P., Stera, P. (eds.) CN 2013. CCIS, vol. 370, pp. 21–30. Springer, Heidelberg (2013). doi:10.1007/978-3-642-38865-1_3
13. Amirante, A., Castaldi, T., Miniero, L., Romano, S.P.: Janus: a general purpose WebRTC gateway. In: 14th Proceedings of the Conference on Principles, Systems and Applications of IP Telecommunications (2014)
14. Kurento documentation, release 6.6.0 (2016). https://www.kurento.org/
15. Fernandez, L.L., Diaz, M.P., Mejias, R.B., Lpez, F.J., Santos, J.A.: Kurento: a media server technology for convergent WWW/mobile real-time multimedia communications supporting WebRTC. In: 14th IEEE International Symposium on A World of Wireless, Mobile and Multimedia Networks (WoWMoM), pp. 1–6 (2013)

Analytical Modelling of Multi-tier Cellular Networks with Traffic Overflow

Mariusz Głąbowski$^{(\boxtimes)}$, Adam Kaliszan, and Maciej Stasiak

Chair of Communication and Computer Networks,
Faculty of Electronics and Telecommunications, Poznan University of Technology,
Poznań, Poland
{mariusz.glabowski,adam.kaliszan,maciej.stasiak}@put.poznan.pl,
http://nss.et.put.poznan.pl/

Abstract. In the paper, a new analytical model of multi-service communication networks with traffic overflow is elaborated. It is assumed that both: primary resources and alternative resources, forming networks with traffic overflow, are distributed, as, e.g. in mobile networks in which traffic from a group of 4G cells is overflowed to a group of 3G cells. The paper proposes also a new method for determining parameters of multi-service traffic that overflows from the distributed primary resources to the distributed alternative resources, as well as a method for occupancy distribution calculation and blocking probability calculation in the alternative resources servicing overflow traffic. The results of the analytical calculations are compared with the results of the simulation experiments of some selected structures of overflow systems with the distributed primary and alternative resources.

1 Introduction

The very dynamic increase in the volume of data traffic in mobile telecommunications networks (HSPA, LTE, CDMA2000 EV-DO, TDSCDMA and Mobile WiMax) seems to be posing a real challenge for mobile operators. The operators of these networks are compelled to introduce mechanisms that would increase the effectiveness the usage of network resources. One of such mechanisms is optimization of cell loads, that can be obtained, among others, due to a concept of self-organising (self-optimizing, self-configuring) networks (SON) [1]. The execution of this concept is possible, i.a., due to the well-known connection handover mechanism that transfers calls from cells with high load to neighbouring cells. Such mechanism is usually executed between cells operating within the same technology only, e.g., GSM.

Over time, the increasing diversity of wireless cellular network technologies led to the emergence of hierarchical (multi-tier) overlay wireless systems, deploying different technologies, e.g. 2G (GSM), 3G (UMTS), and 4G (LTE) in the same area. The availability of different network technologies that co-exist in the same area allowed mobile network operators to apply the well-known optimization technique, the so-called traffic overflow. Applying the concept of traffic

© Springer International Publishing AG 2017
P. Gaj et al. (Eds.): CN 2017, CCIS 718, pp. 256–268, 2017.
DOI: 10.1007/978-3-319-59767-6_21

overflow, the mobile operators usually overflow traffic that cannot be serviced in a group of cells with lower range (e.g., picocells) to a group of cells with higher range (macrocells) [2–5]. In this case, traffic directed to macrocells (upper tier, treated as alternative resources) from picocells (lower tier, treated as primary resources) is the overflow traffic with different characteristics than the traffic that is directly offered to a network by its users [2,4]. As a consequence, the literature distinguishes between two types of handover: horizontal – when handover of connections takes place within the same technology – and vertical – when handover of connections is executed between different technologies (for example, between 2G and 3G, or WiMAX and WiFi).

The traffic overflow mechanism is known in the literature for about 70 years. Its first applications were limited to the single-service (single-rate) hierarchical telecommunications networks with alternate routing [6–15]. Subsequently, this effective optimization technique was applied to the multi-service (multi-rate) networks in order to ensure required quality of service parameters for various traffic classes [16–22], as well as to increase the performance and capability of networked cloud data centers [23].

In all the models presented in literature the assumption was that each of primary resources, to which calls are directly offered, can be considered as a single resource with complete sharing policy and modelled using the so-called full-availability group model [24,25]. Such an assumption is justified in the case of the systems without load balancing between neighbouring cells of the same tier [17]. However, the modern load-balancing mechanisms (self-optimization techniques) treat a group of neighbouring cells, covering a given area, as a single subsystem, and execute horizontal connection handover (within the same tier/technology) from cells with high load to neighbouring cells of the group [26]. Consequently, the area covered by a given group of cells can be treated – from the point of view of traffic engineering – as a multi-service system in which the service of multi-rate calls is performed with the use of "separated" (distributed) resources (cells) that are component parts of a given network and – in consequence – should be modelled using the so-called limited-availability group model [27,28].

The present article considers for the first time a system in which both primary resources and alternative resources of multi-service networks with traffic overflow are composed of a number of separated (distributed) component resources. The term "separation" means that a call will be serviced only when at least one of the primary resources (one of a cell of the group of cells at the lower tier, e.g., picocells) and at least one of the alternative resources (one of a cell of the group of cells at the upper tier, e.g., macrocells) have the appropriate amount of resources that is required for the call to be serviced.

The paper is organized as follows. Section 2 presents a description of a multi-service overflow system with distributed primary and alternative resources. In Sect. 3, the model of the primary and alternative resources of the multi-service overflow system is presented. Section 4 presents a comparison of the results of the analytical calculations with the results of the simulation experiments for some selected structures of systems with distributed primary and alternative resources. Section 5 sums up the paper.

2   Multi-service System with Traffic Overflow

Let us consider a general diagram of an overflow system with distributed primary and alternative resources. In the system, there are r primary resources. Each of

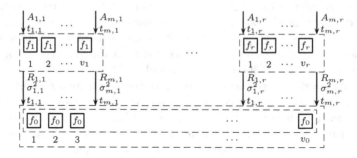

Fig. 1. Traffic overflow system with distributed primary and alternative resources

r primary resources are composed of v_r component resources. Each component resource k $(0 < k \leq v_r)$ has the capacity of f_k BBUs. The total capacity V_j of the primary resources No. j, expressed in the so-called basic bandwidth units – BBUs [29], is then:

$$V_j = v_j f_j \tag{1}$$

The given primary resources can admit a call of a given class for service only when it can be entirely serviced by BBUs of one of the distributed component resources. This means that $t_{i,j}$ BBUs that are necessary to service a call of class i in resource j, cannot be "divided" between different component resources and must be serviced exclusively by the BBUs that belong to one, randomly selected component resource. Each of the primary resources j $(0 < j \leq r)$ is offered a set of m Erlang traffic streams. The intensity of class i $(0 < i \leq m)$, offered to resource j $(0 < j \leq r)$, is denoted as $A_{i,j}$ (Fig. 1).

The traffic that cannot be serviced by the primary resources is not rejected but it is offered to the alternative resources. The alternative resources are composed of v_0 components resources. Each component resource of the alternative resources has the capacity f_0. The total capacity V_0 of the alternative resources system:

$$V_0 = v_0 f_0 \tag{2}$$

As in the case of the primary resources, the alternative resources system can admit a call of a given class for service only when it can be entirely serviced by BBUs of one of the distributed component resources.

The alternative resources are offered traffic streams that overflow from the primary resources. The class i traffic stream offered to the alternative resources, is characterised by two parameters: the average value $R_{i,j}$ of the intensity of traffic of class i $(0 < i \leq m)$ that overflows from resource j $(0 < j \leq r)$, and the variance $\sigma_{i,j}^2$ of the intensity of traffic of class i that overflows from resource j of the primary resources system. The variance of overflow traffic is characterized by values that exceed the average value of this traffic.

3   Analytical Model of Overflow System with Distributed Resources

3.1   Analytical Model of Primary Resources

In the paper it is assumed that the primary resources as well as the alternative resources can be analysed using the model of the limited-availability group (LAG) [27]. The LAG is composed of a number of separated resources with different capacities. In the case of the LAG, $t_{i,j}$ BBUs, required to set up a class i connection in primary group j, cannot be "divided" between different component resources and must be serviced exclusively by the BBUs that belong to one, randomly selected component resource. Thus, the LAG is an example of the so-called state-dependent system in which admission for service of a new call depends on the structure of the system [30–32].

The occupancy distribution in LAG, modelling the primary resources with the capacity V_j, can be approximated by the generalised Kaufman-Roberts distribution [27]:

$$n[P_n]_{V_j} = \sum_{i=1}^m A_{i,j} t_{i,j} \xi_{i,j}(n - t_{i,j})[P_{n-t_{i,j}}]_{V_j} \tag{3}$$

where $\xi_{i,j}(n)$ is the so-called conditional transition probability that defines which part of class i stream offered the primary resources j transfers the service process between the neighbouring states. The conditional transition probability $\xi_{i,j}(n)$ is determined in a combinatorial way as follows:

$$\xi_{i,j}(n) = 1 - \frac{F(V_j - n, v, t_{i,j} - 1)}{F(V_j - n, v, f)} \tag{4}$$

The combinatorial function $F(x, s, f)$ determines the number of arrangements of x free BBUs in s resources with the capacity of f BBUs each:

$$F(x, v, f) = \sum_{u=0}^{\lfloor \frac{x}{f+1} \rfloor} (-1)^u \binom{v}{u} \binom{x + v - u(f + 1) - 1}{v - 1} \tag{5}$$

Having the occupancy distribution (3), the blocking probability for individual multi-service traffic classes offered to the primary resources No. j [27] can be determined:

$$[E_{i,j}]_{V_j} = \sum_{n=0}^{V_j} \{1 - \xi_{i,j}(n)\}[P_n]_{V_j} \tag{6}$$

3.2 Parameters of Overflow Traffic that Overflows from the Primary Resources

Overflow traffic of class i that overflows from the primary resource j is characterised be the three following parameters: the average value $R_{i,j}$, variance $\sigma_{i,j}^2$ and the volume of demands $t_{i,j}$. The average value of this traffic is the part of the traffic $A_{i,j}$ that cannot be serviced in primary by resources. Having the blocking probability $[E_{i,j}]_{V_j}$ (Formula (6)), the average value $R_{i,j}$ for traffic of class i that overflows from the resource j will be determined by the following formula:

$$R_{i,j} = A_{i,j} \cdot [E_{i,j}]_{V_j} \tag{7}$$

To determine the variance of overflow traffic we use the approximate method proposed in [17]. According to this method, a decomposition of each of the primary resources V_j into m fictitious resources with the capacity $V_{i,j}^*$ is to be carried out first. The assumption is that each fictitious group will service exclusively calls of one class, which provides an opportunity to apply Riordan formula [7] to determine the variance of traffic of class i that overflows from the resource j. The capacity of the fictitious resource $V_{i,j}^*$ is defined as this part of the real resource V_j that is not occupied by calls of the remaining classes (different from the class i) [17]. Eventually, we get:

$$V_{i,j}^* = V_j - \sum_{l=1,l\neq i}^{m} Y_{l,j} t_{l,j} \tag{8}$$

where $Y_{l,j}$ is the average value of traffic of class l serviced in the resource j, i.e., the average number of calls of class l serviced in the resource j. According to the definition of traffic intensity, we have:

$$Y_{l,j} = A_{l,j}(1 - [E_{l,j}]_{V_j}) \tag{9}$$

Having the parameters $A_{i,j}$, $R_{i,j}$ and $V_{i,j}^*$ we can determine the variance $\sigma_{i,j}^2$ for individual call streams that overflow from the primary resource j to the system of alternative resources, using the Riordan's formula [7]:

$$\sigma_{i,j}^2 = R_{i,j} \left(\frac{A_{i,j}}{\frac{V_{i,j}^*}{t_{i,j}} + 1 - A_{i,j} + R_{i,j}} + 1 - R_{i,j} \right) \tag{10}$$

In Formula (10), the quotient $V_{i,j}^*/t_{i,j}$ normalizes the system to a single-service case. The operation of this kind is necessary because Riordan Formulas (7) and (10) determine the parameters of overflow traffic in single-service systems. At this point we already have all the parameters that characterize traffic streams offered to the system of alternative resources and we are in position to determine their peakedness coefficients:

$$Z_{i,j} = \sigma_{i,j}^2 / R_{i,j} \tag{11}$$

3.3   Analytical Model of Alternative Resources

In order to model the alternative resources with overflow traffic, we can apply the same approach as in [21]. In [21], the overflow system with state-independent non-separated primary resources and with distributed (separated) alternative resources was considered. To model the state-dependent alternative resources, in [21] the Hayward approach was applied [33]. As a result, the occupancy distribution $[P_n]$ and the blocking probability $E_{i,j}$ for the alternative distributed resources with the capacity of V_0 BBS, can be written as follows [21]:

$$n[P_n]_{V_0/Z_0} = \sum_{j=1}^{r}\sum_{i=1}^{m} \frac{R_{i,j}}{Z_{i,j}} t_{i,j}\xi_{i,j}(n - t_{i,j})[P_{n-t_{i,j}}]_{V_0/Z_0} \tag{12}$$

$$[E_{i,j}]_{V_0/Z_0} = \sum_{n=0}^{V_0/Z_0}\{1 - \xi_{i,j}(n)\}[P_n]_{V_0/Z_0} \tag{13}$$

while the the the transition probability $\xi_{i,j}(n)$ is determined on the basis of Eq. (5), as follows:

$$\xi_{i,j}(n) = 1 - \frac{F\left(\frac{V_0}{Z_0} - n, v_0, t_{i,j} - 1\right)}{F\left(\frac{V_0}{Z_0} - n, v_0, \frac{f_0}{Z_0}\right)} \tag{14}$$

The parameter Z_0 in Formulas (12)–(14) is the so-called aggregated peakedness factor and is calculated as a weighted average of peakedness factors for individual traffic classes $Z_{i,j}$ offered to secondary resources. A weight related to the share of the coefficient $Z_{i,j}$ in the aggregated peakedness coefficient Z_0 is directly proportional to this number of BBUs that is required to service offered traffic $R_{i,j}$:

$$Z_0 = \sum_{j=1}^{r}\sum_{i=1}^{m} Z_{i,j}\frac{R_{i,j}t_{i,j}}{\sum_{u=1}^{r}\sum_{l=1}^{m} R_{l,u}t_{l,u}} \tag{15}$$

Formulas (12)–(14) make it possible to determine all important characteristics of the overflow system with distribution of alternative resources, i.e., the occupancy distribution $[P]_{V_0/Z_0}$ (Formula (12)) and the blocking probabilities $[E_{i,j}]_{V_0/Z_0}$ (Formula (13)) for particular traffic classes.

4   Numerical Results

The proposed method of modelling the overflow systems with distributed both primary and alternative resources is an approximate one. In order to evaluate the accuracy of the proposed method, the results of the analytical calculations were compared with the results of the simulation experiments for a number of selected overflow system.

To evaluate the accuracy of the proposed model the results of the analytical calculations were compared with the data provided by the simulation experiments. For this particular purpose, a dedicated simulator of the considered networks at the call level was constructed [34]. The approach to modelling of multiservice systems with traffic overflow at the call level adopted in this article makes it possible to use it to determine traffic characteristics, e.g. hierarchical wireless networks [31]. The results of the simulation are presented in Figs. 2, 3, 4, 5 and 6 in the from of appropriately marked points with 99-percent confidence interval calculated after the t-Student distribution for the number of ten series.

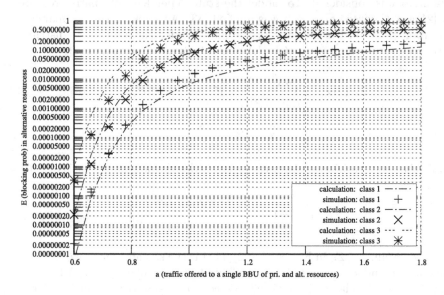

Fig. 2. Blocking probability in System 1

The parameters of the considered systems were as follows.

Parameters of the analyzed systems No. 1

1. Primary resources
   - The number of primary resources $r = 3$
   - Resources 1, 2 and 3 (all primary resources have the same structure):
     • $f_1 = 20$ BBUs, $v_1 = 3$, $V_1 = V_2 = V_3 = 3 \times 20 = 60$ BBUs,
     • offered traffic: $t_{1,*} = 1$ BBUs, $t_{2,*} = 2$ BBUs, $t_{3,*} = 4$ BBUs,
     • proportion of offered traffic: $A_{1,1}t_{1,1} : A_{2,1}t_{2,1} : A_{3,1}t_{3,1} = 1 : 1 : 1$; $A_{1,2}t_{1,2} : A_{2,2}t_{2,2} : A_{3,2}t_{3,2} = 1 : 1 : 1$; $A_{1,3}t_{1,3} : A_{2,3}t_{2,3} : A_{3,3}t_{3,3} = 1 : 1 : 1$;
2. Secondary resources
   - $f_0 = 20$ BBUs, $v_0 = 6$, $V_0 = 120$ BBUs.

Parameters of the analyzed systems No. 2

1. Primary resources
   - The number of primary resources $r = 3$
   - Resources 1:
     • $f_1 = 20$ BBUs, $v_1 = 3$, $V_1 = 3 \times 20 = 60$ BBUs,

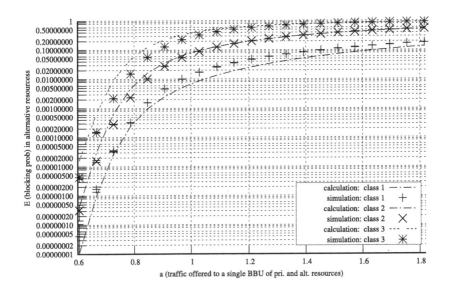

Fig. 3. Blocking probability in System 2

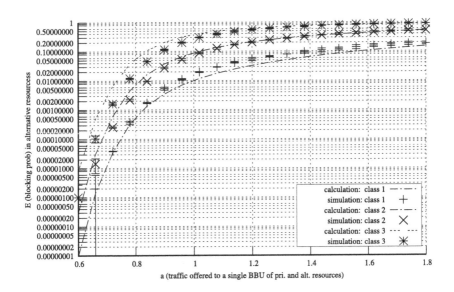

Fig. 4. Blocking probability in System 3

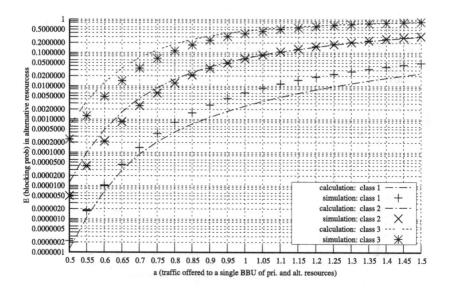

Fig. 5. Blocking probability in System 4

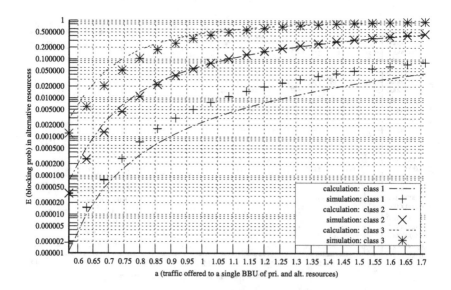

Fig. 6. Blocking probability in System 5

- offered traffic: $t_{1,1} = 1$ BBUs, $t_{2,1} = 2$ BBUs, $t_{3,1} = 4$ BBUs,
- proportion of offered traffic: $A_{1,1}t_{1,1} : A_{2,1}t_{2,1} : A_{3,1}t_{3,1} = 1 : 1 : 1$;
- Resources 2:
  - $f_2 = 30$ BBUs, $v_2 = 2$, $V_2 = 2 \times 30 = 60$ BBUs,
  - offered traffic: $t_{1,2} = 1$ BBUs, $t_{2,2} = 2$ BBUs, $t_{3,2} = 4$ BBUs,
  - proportion of offered traffic: $A_{1,2}t_{1,2} : A_{2,2}t_{2,2} : A_{3,2}t_{3,2} = 1 : 1 : 1$;
- Resources 3:
  - $f_3 = 16$ BBUs, $v_3 = 4$, $V_3 = 4 \times 16 = 64$ BBUs,
  - offered traffic: $t_{1,3} = 1$ BBUs, $t_{2,3} = 2$ BBUs, $t_{3,3} = 4$ BBUs,
  - proportion of offered traffic: $A_{1,3}t_{1,3} : A_{2,3}t_{2,3} : A_{3,3}t_{3,3} = 1 : 1 : 1$;
2. Secondary resources
   - $f_0 = 20$ BBUs, $v_0 = 6$, $V_0 = 120$ BBUs.

Parameters of the analyzed systems No. 3

1. Primary resources
   - The number of primary resources $r = 3$
   - Resources 1, 2 and 3 (all primary resources have the same structure):
     - $f_1 = 20$ BBUs, $v_1 = 3$, $V_1 = V_2 = V_3 = 3 \times 20 = 60$ BBUs,
     - offered traffic: $t_{1,*} = 1$ BBUs, $t_{2,*} = 2$ BBUs, $t_{3,*} = 4$ BBUs,
     - proportion of offered traffic: $A_{1,1}t_{1,1} : A_{2,1}t_{2,1} : A_{3,1}t_{3,1} = 1 : 1 : 1$;
       $A_{1,2}t_{1,2} : A_{2,2}t_{2,2} : A_{3,2}t_{3,2} = 1 : 1 : 1$; $A_{1,3}t_{1,3} : A_{2,3}t_{2,3} : A_{3,3}t_{3,3} = 1 : 1 : 1$;
2. Secondary resources
   - $f_{0,1} = 20$ BBUs, $v_{0,1} = 3$, $f_{0,2} = 30$ BBUs, $v_{0,2} = 2$, $V_0 = 3 \cdot 20 + 2 \cdot 30 = 120$ BBUs.

Parameters of the analyzed systems No. 4

1. Primary resources
   - The number of primary resources $r = 3$
   - Resources 1, 2 and 3 (all primary resources have the same structure):
     - $f_1 = 20$ BBUs, $v_1 = 2$, $V_1 = V_2 = V_3 = 2 \times 20 = 40$ BBUs,
     - offered traffic: $t_{1,*} = 1$ BBUs, $t_{2,*} = 3$ BBUs, $t_{3,*} = 8$ BBUs,
     - proportion of offered traffic: $A_{1,1}t_{1,1} : A_{2,1}t_{2,1} : A_{3,1}t_{3,1} = 1 : 1 : 1$;
       $A_{1,2}t_{1,2} : A_{2,2}t_{2,2} : A_{3,2}t_{3,2} = 1 : 1 : 1$; $A_{1,3}t_{1,3} : A_{2,3}t_{2,3} : A_{3,3}t_{3,3} = 1 : 1 : 1$;
2. Secondary resources
   - $f_0 = 30$ BBUs, $v_0 = 4$, $V_0 = 120$ BBUs.

Parameters of the analyzed systems No. 5

1. Primary resources
   - The number of primary resources $r = 3$
   - Resources 1:
     - $f_1 = 20$ BBUs, $v_1 = 2$, $V_1 = 2 \times 20 = 40$ BBUs,
     - offered traffic: $t_{1,1} = 1$ BBUs, $t_{2,1} = 3$ BBUs, $t_{3,1} = 8$ BBUs,

- proportion of offered traffic: $A_{1,1}t_{1,1} : A_{2,1}t_{2,1} : A_{3,1}t_{3,1} = 1 : 1 : 1$;
  - Resources 2:
    - $f_2 = 30$ BBUs, $v_2 = 2$, $V_2 = 2 \times 30 = 60$ BBUs
    - offered traffic: $t_{1,2} = 1$ BBUs, $t_{2,2} = 3$ BBUs, $t_{3,2} = 8$ BBUs,
    - proportion of offered traffic: $A_{1,2}t_{1,2} : A_{2,2}t_{2,2} : A_{3,2}t_{3,2} = 1 : 1 : 1$;
  - Resources 3:
    - $f_3 = 20$ BBUs, $v_3 = 3$, $V_3 = 3 \times 20 = 60$ BBUs,
    - offered traffic: $t_{1,3} = 1$ BBUs, $t_{2,3} = 3$ BBUs, $t_{3,3} = 8$ BBUs,
    - proportion of offered traffic: $A_{1,3}t_{1,3} : A_{2,3}t_{2,3} : A_{3,3}t_{3,3} = 1 : 1 : 1$;
2. Secondary resources
  - $f_0 = 30$ BBUs, $v_0 = 4$, $V_0 = 120$ BBUs.

The analytical and simulation results of the blocking probability in the alternative resources of the overflow system with distributed primary resources are presented in Figs. 2, 3, 4, 5 and 6.

The results of the present study clearly demonstrate that the accuracy of the method is not dependent on either the number and structure of LAGs that compose primary groups or LAGs that form secondary groups. In the same way, a differentiation in demands of individual classes (the parameter t) has no influence on the accuracy of the proposed method which is comparable with the accuracy of calculations for a single LAG.

The results of the blocking probability for the traffic classes with the highest demands obtained on the basis of the simulation are marginally higher than the results obtained on the basis of the proposed analytical method. For classes with the lowest demands, the situation is reversed – the results of the simulation slightly exceed the results of the calculations. This particular dependence between the results of the analytical modeling and the results of simulation experiments is, however, convenient from the practical point of view. The same degree of overestimation applies to and involves those traffic classes that are characterized by the highest blocking probability and, in consequence have the highest influence on the operation of a system. Engineering rules for designing network systems are consistent with the principle of dimensioning that is based on the "worst case" scenario. Within this context, a slight overestimation of the QoS parameters for the traffic classes that have the highest impact on the operation of a system fits well into the principle.

5   Conclusion

The paper proposes a new analytical model of multi-service overflow systems with distributed primary and alternative resources. The results of analytical modelling have been compared with the simulation results. The accuracy of the proposed method is comparable with the accuracy of the classical models elaborated for modelling the overflow systems with fully available primary resources. The model can be applied for analytical modelling of multi-service multi-tier cellular network.

References

1. Nagesh, C.B., Jayaramaiah, D.: A survey of self-organizing networks (son). Int. J. Innov. Res. Comput. Commun. Eng. **3**(4), 2916–2923 (2015). http://www.rroij.com/abstract.php?abstract_id=56994
2. Tripathi, N., Reed, J., VanLandinoham, H.: Handoff in cellular systems. IEEE Pers. Commun. **5**(6), 26–37 (1998)
3. Fernandes, S., Karmouch, A.: Vertical mobility management architectures in wireless networks: a comprehensive survey and future directions. IEEE Commun. Surv. Tutorials **14**(1), 45–63 (2012)
4. Lin, Y.-B., Chang, L.-F., Noerpel, A.: Modeling hierarchical microcell/macrocell PCS architecture. In: 1995 IEEE International Conference on Communications, ICC 1995, Seattle. Gateway to Globalization, vol. 1, pp. 405–409, June 1995
5. Li, S., Grace, D., Wei, J., Ma, D.: Guaranteed handover schemes for a multilayer cellular system. In: 7th International Symposium on Wireless Communication Systems, pp. 300–304, September 2010
6. Bretschneider, G.: Die Berechnung von Leitungsgruppen für berfließenden Verkehr in Fernsprechwählanlagen. Nachrichtentechnische Zeitung (NTZ) (11), 533–540 (1956)
7. Wilkinson, R.: Theories of toll traffic engineering in the USA. Bell Syst. Techn. J. **40**, 421–514 (1956)
8. Matsumoto, J., Watanabe, Y.: Theoretical method for the analysis of queueing system with overflow traffic. Electron. Commun. Japan (Part I: Commun.) **64**(6), 74–83 (1981). http://dx.doi.org/10.1002/ecja.4410640610
9. Rapp, Y.: Planning of junction network in a multi-exchange area. In: Proceedings of 4th International Teletraffic Congress, London, p. 4 (1964)
10. Herzog, U.: Die exakte Berechnung des Streuwertes von überlaufverkehr hinter Koppelanordnungen beliebiger Stufenzahl mit vollkommener bzw. unvollkommener Erreichbarkeit. AEÜ **20**(3) (1966)
11. Schehrer, R.: On the exact calculation of overflow systems. In: Proceedings of 6th International Teletraffic Congress, Munich, pp. 147/1–147/8, September 1970. https://itc-conference.org/_Resources/Persistent/21594392c496c4c0193b4864ec86 0502ebdec2bc/schehrer70.pdf
12. Bretschneider, G.: Extension of the equivalent random method to smooth traffics. In: Proceedings of 7th International Teletraffic Congress, Stockholm (1973)
13. Schehrer, R.: On the calculation of overflow systems with a finite number of sources and full availiable groups. IEEE Trans. Commun. **26**(1), 75–82 (1978)
14. Fredericks, A.: Congestion in blocking systems - a simple approximation technique. Bell Syst. Techn. J. **59**(6), 805–827 (1980). http://onlinelibrary.wiley.com/doi/10.1002/j.1538-7305.1980.tb03034.x/abstract
15. Shortle, J.F.: An equivalent random method with hyper-exponential service. J. Perform. Eval. **57**(3), 409–422 (2004)
16. Głąbowski, M., Hanczewski, S., Stasiak, M.: Erlang's ideal grading in diffserv modelling. In: Proceedings of IEEE Africon 2011, pp. 1–6. IEEE, Livingstone, September 2011. http://dx.doi.org/10.1109/AFRCON.2011.6072139
17. Głąbowski, M., Kubasik, K., Stasiak, M.: Modeling of systems with overflow multirate traffic. Telecommun. Syst. **37**(1–3), 85–96 (2008)

18. Głąbowski, M., Kubasik, K., Stasiak, M.: Modelling of systems with overflow multi-rate traffic and finite number of traffic sources. In: Proceedings of 6th International Symposium on Communication Systems, Networks and Digital Signal Processing 2008, Graz, pp. 196–199, July 2008. http://dx.doi.org/10.1109/CSNDSP.2008.4610800

19. Huang, Q., Ko, K.-T., Iversen, V.B.: Approximation of loss calculation for hierarchical networks with multiservice overflows. IEEE Trans. Commun. **56**(3), 466–473 (2008)

20. Głąbowski, M., Kaliszan, A., Stasiak, M.: Two-dimensional convolution algorithm for modelling multiservice networks with overflow traffic. Math. Prob. Eng. **2013**, 18 (2013). Article ID 852082. http://dx.doi.org/10.1155/2013/852082

21. Głąbowski, M., Kaliszan, A., Stasiak, M.: Modelling overflow systems with distributed secondary resources. Comput. Netw. **108**, 171–183, (2016). http://www.sciencedirect.com/science/article/pii/S1389128616302675

22. Głąbowski, M., Hanczewski, S., Stasiak, M.: Modelling of cellular networks with traffic overflow. Math. Prob. Eng. **2015**, 15 (2015). Article ID 286490. http://dx.doi.org/10.1155/2015/286490

23. Kühn, P., Mashaly, M.E.: Multi-server, finite capacity queuing system with mutual overflow. In: Fiedler, M. (ed.) Proceedings of 2nd European Teletraffic Seminar, pp. 1–6. Karlskrona, September 2013. paper S1_2

24. Kaufman, J.: Blocking in a shared resource environment. IEEE Trans. Commun. **29**(10), 1474–1481 (1981)

25. Roberts, J.: A service system with heterogeneous user requirements – application to multi-service telecommunications systems. In: Pujolle, G. (ed.) Proceedings of Performance of Data Communications Systems and their Applications, pp. 423–431. North Holland, Amsterdam (1981)

26. Głąbowski, M., Hanczewski, S., Stasiak, M.: Modelling load balancing mechanisms in self-optimising 4G mobile networks with elastic and adaptive traffic. IEICE Trans. Commun. **E99-B**(8), August 2016

27. Stasiak, M.: Blocking probability in a limited-availability group carrying mixture of different multichannel traffic streams. Ann. Télécommun. **48**(1–2), 71–76 (1993)

28. Głąbowski, M., Stasiak, M.: Multi-rate model of the group of separated transmission links of various capacities. In: Souza, J.N., Dini, P., Lorenz, P. (eds.) ICT 2004. LNCS, vol. 3124, pp. 1101–1106. Springer, Heidelberg (2004). doi:10.1007/978-3-540-27824-5_143

29. Roberts, J., Mocci, V., Virtamo, I. (eds.): Broadband Network Teletraffic, Final Report of Action COST 242. Commission of the European Communities. Springer, Berlin (1996)

30. Głąbowski, M., Kaliszan, A., Stasiak, M.: Modeling product-form state-dependent systems with BPP traffic. Perform. Eval. **67**, 174–197 (2010)

31. Stasiak, M., Głąbowski, M., Wiśniewski, A., Zwierzykowski, P.: Modeling and Dimensioning of Mobile Networks. Wiley, Chichester (2011)

32. Głabowski, M., Hanczewski, S., Stasiak, M., Weissenberg, J.: Modeling Erlang's ideal grading with multi-rate BPP traffic. Math. Prob. Eng. **2012**, 35 (2012). Article ID 456910

33. Ershova, E., Ershov, V.: Digital Systems for Information Distribution. Radio and Communications, Moscow (1983). in Russian

34. Głąbowski, M., Kaliszan, A.: Simulator of full-availability group with bandwidth reservation and multi-rate Bernoulli-Poisson-Pascal traffic streams. In: Proceedings of Eurocon 2007, Warszawa, pp. 2271–2277, September 2007. http://dx.doi.org/10.1109/EURCON.2007.4400605

New Technologies

Multi-level Stateful Firewall Mechanism
for Software Defined Networks

Fahad Nife[1,2(✉)] and Zbigniew Kotulski[1]

[1] Faculty of Electronics and Information Technology,
Warsaw University of Technology, Warsaw, Poland
fahadnaim114@gmail.com, zkotulsk@tele.pw.edu.pl
[2] Faculty of Science, Al-Muthanna University, Muthanna, Iraq

Abstract. Traditional networks are often quite static, slow to modify, dedicated for a single service and very difficult to scale, what is typical for a large number of different network devices (such as switches, routers, firewalls, and so on), with many complex protocols implemented or embedded on them. Software Defined Network (SDN) is a new technology in communication industry that promises to provide new approach attempting to overcome this weakness of the current network paradigm. The SDN provides a highly scalable and centralized control architecture in which the data plane is decoupled from the control plane; this abstraction gives more flexible, programmable and innovative network architecture. However, centralization of the control plane and ability of programming the network are very critical and challenging tasks causing security problems. In this paper we propose a framework for securing the SDN by introducing an application as an extension to the controller to make it able to check every specific flow in the network and to push the security instructions in real-time down to the network. We also compare our proposal with other existing SDN-based security solutions.

Keywords: Software Defined Network · Stateful-Packet filtering · Flow table · OpenFlow protocol

1   Introduction

In computer networking, the emerging network architecture that decouples the control plane from the data plane, logically centralizes the network intelligence and state, and abstracts the underlying network infrastructure from the applications is known as Software Defined Network [1]. SDN is a new framework, where the control plane of all devices in the network is migrated from the forwarding plane and consolidated into a software-based device called the controller. This controller is responsible for making and forwarding decisions for the entire network, while the switches and routers just perform basic packets forwarding. The controllers have full view of the network and consider the totality of the network before making any decision. Since SDN relies more on software and, because of

P. Gaj et al. (Eds.): CN 2017, CCIS 718, pp. 271–286, 2017.
DOI: 10.1007/978-3-319-59767-6_22

its network intelligence, SDN is logically centralized [2], it supports rapid innovation and provides new ways and opportunities to enhance and protect the network.

In this paper we introduce a stateful firewall solution for securing an open network using SDN technology. This proof of concept proposal aims to build a reactive Stateful SDN-Based Traffic Filtering framework for SDN networks by getting the advantages of using the OpenFlow protocol messages, which are exchanged between the controller and the OpenFlow-enabled switches. It extends the controller using a software which makes use of some tables used to keep track of each flow in the network to provide stateful packet filtering. The method relies on dividing the network into planes to create a multi-plane system of defense and to make it easy protecting the network from external attacks as well as from internal malicious users. The network security policy will be centralized in the controller, where there is the place of making decision regarding how the switches should handle the packets, while a switch will only enforce this decision. The controller can at any time reconfigure the security rules and redeploy them at any device under its control. Thus, we have more efficient, flexible and centralized point of security management represented by the controller, which in turn spreads the network security policy dynamically across the distributed checkpoints.

The rest of this paper is organized as follows. Section 2 describes the theoretical background of SDN, OpenFlow and firewalls. Section 3 presents state of the art of current SDN security solutions. The overview of the proposed system's architecture is specified in Sect. 4. The analysis of the new prototype solution and its comparison to other known SDN-based firewalls are presented in Sect. 5, while Sect. 6 concludes the paper with a summary of its content and future enhancements.

2   Background

To relate our proposal to other existing network solutions and to give the background of the suggested method along with its functionality and components, we start from a brief description of the SDN architecture, the OpenFlow protocol and firewalls.

2.1   SDN Architecture

The OpenFlow-based SDN architecture mainly consists of two planes, see Fig. 1, the control plane and the data plane [3]. The control plane strives to act as a network operating system and it is responsible for programming and managing the data plane. It makes use of the information provided by the data plane and defines network operation and routing [4]. The other plane, which is the data plane, comprises the forwarding network elements that are connected to each other. It is responsible for data forwarding, as well as monitoring the network by gathering local information and statistics [5]. For more eligibility and scalability in SDN networks, network services provided by the network applications are abstracted as a separate plane, which may introduce new network features, such as security, QoS, load balance, bandwidth management, and so on. Two interfaces that must be implemented by the control plane are the Northbound interface (NBI) and the Southbound Interface (SBI). The Northbound Interface presents a programmable API used by the application plane to receive an abstracted and global view of the network from the controllers and then use that information to provide an appropriate guidance back to the control plane. Typically, control software provides its own API to applications as there is no standardized API today [6]. The Southbound Interface (e.g. OpenFlow) is used for the communication between the controller and switches. The data plane elements use the SBI to reveal their capabilities toward the controller (Control Plane). It is used to provide grammatic control, statistics reporting and events notifications.

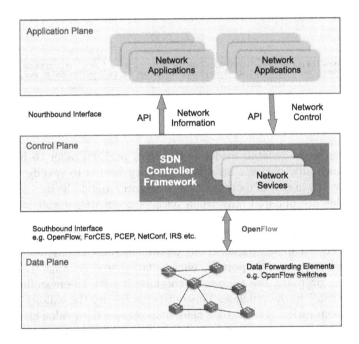

Fig. 1. Software-Defined Network abstraction

2.2 OpenFlow

The most commonly used protocol for the Southbound interface of SDN is the OpenFlow protocol [7]. It is standardized by the Open Networking Foundation

(ONF) and it is typically responsible for establishing the communication through secure control channel between the controller and switches under its control. An OpenFlow-enable switch is a basic forwarding device that forwards packets according to its flow table. The flow table contains a set of flow entries (also called flow rules) each of which consists of headers fields, activity counters, priority, timeout, actions, cookie and flags, see Fig. 2.

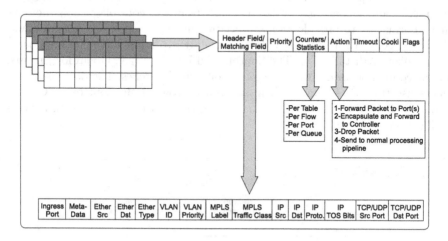

Fig. 2. OpenFlow table

The header fields (Match fields) is 15-tuples packet header to be matched against the incoming packet's header. A Priority field is to specify which flow should be selected in case of more than one flow entry match the incoming packet. Counters fields are updated only when packets are matched and used to keep track of each flow, table and port and provide statistics to the Controller. The action fields contain data, where there is a set of zero or more actions that should be applied to the matched packets. The possible actions include forwarding the incoming packet to specific port(s), encapsulating and sending the packet to the controller, dropping the packet, or checking it with the next flow table or with specific table in the pipeline. Timeout fields specify the validity of the flow entry before expiration. The Cookie field is an opaque data value chosen by the controller. It can be used by the controller to filter the flow entries that affected by flow statistics, modification and deletion requests. Cookie is not used when processing packets. Flags fields alter the way flow entries are managed. The flow tables of the switch are manipulated by the controller, where it can modify the behavior of the switches with regard to forwarding by inserting, modifying and removing the switch flow entries.

Every packet received by the switch, the lookup process begin with the first flow table entries till the last flow table in the pipeline, the process end either with match and execute the corresponding action for the matched entry or with miss (no rule match the packet) and execute the default rule which is either send the packet to the controller or drop the packet. If matching occurs, the action field for that entry is applied on that packet, While in case of no matching, the default rule will applied which may encapsulate the packet (or part of the packet, it is common the packet header and the number of bytes defaults to 128 if the switch is capable of buffering) with Packet-in message and send it to the controller. The controller according to its policy rules will take a decision of how should that switch handle the packet and forward it back to the requested switch. According to that decision, the switch will apply it on the packet and modify its flow table by adding new entry to help handling next similar packets [8].

2.3   Firewall

Firewalls are fundamental security mechanisms for any network defense systems. Firewall is usually placed between an authorized system and external environments to act as a gateway through which all inbound and outbound network traffic must pass [9]. It is checkpoint that compares each incoming packet with its policy rules to decide whether to allow or deny the packet from pass through it. Each firewall have its policy rules database which define how an firewall should handle incoming and outgoing network traffic for some specific IP addresses and range of addresses, protocols, applications, and packet content types based on the organization's security policies [10]. Firewalls can be software or hardware-based and can be categorized according to deferent criteria such as their functionality or the layer (ISO/OSI Reference Model and TCP/IP model) they operate: the Network, Session, Transport or Application layer. The most common firewalls that work on the network and transport layers are Stateless Firewalls and Stateful Firewalls.

Stateless Firewall. It enforces security policies within a network by employ a filtering process for incoming and outgoing packets. It is looks at the IP address of the packet and compare it against predetermined Rule set defined by the network administrator to decide whether it should be allowed through or not [11]. Each packet maintained separately, so it does not need to save the state or any information about the packet to be used for processing the next packet in the same flow, see Fig. 3.

Stateful Firewall. It attempts to track the state of network connections when filtering packets. It often makes use of a table called state table to store information about each traffic flow passing through. This state table holds entries that represent all the communication sessions of which the device is aware. Every entry holds specific information (e.g. source and destination IP addresses, Port numbers, flags, sequence and acknowledgment numbers) that uniquely identifies the communication session it represents [12]. For each connection started out through the stateful device, a state table entry is created. When the inspected packet matches an existing firewall rule that permits it, the packet is passed and new entry is added to the state table, see Fig. 4. Then, when traffic returns, the

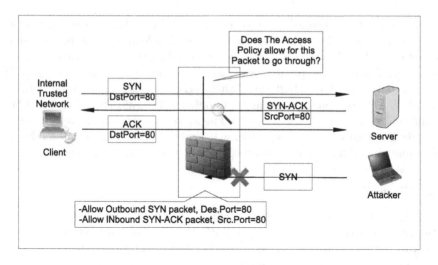

Fig. 3. Stateless firewall

packet's information compared against the state table information to determine whether it is part of a currently logged communication session. The packet is allowed to pass only if it is related to a current table entry.

3   Related Works

Several recent efforts in the field of SDN security have been presented in many previous contexts, such as scanning attack prevention [13], DDoS attack detection [14], IDS [15], IPS [16], Stateless firewall and Stateful firewall.

FortNox [17] is an extension to the open source NOX controller. The security policy enforcement kernel implement a novel analysis algorithm to prevent any of application attempts to insert flow rules that may change flow rules enforced by security policy. FortNox is a real-time rule conflict detection engine, where after detecting any conflict, it is either accept or refuse the new rule based on rule-based authorization. FRESCO [18] introduce security application framework consists of several reusable modular libraries that can be connected together to build more sophisticated security applications. FRESCO developers can easily connect the necessary modules including its libraries and assigning values to its interfaces to produce the required security functions, like firewalls, IDS, or scan detections. FLOWGUARD [19], represents a comprehensive SDN framework for facilitate accurate detection and flexible resolution of firewall policy violations in dynamic OpenFlow networks. FLOWGUARD provide an automatic and real-time violation resolution mechanism, when the network states are updated or the configurations are dynamically changed, it detect firewall policy violations by checks the flow path spaces and compare it against the specified authorized space in the firewall.

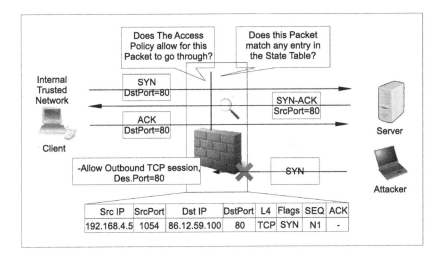

Fig. 4. Stateful firewall

FleXam [20,21], proposes a sampling method that make the controller able to access packet-level information that is necessary for different security applications. With FleXam, the controller can decide which part of the packet (e.g. only headers or, maybe, only payload) and where they should be sent. Moreover, the packet can be sampled either stochastically based on predetermined probability or deterministically based on pattern. The authors of [22] propose a prototype in which the security rules are specified in the flow table in both controller and switches. The switch acts as simple packet-pusher based on its flow table and any unknown traffics are sent to the controller for inspection. The prototype proposed in [23] provides orchestration services for security policy management, reactive application that processing the state of the connection, the security are expressed with OpenFlow rules and in the data plane and the generic algorithm is introduced for processing the Finite State Machine (FSM) of the TCP protocol. Proactive Security Mechanism has been introduced in [24]. In that work, fuzzy based system was designed to evaluate the threat level of identified threats. The authors of [25] introduce an SDN-based mitigation system for ransomware threats. Their solution is based on CryptWall findings and the two designed SDN1 and SDN2 mitigation ransomware methods, which rely on utilizing a dynamic blacklist database of up-to-date list of known proxy servers that used for the purpose of ransomware. Both methods try to break the communication between the victim and the infected C&C server, so no public key will retrieve and as a result the encryption process will not happen.

4   System Architecture

4.1   Overview

The diagram presented in Fig. 5 shows that the system's architecture is compatible with the SDN structure: it consists of three planes distributed into the SDN planes. The highest plane involves the system functionality, which is executed by different modules, responsible for maintaining the tables in the other planes and providing the required security functions to protect the SDN network. The second firewall plane consists of the set of tables that are built in the control plane. These tables are used to improves the overall performance by reducing the lookup process time needed to handle a packet. The most down plane provides the flow state-aware using tables placed in the OpenFlow-enabled switches and holds entries used to keep track for each specific flow.

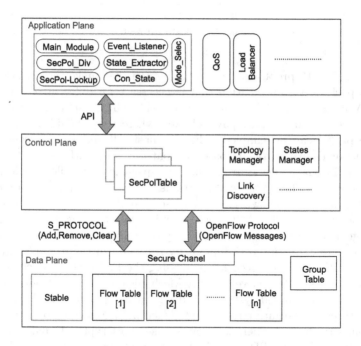

Fig. 5. System architecture overview

The functionality of the firewall application that run on a top of the controller are executed by 7 modules, which in turn maintain tables in, both, the controller and the OpenFlow-enabled switches. The first module is Main_Module. It is the control unit that play the coordinator and manager roles for other units. The other module is Event_Listener which is triggered by some predefined events and which responds with an action corresponding to that event. The State_Extractor module is responsible for extracting some information from the

packet. The SecPol_Div module is initiating SecPolTable for each switch joins
to the controller. Each entry in the SecPolTable table represents the security
policy for that switch prepared by the network administrator. The other table
is STable. It is maintained in the switch and used to keep track of each traffic
flow pass through the switch. SecPol_Lookup module is responsible for checking
out the SecPolTable for all switches in the whole path for that packet from the
source to the destination if it is inside the network or to the gate way switch if
the final destination is outside the network. Con_State module is used to install
new entry in STable of all switches in the path using S_PROTOCOLE protocol,
this approach is to avoid additional delays (switch-controller communication)
when the first packet arrives for next switches in the path. The last module is
Mode_Selec, which is responsible of switching between the tow control behavior
modes.

Now, before explaining the main components and tables of the proposed
stateful application, let us discuss interactions among OpenFlow switches, controllers and the stateful firewall application. Every time an OpenFlow-enabled
switch receives a packet, it parses the packet's header, which is matched against
the STable entries before the flow table pipeline for any match. If the match
occurs, the packet will forwarded meaning that the packet is part of previous
flow, otherwise no matching occurs, The matching will be against the flow table
entries to do the action related with matched entry, or if no matching entry, the
packet will encapsulated with Packet-in message and forwarded for the controller
to decide how to handle the packet, see Fig. 6.

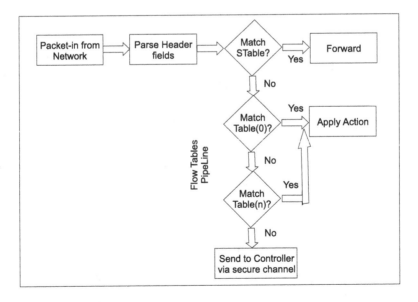

Fig. 6. Packet flow in stable enabled open flow switch

Once the controller receives the Packet-in notification, it identifies the correct action for the packet according to the security policy stored in the SecPolTable corresponding for that switch and then using Packet-out messages installs one or more appropriate entries in the flow table of requesting switch. The event of the controller sends the Packet-out message that will trigger the Event_Listener object to Call other three functions, State_Extractore object, which in turn extracts some information and attributes that make up the communication session (e.g. IP addresses, Port numbers, Sequence and Acknowledgment numbers, and flags) which can all combined to be used as a fingerprint for an individual connection. The other object is SecPol_Lookup, which is responsible for checking out the SecPolTable for all switches in the whole path for that packet to be sure it does not violate the security policy of the network, if so, and finally the Con_State module to install entry in STable of corresponding switches using S_PROTOCOLE Add message and all switches in the path, this approach is to avoid additional delays (switch-controller communication) when the first packet arrives for next switches in the path.

One more important introduced mechanism done by the SecPol_Div, which for each data plane device join to the controller, this SecPol_Div object will initiate SecPolTable for that new connected switch, this mechanism will improve the overall performance by reducing time needed by the controller to decide how should that switch handle the packet if it does not match any entry in its flow table. When security policy updated by the network administrator, the SecPol_Div will rebuild all the SecPolTable and reinforce and dynamically propagate the security policies for every data plane.

To specify the controller behavior on the flow and STable entries, 2 modes Norm_Mode and Critic_Mode are implemented. The Mode_Selec module is responsible of switching between this modes. The Norm_Mode is the default mode in which the Timeout field in the STable is started with 60 s and there is no limit on the number of entries of STable, once the S_PROTOCOLE enabled switch exposure to onslaught or its Stable nearly to full (exceed the threshold) it switch will send S_PROTOCOLE Critical message to the controller and accordingly the Mode_Selec module will implement the Critic_Mode in which rather than start with 60 s as timeout for establishing the flow, it will to be 15 s, and put limit on the number of entries of the STable per source and destination IP, per source and destination ports, or even per rule policy to help reducing the risk of fulling the STable. When the STable entries decreased to some reasonable size the switch will send S_PROTOCOLE Normal message to the controller to inform the Mode_Selec module to implement the Norm_Mode.

4.2   STable Table and SecPolTable Table

STable is the table generally implemented as a hash table, which typically resides in the OpenFlow-enabled Switch RAM and is used to keep track of each valid connection. STable offers stateful inspection capability by records the state of a connection based on protocol specific information. Each new connection requires adding new entry in the STable for the connection. The STable entry includes the following information: Protocols, Source IP addresses and Destination IP addresses, Source port and Destination port, Reversed Source and destination IP addresses and ports (to represent response traffic), the protocol being used for the connection, the time specifies the validity of rule before removing it, the TCP state of the connection (in a case TCP is used), the sequence number, and the acknowledgment number, see Table 1. The stateful firewall application maintains this table using Add, Remove, and Clear messages of the designated protocol called S_PROTOCOLE, see Sect. 4.3.

The other table is SecPolTable table. It is simple table consists of rows and columns initiated by the SecPol_Div module for each specific switch join to the controller and used to holds Security Policy related for that switch. This table will improve the overall performance by reducing time needed by the controller to decide how that switch should handle the packet if it is not match any entry in its flow table. Any change of the network security by the network administrator will enforce the SecPol_Div to rebuild all the SecPolTable tables and dynamically propagate the security policies for every data plane.

Table 1. Example of STable Table's entries

Prtcl	Src IP	Dst IP	SrcPrt	DstPrt	State	SeqNo	AckNo	Timout
TCP	192.168.10.9	172.16.2.23	1054	21	SYN_SENT	0	0	93
TCP	172.16.2.23	192.168.10.9	21	1054	ESTABLISHED	1	726	2845
...	....	....	..	...	....	..	...	...

4.3 S_PROTOCOLE

It is designated communication protocol between the controller and the OpenFlow-enabled switch. The protocol S_PROTOCOLE defines five messages. Three of them: S_PROTOCOLE Add, S_PROTOCOLE Remove and S_PROTOCOLE Clear are to enable state-based connection monitoring and to help the ConState module to maintain the STable by adding, removing and clearing the STable entries of the data plane. The two other messages: S_PROTOCOLE Normal and S_PROTOCOLE Critical are to help the Mode_Selec restricting the controller behavior. S_PROTOCOLE with its messages is used only by the ConState module and it is not affected by other SDN Modules and applications. Table 2 shows the direction of the messages flow along with descriptions of their functionality.

5 Analysis

We now estimate effectiveness of the presented solution, such performance analysis calculated for memory consumption, processing times and throughputs and vulnerabilities mitigations.

Table 2. S_PROTOCOLE messages

Message name	Direction	Description
S_PROTOCOLE Add	Controller_to_Switch	Add new STable entry for the new traffic
S_PROTOCOLE Remove	Controller_to_Switch	Delete specific entry from the STable
S_PROTOCOLE Clear	Controller_to_Switch	Empty the STable
S_PROTOCOLE Normal	Switch_to_Controller	Enforce the Mode_Selec module to implement the Norm_Mode
S_PROTOCOLE Critical	Switch_to_Controller	Enforce the Mode_Selec module to implement the Critic_Mode

In order to compute memory consumption required by our approach, we need to focus on the size of STable which run in OpenFlow-enabled switch. As explained in Sect. 4.2 each STable entry is 26-byte, so for STable with 10000 entries the required amount of memory is 26 MB, while for STable with 100000 entries the required amount of memory is 260 MB, which can be considered a reasonable amount of memory. The other data structure used are SecPolTable tables does not require large memory as it divide the rule table into small tables, one main table for shared rules and other tables for every joined switch.

Optimization in our proposed prototype happens in different places: (1) using the STable, that will help to increase performance and security by associating the individual packets with their respective flows, so any match between the incoming packet with any Stable entry lead to forward the packet to its destination through the appropriate port(s) without requiring any extra computation or controller communication overhead; (2) by rule base checking, where in SDN, every Packet-in arrives some algorithm must be used to check the rules and decide how to handle the packet. An essayist approach is to scan every rule in priority order until a match is found or the rule base is exhausted. This linear search through the rule base is convenient for small rule policies, but it is inefficient for larger policies, particularly when match done with the default rule or the last rule entry of the policy. So the worst-case, per packet running time (for each Packet-in from each switch) is $O(n)$, where n is the number of entries in the rule base.

In contrast, let us calculate it for our proposed solution, for ease of measurement we break up the process into two parts: the first part searches the rule policy and builds up SecPolTable and the second part is scanning the SecPolTable policy rules to find the match. Let j be the number of switches connected to the SDN controller and n the number of policy rules. Building each SecPolTable table requires scanning through the policy file or just the remaining of it to determine which entry belongs to it. By doing so, the first joined switch requires scanning the whole policy file with $O(n)$ processing time, producing first SecPolTable with, say, n_1 entries. Now for the next joined switch,

Table 3. Comparison of the proposed approach with other SDN security solutions

Publication	Problem approached	Proposed solution	Behavior	Extra memory	Capabilities exploited	Performance
Collings, J. et al. [22]	Dynamic packet filtering	Stateful H/W firewall	Reactive	No	Centralized control	High communication overhead
Zerkane, S. et al. [23]	DoS attack mitigation	Stateful/ Stateless	Reactive	Yes	Centralized control	High communication overhead
Suh, M. et al. [26]	Block unnecessary mitigation	Stateless firewall	Reactive	No	Centralized control	High performance
Pena, J. G. et al. [27]	Packet filtering	Stateless firewall	Proactive	Yes	Distributed control	High performance
Vasudevan, S. et al. [28]	Prevent unauthorized access	Stateless firewall	Reactive	Yes	Distributed control	-
Jeong C. et al. [15]	Malicious traffic inspection	IDS	Proactive	Yes	Centralized control	High performance
Proposed solution	Dynamic packet filtering	Stateful firewall	Reactive	Yes	Centralized control	High performance

building up the SecPolTable requires only to search the remaining $(n - n_1)$ entries with $O(n - n_1)$ processing time. We can observe that each time a new switch joins the network, the processing time needed to build SecPolTable is decreasing. Fortunately this process is done only one time for each switch or, additionally, when the policy file is updated. Now on, each Packet-in arrives (for example from switch 1) only we need to search the SecPolTable related to that switch with $O(n_1)$ processing time that is lease than $O(n)$, which in result will increase the overall throughputs.

Regarding the protection level provided by a network security solution, a simple stateful firewall, which is a good security system to protect the network still is vulnerable to some attacks like DDoS or DoS. Such attacks adapt at exploiting the heart of most stateful firewalls STable and try to fall it to capacity and block new incoming connections. Our suggested prototype uses mechanisms and implements modes that restrict the timeout for establishing connection and limits the number of the entries in the STable, which in turn can prevent or at least mitigate possibilities of such attacks in the network.

To summarize this draft analysis of security functions and performance let us look at Table 3. We compare the proposed approach with SDN-dedicated security solutions known from the literature. It is seen that some of the systems are only simple packet filtering firewalls while some other additionally provide dynamic packet checkup. This requires new tables in the controller side which in turn leads to high communication overhead. To overcome this limitations of existing approaches, we make use of a new table in the data plan side that provides the state-awareness and decreases the switch controller communication overhead. Another advantage of our approach is application of a multi-level structure of the firewall with the levels associated with SDN planes, distributed sensing and modular structure. Such an approach makes it possible easy extensions of the firewall functions and integration with other security systems. For instance, our firewall is compatible with the ISO/OSI Layer 2 firewall solution proposed in [29], which consists of two basic elements: Ethernet frames filtering in Layer 2 and a three-tier reputation system of critical data acquisition and analysis for the firewall rules configuration. Such a distributed multi-tier reputation system modulus could slightly improve threats identification and enforce SDN protection.

6 Conclusion

In this paper we introduced a reactive stateful firewall approach for securing SDN-based networks. Our under construction suggested solution does not enforce any change to the current SDN paradigm in terms of its design and behavior. Moreover, it attempts to enhance the security of the programmable network and simplify security management. This proof-of-concept solution tries to overcome the limitations of the OpenFlow implementation that relies on a simple "match-action" paradigm. Such an OpenFlow approach causes lacks of stateful processing function in the SDN data plane, where it makes use of specified tables in the data plane devices to keep track of TCP connection or pseudo-connection. The security policy of SDN proposed in our approach is enforced by the firewall application that is centralized on the top of the controller, while the data plane device is considered as a distributed checkpoint. Thus, instead of one firewall placed at the edge of the network's boundary and providing a single point of failure, we can have a plane-like firewall that creates additional boundaries within the network and provides "defense-in-depth", which is considered as a good solution for networks with different levels of trust. The behavior of our solution is reactive, which gives an advantage of keeping the flow tables in the data plane small and which helps not to fill them with unmatched rules. For the next enhancements, we strive to improve the expected performance of the proposed prototype. We will focus on implementing the application with proactive approach to improve the system's overall performance by avoiding redundant times caused by communication between the switch and the controller for undefined packets. Also we plan to deploy the prototype in a real network environment to confirm its strength against active attackers. As a next step we will integrate the system with a soft methods-based security tool (a multi-tier trust and reputation system) what is a reasonable enhancement of a firewall in widespread and untrusted networks.

References

1. Adrian, L., Kolasani, A., Ramamurthy, B.: Network innovation using openflow: a survey. IEEE Commun. Surv. Tutorials **16**, 493–512 (2013)
2. Azodolmolky, S.: Software Defined Networking with OpenFlow Get Hands-on with the Platforms and Development Tools Used to Build OpenFlow Network Applications. Packt Publishing Ltd., Birmingham (2013)
3. Pujolle, G.: Software Networks Virtualization, SDN, 5G and Security. ISTE Ltd. and Wiley, Great Britain, United States (2015)
4. Kreutz, D., Ramos, F.M.V., Verissimo, P.E., Rothenberg, C.E., Azodolmolky, S., Uhlig, S.: Software-defined networking: a comprehensive survey. Proc. IEEE **103**, 14–76 (2015)
5. Underdahl, B., Kinghorn, G.: Software Defined Networking for Dummies. Wiley, Hoboken (2015)
6. Jain, R., Paul, S.: Network virtualization and software defined networking for cloud computing: a survey. IEEE Commun. Mag. **51**, 24–31 (2013)
7. The Open Networking Foundation, OpenFlow Switch Specification (2014)
8. Astuto, B.N., Mendonca, M., Nguyen, X.N., Obraczka, K., Turletti, T.: A survey of software-defined networking: past, present, and future of programmable networks. IEEE Commun. Surv. Tutorials **16**, 1617–1634 (2014)
9. Sharma, R.K., Kalita, H.K., Issac, B.: Different firewall techniques: a survey. In: 5th IEEE International Conference on Computing. Communications and Networking Technologies (ICCCNT), Hefei, Anhui, China, pp. 1–6 (2014)
10. Scarfone, K., Hoffman, P.: Guidelines on Firewalls and Firewall Policy, Gaithersburg (2009)
11. Duan, Q., Al-Shaer, E.: Traffic-aware dynamic firewall policy management: techniques and applications. IEEE Commun. Mag. **51**, 73–79 (2013)
12. Trabelsi, Z.: Teaching stateless and stateful firewall packet filtering: a hands-on approach. In: 16th Colloquium for Information Systems Security Education, Lake Buena Vista, Florida, pp. 95–102 (2012)
13. Mehdi, S.A., Khalid, J., Khayam, S.A.: Revisiting traffic anomaly detection using software defined networking. In: Sommer, R., Balzarotti, D., Maier, G. (eds.) RAID 2011. LNCS, vol. 6961, pp. 161–180. Springer, Heidelberg (2011). doi:10.1007/978-3-642-23644-0_9
14. Braga, R., Mota, E., Passito, A.: Lightweight DDoS flooding attack detection using NOX/OpenFlow. In: 35th Annual IEEE Conference on Local Computer Networks, LCN, Denver, Colorado, pp. 408–415 (2010)
15. Jeong, C., Ha, T., Narantuya, J., Lim, H., Kim, J.: Scalable network intrusion detection on virtual SDN environment. In: 2014 IEEE 3rd International Conference on Cloud Networking (CloudNet), Luxembourg, pp. 264–265 (2014)
16. Francois, J., Aib, I., Boutaba, R.: Firecol: a collaborative protection network for the detection of flooding DDoS attacks. IEEE/ACM Trans. Networking (TON) **20**, 1828–1841 (2012)
17. Porras, Ph., Shin, S., Yegneswaran, V., Fong, M., Tyson, M., Gu, G.: A security enforcement kernel for openflow networks. In: HotSDN 2012 ACM Special Interest Group on Data Communication SIGCOMM, pp. 121–126. ACM, Helsinki (2012)
18. Shin, S., Porras, P., Yegneswaran, V., Fong, M., Gu, G., Tyson, M.: FRESCO: Modular composable security services for software-defined networks. In: NDSS 2013 Network and Distributed System Security Symposium, San Diego, CA, pp. 1–16 (2013)

19. Hu, H., Han, W., Ahn, G.J., Zhao, Z.: FLOWGUARD: building robust firewalls for software-defined networks. In: HotSDN 2014, pp. 97–102. ACM, Chicago (2014)
20. Shirali-Shahreza, S., Ganjali, Y.: Flexam: flexible sampling extension for monitoring and security applications in OpenFlow. In: HotSDN 2013, Hong Kong, China, pp. 167–168. ACM, New York (2013)
21. Shirali-Shahreza, S., Ganjali, Y.: Efficient implementation of security applications in openflow controller with flexam. In: IEEE 21st Annual Symposium on High-Performance Interconnects, San Jose, CA, USA, pp. 49–54 (2013)
22. Collings, J., Liu, J.: An openflow-based prototype of SDN-oriented stateful hardware firewalls. In: IEEE 22nd International Conference on Network Protocols, Raleigh, NC, USA, pp. 525–528 (2014)
23. Zerkane, S., Espes, D., Le Parc, P., Cuppens, F.: Software defined networking reactive stateful firewall. In: Hoepman, J.-H., Katzenbeisser, S. (eds.) SEC 2016. IFIP AICT, vol. 471, pp. 119–132. Springer, Cham (2016). doi:10.1007/978-3-319-33630-5_9
24. Lar, S., Liao, X., ur Rehman, A., Ma, Q.: Proactive security mechanism and design. J. Inf. Secur. 2, 122–130 (2011)
25. Cabaj, K., Mazurczyk, W.: Using software-defined networking for ransomware mitigation: the case of cryptowall. IEEE Netw. Mag. Global Internetworking 30, 14–20 (2016)
26. Suh, M., Park, S.H., Lee, B., Yang, S.: Building firewall over the software-defined network controller. In: The 16th IEEE International Conference on Advanced Communications Technology, Daejeon, South Korea, pp. 744–748 (2014)
27. Pena, J.G., Yu, W.E.: Development of a distributed firewall using software defined networking technology. In: 4th IEEE International Conference on Information Science and Technology, Shenzhen, China, pp. 449–452 (2014)
28. Vasudevan, S.: Firewall a new approach to solve issues in software define networking. In: 6th International Conference on Emerging Trends in Engineering and Technology, pp. 14–19 (2016)
29. Konorski, J., Pacyna, P., Kolaczek, G., Kotulski, Z., Cabaj, K., Szalachowski, P.: A virtualization-level future internet defense-in-depth architecture. In: Thampi, S.M., Zomaya, A.Y., Strufe, T., Alcaraz Calero, J.M., Thomas, T. (eds.) SNDS 2012. CCIS, vol. 335, pp. 283–292. Springer, Heidelberg (2012). doi:10.1007/978-3-642-34135-9_29

Quantum Direct Communication Wiretapping

Piotr Zawadzki[(⊠)]

Institute of Electronics, Silesian University of Technology,
Akademicka 16, 44-100 Gliwice, Poland
Piotr.Zawadzki@polsl.pl

Abstract. The analyses of the ping-pong communication paradigm anticipated possibility to undetectably wiretap some variants of quantum direct communication. The only proposed so far instantiation of attacks of this type is formulated for the qubit based version of the protocol and it implicitly assumes the existence of losses. The essential features of undetectable attack transformations are identified in the study and the new generic eavesdropping scheme is proposed. The scheme does not refer to the properties of the vacuum state, so it is fully consistent with the absence of losses assumption. It is formulated for the space of any dimension and it can be used to design the family of circuits that enable undetectable eavesdropping.

Keywords: Ping-Pong protocol · Quantum direct communication

1 Introduction

The ping-pong communication paradigm is frequently used in quantum cryptography to realize the tasks impossible in the classical approach. The most notable areas of its application are quantum key distribution (QKD) and quantum direct communication (QDC) that aim at secure key agreement and confidential communication without encryption in open communication channels, respectively. The first applications of this quantum communication technique to QKD and QDC should be credited to Long et al. [1] and Deng et al. [2]. However, in these first proposals it was assumed that communicating parties possess long term quantum memories. Unfortunately, this strong requirement excluded their practical implementation. The first practically feasible QDC protocol that exploits the properties of the ping-pong communication paradigm applied to Einstein-Podolsky-Rosen (EPR) pairs has been proposed by Boström et al. [3,4]. Although the Boström's QDC protocol, frequently referred as the ping-pong protocol, is "only" quasi secure in perfect channels [5], it can be used as an engine for unconditionally secure QKD. The privacy amplification step applied to the data received from the QDC core protocol can reduce eavesdropper's knowledge on the final key to the arbitrary small value provided that his information gain is less than the mutual information of legitimate parties. Protocols of this type are referred as deterministic QKD and some of them have been recently experimentally demonstrated [6,7].

© Springer International Publishing AG 2017
P. Gaj et al. (Eds.): CN 2017, CCIS 718, pp. 287–294, 2017.
DOI: 10.1007/978-3-319-59767-6_23

Due to its conceptual simplicity, the idea of ping-pong communication paradigm is a subject of the ongoing research – it is adapted to higher dimensional systems [8–11] and/or modified to enhance capacity via dense coding [12,13]. Recently, Pavičić [14] proposed the attack that demonstrates inability to double the capacity of the protocol via dense coding. The attack targets only seminal version of the protocol utilizing qubits as information carriers. The custom control modes that detect the malevolent circuit have been proposed [15,16]. However, it was not known whether the circuit can be generalized to higher dimensions. In this contribution we propose such generalization.

The paper is organized as follows. In Sect. 2, we introduce notation and summarize the operation of the Pavičić's attack. Section 3 presents the main contribution. In particular, it is shown that Pavičić's circuit can be described as CNOT gate in the state space supplemented with the vacuum state. Then, on the basis of this observation, the attack is generalized to the qudit based systems. The impact of the obtained results is summarized in Sect. 4.

2 Analysis

Personification rules are used to simplify the description of the protocol – Alice and Bob are the names of the legitimate parties while the malevolent eavesdropper is referred as Eve. Communication protocol described below is a ping-pong paradigm variant analysed in [14]. Compared to the seminal version [3], it differs only in the encoding operation – the sender uses dense coding instead of phase flips and the remaining elements of the communication scenario are left intact.

The communication process is started by Bob. He creates EPR pair

$$|\Psi^-\rangle = (|0_h\rangle|1_t\rangle - |1_h\rangle|0_t\rangle) /\sqrt{2}. \tag{1}$$

and sends one of the qubits to Alice. The sent qubit is further referred as the signal/travel particle. Alice encodes two classic bits μ, ν using unitary transformation of the form $\mathcal{A}_{\mu,\nu} = \mathcal{X}^\mu \mathcal{Z}^\nu$ where $\mathcal{X} = |1\rangle\langle 0| + |0\rangle\langle 1|$, $\mathcal{Z} = |0\rangle\langle 0| - |1\rangle\langle 1|$ are bit-flip and phase-flip operations, respectively. After encoding, depending on the values of the information bits, the system state lands in one of the four EPR pairs

$$|\Psi_B\rangle = (|0_h\rangle|(1 \oplus \mu)_t\rangle - (-1)^\nu |1_h\rangle|(0 \oplus \mu)_t\rangle) /\sqrt{2}, \tag{2}$$

where $\oplus$ denotes summation modulo 2. Alice sends signal particle back to Bob, who detects applied transformation by collective measurement of both qubits.

Passive eavesdropping is impossible, but Eve can mount man in the middle (MITM) attack. As a countermeasure, Alice and Bob have to verify whether Alice received a genuine qubit. In some randomly selected protocol cycles, instead of the encoding step, she measures the travel qubit and signals that fact to Bob. Bob also measures the possessed/home qubit. Now, the legitimate parties are in position to verify the expected correlation of outcomes, which is preserved only if they share a genuine entangled state. They can use public classical channel for this purpose. This way Alice and Bob can convince themselves that the quantum channel is not spoofed with a confidence approaching certainty, provided that they have executed a sufficient number of control cycles.

An incoherent attack is yet another way of active information interception. The qubit that travels forth and back between legitimate parties is the subject of some quantum action $\mathcal{Q}$ introduced by Eve (Fig. 1). This malevolent activity can be described as unitary transformation in the space extended with two additional qubits. Consequently, Eve can implement the most generic attack by using ancilla system composed of two qubit registers.

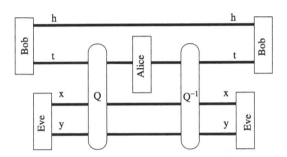

Fig. 1. The incoherent attack.

The circuit $\mathcal{P}$ from Fig. 2 has been proposed by Pavičić [14] as the quantum action $\mathcal{Q}$ able to detect Alice's bit-flip actions. It is composed of two Hadamard gates followed by the controlled polarization beam splitter ($CPBS$), which is a generalization of the polarization beam splitter (PBS) concept. The PBS is a two port gate that swaps horizontally polarized photons $|0_x\rangle$ ($|0_y\rangle$) entering its input to the other port $|0_y\rangle$ ($|0_x\rangle$) on output while vertically polarized ones $|1_x\rangle$ ($|1_y\rangle$) retain in their port $|1_x\rangle$ ($|1_y\rangle$) i.e.:

$$PBS|v_x\rangle|0_y\rangle = |0_x\rangle|v_y\rangle, \qquad\qquad PBS|v_x\rangle|1_y\rangle = |v_x\rangle|1_y\rangle, \qquad (3a)$$
$$PBS|0_x\rangle|v_y\rangle = |v_x\rangle|0_y\rangle, \qquad\qquad PBS|1_x\rangle|v_y\rangle = |1_x\rangle|v_y\rangle, \qquad (3b)$$

where $|v\rangle$ denotes the vacuum state. The $CPBS$ behaves as normal PBS if the control qubit is set to $|0_t\rangle$. The roles of horizontal and vertical polarization are exchanged for control qubit set to $|1_t\rangle$. The circuit $\mathcal{P}$ implements the following actions

$$\mathcal{P}_{txy}|0_t\rangle|\chi_0\rangle = |0_t\rangle|a_E\rangle, \qquad\qquad \mathcal{P}_{txy}|1_t\rangle|\chi_0\rangle = |1_t\rangle|d_E\rangle, \qquad (4a)$$
$$\mathcal{P}_{txy}|0_t\rangle|\chi_1\rangle = |0_t\rangle|d_E\rangle, \qquad\qquad \mathcal{P}_{txy}|1_t\rangle|\chi_1\rangle = |1_t\rangle|a_E\rangle, \qquad (4b)$$

where

$$|\chi_0\rangle = |v_x\rangle|0_y\rangle, \quad |\chi_1\rangle = |0_x\rangle|v_y\rangle$$

and

$$|a_E\rangle = \left(|0_x\rangle|v_y\rangle + |v_x\rangle|1_y\rangle\right)/\sqrt{2}, \quad |d_E\rangle = \left(|v_x\rangle|0_y\rangle + |1_x\rangle|v_y\rangle\right)/\sqrt{2}.$$

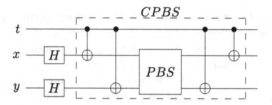

Fig. 2. The $\mathcal{P}$ circuit from [14].

Eve's subsystem is initially decoupled, so without loss of generality it may be assumed that the system is in the state

$$|\psi_{\text{init}}\rangle = |\Psi^-\rangle|\chi_0\rangle. \tag{5}$$

Under attack, the travel qubit, in its way to Alice, is entangled with the ancilla:

$$|\psi_{htE}\rangle = (|0_h\rangle|1_t\rangle|d_E\rangle - |1_h\rangle|0_t\rangle|a_E\rangle)/\sqrt{2}. \tag{6}$$

It is clear from (6) that attack does not introduce errors nor losses in control mode and the expected correlation of outcomes is preserved. Let us consider bit-flip encoding. It transforms the state of the system to

$$|\psi_{\text{bit}}\rangle = (\mathcal{I}_h \otimes \mathcal{X}_t)|\psi_{htE}\rangle = \frac{1}{\sqrt{2}}(|0_h\rangle|0_t\rangle|d_E\rangle - |1_h\rangle|1_t\rangle|a_E\rangle). \tag{7}$$

In its way back to Bob, the travel qubit is affected by the disentangling transformation $\mathcal{P}^{-1}$. It follows from (4), that

$$\mathcal{P}_{txy}^{-1}|0_t\rangle|d_E\rangle = |0_t\rangle|\chi_1\rangle, \quad \mathcal{P}_{txy}^{-1}|1_t\rangle|a_E\rangle = |1_t\rangle|\chi_1\rangle.$$

so the resulting state takes the form

$$|\phi_{\text{bit}}\rangle = \mathcal{P}_{txy}^{-1}|\psi_{\text{bit}}\rangle = \frac{|0_h\rangle|0_t\rangle - |1_h\rangle|1_t\rangle}{\sqrt{2}} \otimes |\chi_1\rangle = |\Phi^-\rangle \otimes |\chi_1\rangle. \tag{8}$$

Bob's decoding is limited to the ht space. His part of the system looks like as if there was no wiretapping device on the line and he indeed correctly detects Alice's bit-flip action. Eve's observations take place in xy space. She observes that the state $|\chi_0\rangle$ has been flipped to $|\chi_1\rangle$, so she is able to infer the value of bit μ from her observations. Similar analysis can be conducted for the phase flip encoding

$$|\phi_{\text{phase}}\rangle = \mathcal{P}_{txy}^{-1}(\mathcal{I}_h \otimes \mathcal{Z}_t)|\psi_{htE}\rangle = -|\Psi^+\rangle \otimes |\chi_0\rangle, \tag{9}$$

but this time Eve observes no change in her ancilla and Bob also correctly decodes phase flip operation. It is clear that $\mathcal{P}$ circuit enables detection of bit-flip operation without (a) disturbing expected correlation in the control mode, (b) introducing errors in transmission mode. In other words, it permits undetectable protocol wiretap.

3   Results

The in depth analysis of the $\mathcal{P}$-circuit operation reveals that properties described by expressions (4) determine the success of the attack: property (4a) provides entanglement undetectable by the control mode and property (4b) permits decoupling without disturbing information encoded by Alice. In fact, any map $\mathcal{Q}$ from Fig. 1 that satisfies

$$\mathcal{Q}|0_t\rangle|\chi_E\rangle \rightarrow |0_t\rangle|\alpha_E\rangle, \qquad\qquad \mathcal{Q}|1_t\rangle|\chi_E\rangle \rightarrow |1_t\rangle|\delta_E\rangle, \qquad (10\text{a})$$
$$\mathcal{Q}|0_t\rangle|\phi_E\rangle \rightarrow |0_t\rangle|\delta_E\rangle, \qquad\qquad \mathcal{Q}|1_t\rangle|\phi_E\rangle \rightarrow |1_t\rangle|\alpha_E\rangle \qquad (10\text{b})$$

can be used to attack the protocol, provided that probe states $|\chi_E\rangle$, $|\phi_E\rangle$ are perfectly distinguishable and $|\alpha_E\rangle \neq |\delta_E\rangle$ are some ancilla states. The operation of the protocol under attack is then described by

$$\left[\mathcal{Q}^{-1}\left(\mathcal{I}_h \otimes \mathcal{I}_t \otimes \mathcal{I}_E\right)\mathcal{Q}\right]|\Psi^-\rangle|\chi_E\rangle = |\Psi^-\rangle|\chi_E\rangle, \qquad (11\text{a})$$
$$\left[\mathcal{Q}^{-1}\left(\mathcal{I}_h \otimes \mathcal{X}_t \otimes \mathcal{I}_E\right)\mathcal{Q}\right]|\Psi^-\rangle|\chi_E\rangle = |\Phi^-\rangle|\phi_E\rangle, \qquad (11\text{b})$$
$$\left[\mathcal{Q}^{-1}\left(\mathcal{I}_h \otimes \mathcal{Z}_t \otimes \mathcal{I}_E\right)\mathcal{Q}\right]|\Psi^-\rangle|\chi_E\rangle = -|\Psi^+\rangle|\chi_E\rangle, \qquad (11\text{c})$$
$$\left[\mathcal{Q}^{-1}\left(\mathcal{I}_h \otimes \mathcal{X}_t\mathcal{Z}_t \otimes \mathcal{I}_E\right)\mathcal{Q}\right]|\Psi^-\rangle|\chi_E\rangle = -|\Phi^+\rangle|\phi_E\rangle. \qquad (11\text{d})$$

It follows from (11) that the registers used for communication are left untouched and decoupled but the Eve's ancilla state is flipped from $|\chi_E\rangle$ to $|\phi_E\rangle$ when Alice applies bit-flip operation. In consequence, Eve can successfully decode a half of the message content as long as the detection states $|\chi_E\rangle$, $|\phi_E\rangle$ are perfectly distinguishable. The map $\mathcal{Q}$ that satisfies (10) can be considered as a generalization of the $\mathcal{P}$-circuit.

The simpler equivalents of the $\mathcal{P}$-circuit can be found on a basis of the introduced generalization. Let us consider map $\mathcal{Q}$ of the form

$$\mathcal{Q}|0_t\rangle|0_x\rangle|0_y\rangle \rightarrow |0_t\rangle|0_x\rangle|0_y\rangle, \qquad\qquad \mathcal{Q}|1_t\rangle|0_x\rangle|0_y\rangle \rightarrow |1_t\rangle|1_x\rangle|0_y\rangle, \qquad (12\text{a})$$
$$\mathcal{Q}|0_t\rangle|1_x\rangle|0_y\rangle \rightarrow |0_t\rangle|1_x\rangle|0_y\rangle, \qquad\qquad \mathcal{Q}|1_t\rangle|1_x\rangle|0_y\rangle \rightarrow |1_t\rangle|0_x\rangle|0_y\rangle \qquad (12\text{b})$$

The map $\mathcal{Q}$ defined in (12) satisfies (10) in obvious way. But the map from (12) acts as $\mathcal{CNOT}$ gate with travel qubit on its control input and x register as a target. Such version of $\mathcal{Q}$ is also practically feasible as the attacks involving probes entangled via $\mathcal{CNOT}$ operation have been already proposed in QKD context [17,18]. As a result, both, $\mathcal{CNOT}$ gate and $\mathcal{P}$-circuit are equivalent in terms of provided information gain, detectability and practical feasibility. Consequently, in spite of the $\mathcal{P}$-circuit superficial otherness, there is no need to design control modes that address that circuit in a special way [16].

The explicit form of the $\mathcal{P}$-circuit from Fig. 2 does not provide hints how it can be generalized to protocols formulated for qutrits [9] or qudits [10]. However, the key properties identified in (10) can be easily transferred to systems of higher dimension. The conditions (10) can be rewritten in the form

$$\mathcal{Q}|k_t\rangle|\alpha_E^{(m)}\rangle \rightarrow |k_t\rangle|\delta_E^{(m\oplus k)}\rangle, \qquad k, m = 0, \ldots, D - 1 \qquad (13)$$

where $\oplus$ denotes addition modulo D – the dimension of the particles used for signalling and $|\alpha_E^{(k)}\rangle$ and $|\delta_E^{(k)}\rangle$ are the sets of D distinguishable states in the ancilla system. The inverse action can be written as

$$Q^{-1}|k_t\rangle|\delta_E^{(m)}\rangle \rightarrow |k_t\rangle|\alpha_E^{(m\ominus k)}\rangle, \qquad k, m = 0, \ldots, D-1. \tag{14}$$

The expressions (13) and (14) are in fact the main contribution of the paper – they define sufficient conditions of undetectable eavesdropping.

Let the system of legitimate parties and the ancilla be initially in the decoupled state

$$|\psi_{htE}\rangle = \left(\frac{1}{\sqrt{D}} \sum_{k=0}^{D-1} |k_h\rangle|k_t\rangle\right) \otimes |\alpha_E^{(0)}\rangle. \tag{15}$$

When the travel qudit arrives to Alice, the state of the system under attack is given as

$$Q|\psi_{htE}\rangle = \frac{1}{\sqrt{D}} \sum_{k=0}^{D-1} |k_h\rangle|k_t\rangle|\delta_E^{(k)}\rangle. \tag{16}$$

The correlation of the outcomes of the control measurements is still preserved. Alice uses the following generalizations of phase-flip and bit-flip operators

$$\mathcal{Z} = \sum_{k=0}^{D-1} \omega^k |k\rangle\langle k|, \qquad \mathcal{X} = \sum_{k=0}^{D-1} |k \oplus 1\rangle\langle k|, \qquad \omega = e^{j2\pi/D}. \tag{17}$$

This way she can encode two classic μ, ν "cdits" (i.e. symbols from $0, \ldots, D-1$ alphabet) in a single protocol transaction. Information encoding transforms the state of the system to

$$\mathcal{X}_t^\mu \mathcal{Z}_t^\nu Q|\psi_{htE}\rangle = \frac{1}{\sqrt{D}} \sum_{k=0}^{D-1} \omega^{k\nu}|k_h\rangle|(k \oplus \mu)_t\rangle|\delta_E^{(k)}\rangle. \tag{18}$$

In its way back to Bob, the travel qudit is disentangled from the ancilla according to the Formula (14)

$$Q^{-1}\mathcal{X}_t^\mu \mathcal{Z}_t^\nu Q|\psi_{htE}\rangle = \frac{1}{\sqrt{D}} \sum_{k=0}^{D-1} \omega^{k\nu}|k_h\rangle Q^{-1}|(k \oplus \mu)_t\rangle|\delta_E^{(k)}\rangle \tag{19}$$

$$= \left(\frac{1}{\sqrt{D}} \sum_{k=0}^{D-1} \omega^{k\nu}|k_h\rangle|(k \oplus \mu)_t\rangle\right) \otimes |\alpha_E^{(0\ominus\mu)}\rangle. \tag{20}$$

Let us note that expression in braces describes the state of signal particles that legitimate parties expect to find. Thus the introduced coupling is not detectable in the control mode nor it introduces errors in the transmission mode. On the other hand, Eve can unambiguously detect the value of the cdit μ provided that the states $|\alpha_E^{(k)}\rangle$ are properly selected.

4   Conclusion

A generic attack that provides undetectable eavesdropping of dense coded information in the ping-pong protocol is proposed. It can be considered as a generalization of the $\mathcal{P}$-circuit [14]. In contrast to $\mathcal{P}$-circuit, the introduced scheme does not refer to the vacuum state, so it can be applied to the protocol working under perfect quantum channel assumption. The provided analysis revealed that the trivial $\mathcal{CNOT}$ and quite complicated $\mathcal{P}$-circuit are equivalent in terms of provided information gain, detectability and practical feasibility. Consequently, there is no need to design control modes that address the $\mathcal{P}$-circuit in a special way [16] in spite of its superficial specificity. The identification of essential properties of the attack permitted its generalization to systems of any dimension. The existence of attacks with similar properties has been already forecast in relation to qubit [2], qutrit [9] and qudit [11] based protocol. However, no explicit form of the attack transformation has been given. Presented result can be considered as a constructive proof of their existence.

Acknowledgements. Author acknowledges support by the Ministry of Science and Higher Education funding for statutory activities and Rector of Silesian University of Technology grant number 02/030/RGJ17/0025 in the area of research and development.

References

1. Long, G.L., Liu, X.S.: Theoretically efficient high-capacity quantum-key-distribution scheme. Phys. Rev. A **65**, 032302 (2002)
2. Deng, F.G., Long, G.L., Liu, X.S.: Two-step quantum direct communication protocol using the Einstein-Podolsky-Rosen pair block. Phys. Rev. A **68**, 042317 (2003)
3. Boström, K., Felbinger, T.: Deterministic secure direct communication using entanglement. Phys. Rev. Lett. **89**(18), 187902 (2002)
4. Ostermeyer, M., Walenta, N.: On the implementation of a deterministic secure coding protocol using polarization entangled photons. Opt. Commun. **281**(17), 4540–4544 (2008)
5. Boström, K., Felbinger, T.: On the security of the ping-pong protocol. Phys. Lett. A **372**(22), 3953–3956 (2008)
6. Cerè, A., Lucamarini, M., Di Giuseppe, G., Tombesi, P.: Experimental test of two-way quantum key distribution in the presence of controlled noise. Phys. Rev. Lett. **96**, 200501 (2006)
7. Chen, H., Zhou, Z.Y., Zangana, A.J.J., Yin, Z.Q., Wu, J., Han, Y.G., Wang, S., Li, H.W., He, D.Y., Tawfeeq, S.K., Shi, B.S., Guo, G.C., Chen, W., Han, Z.F.: Experimental demonstration on the deterministic quantum key distribution based on entangled photons. Sci. Rep. **6**, 20962 (2016)
8. Wang, C., Deng, F.G., Long, G.L.: Multi-step quantum secure direct communication using multi-particle Green-Horne-Zeilinger state. Opt. Commun. **253**(13), 15–20 (2005)
9. Vasiliu, E.V.: Non-coherent attack on the ping-pong protocol with completely entangled pairs of qutrits. Quant. Inf. Process. **10**(2), 189–202 (2010)
10. Zawadzki, P.: Security of Ping-Pong protocol based on pairs of completely entangled qudits. Quant. Inf. Process. **11**(6), 1419–1430 (2012)

11. Zawadzki, P., Puchała, Z., Miszczak, J.: Increasing the security of the ping-pong protocol by using many mutually unbiased bases. Quant. Inf. Process. **12**(1), 569–575 (2013)

12. Cai, Q., Li, B.: Improving the capacity of the Boström-Felbinger protocol. Phys. Rev. A **69**, 054301 (2004)

13. Wang, C., Deng, F.G., Li, Y.S., Liu, X.S., Long, G.L.: Quantum secure direct communication with high-dimension quantum superdense coding. Phys. Rev. A **71**, 044305 (2005)

14. Pavičić, M.: In quantum direct communication an undetectable eavesdropper can always tell ψ from ϕ Bell states in the message mode. Phys. Rev. A **87**, 042326 (2013)

15. Zawadzki, P.: An improved control mode for the ping-pong protocol operation in imperfect quantum channels. Quant. Inf. Process. **14**(7), 2589–2598 (2015)

16. Zhang, B., Shi, W.X., Wang, J., Tang, C.J.: Quantum direct communication protocol strengthening against Pavičić's attack. Int. J. Quant. Inf. **13**(07), 1550052 (2015)

17. Brandt, H.E.: Entangled eavesdropping in quantum key distribution. J. Mod. Opt. **53**(16–17), 2251–2257 (2006)

18. Shapiro, J.H.: Performance analysis for Brandts conclusive entangling probe. Quant. Inf. Process. **5**(1), 11–24 (2006)

A Qutrit Switch for Quantum Networks

Joanna Wiśniewska[1(✉)] and Marek Sawerwain[2]

[1] Institute of Information Systems, Faculty of Cybernetics,
Military University of Technology, Kaliskiego 2, 00-908 Warsaw, Poland
jwisniewska@wat.edu.pl
[2] Institute of Control and Computation Engineering,
University of Zielona Góra, Licealna 9, Zielona Góra 65-417, Poland
M.Sawerwain@issi.uz.zgora.pl

Abstract. The chapter contains information about a construction of quantum switch for qutrits. It is an expansion of the idea of quantum switch for qubits. For the networks based on quantum effects the switch could realize an operation of changing the direction of transferred information. A presented definition of quantum switch may be easily generalized to qudits. The chapter contains also the figures representing circuits realizing the quantum switch and a proof of its correctness. Additionally, there is an analysis of entanglement level in the circuit which also characterizes the correct operating of the switch.

Keywords: Quantum information transfer · Quantum switch · Qutrits

1 Introduction

The features and properties of quantum information – e.g. biological [10] and quantum model [8,9], non-cloning theorem – cause that some operations used nowadays as parts of the algorithms or performed in computer networks need new definitions. One of these issues is a switch swapping the signals in an electronic circuit.

In [16] a definition and construction of quantum switch operating on qubits were presented. The constant development and research of the quantum computing covers also the issue of information units with freedom level greater than two [5,6,11]. That causes the need to define a quantum switch for qutrits and generally for qudits, which could be used in the computer networks of the future, but also in processors and in logic arrays of quantum circuits [12,13].

In this chapter we discus one of the possible realizations of quantum switch for qutrits, i.e. qudits with three levels of freedom. Just like previously known switch for qubits, the quantum switch for qutrits swaps the information between inputs A and B when the third, so-called, control signal is $|2\rangle$. When the control signals are equal to $|0\rangle$ or $|1\rangle$ then the operation of swapping is not performed. Moreover, the problem of discrete-time switched dynamical system [1] is still discussed in many areas of application for quantum and classical systems.

© Springer International Publishing AG 2017
P. Gaj et al. (Eds.): CN 2017, CCIS 718, pp. 295–304, 2017.
DOI: 10.1007/978-3-319-59767-6_24

The chapter's reminder is organized in the following way: in Sect. 2 we introduce the basic definitions of gates used in the construction of quantum switch for qutrits. Section 3 contains the circuits implementing the switch and a proof of their correct operating. In Sect. 4 the evolution of entanglement's level for exemplary quantum states is presented. The values were calculated for each computational step during the operating process of the switch.

A summary and a draft of further works were presented in Sect. 5. The last element of the chapter is a references section.

2 Preliminary Definitions

In this section we present some basic information and definitions concerning the quantum computing. The more detailed introduction may be found in many textbooks [4,15].

One of the most important notions in quantum computing is a qubit which is a normalized vector in Hilbert space $\mathcal{H}_2$. The definition of qubit, more precisely the definition of qubit's state expressed in Dirac notation as $|\psi\rangle$, is a superposition of two vectors:

$$|\psi\rangle = \alpha|0\rangle + \beta|1\rangle, \tag{1}$$

where $\alpha, \beta \in \mathcal{C}$ and $|\alpha_0|^2 + |\alpha_1|^2 = 1$. The normalized vectors $|0\rangle, |1\rangle$ may be presented as:

$$|0\rangle = \begin{bmatrix} 1 \\ 0 \end{bmatrix}, \quad |1\rangle = \begin{bmatrix} 0 \\ 1 \end{bmatrix}. \tag{2}$$

The vectors $|0\rangle, |1\rangle$ constitute so-called standard computational base, but a computational base for qubits may be set on any two orthonormal vectors.

However, in this chapter we use a qudit as a unit of quantum information. A qudit is a generalization of qubit. The qudit state $|\phi\rangle$ we express as a superposition of d orthonormal vectors:

$$|\phi\rangle = \sum_{i=0}^{d-1} \alpha_i|i\rangle, \tag{3}$$

where $\sum_{i=0}^{d-1} |\alpha_i|^2 = 1$ and $|i\rangle$ denotes the i-th state of a chosen computational base.

Just like for classical bits the qudits (e.g. qubits) may be joined into so-called quantum register. The n-qudit state/register $|\Psi\rangle$ may be presented as a tensor product:

$$|\Psi\rangle = |\psi_0\rangle \otimes |\psi_1\rangle \otimes |\psi_2\rangle \otimes ... \otimes |\psi_{n-1}\rangle. \tag{4}$$

Usually the quantum register consists of qudits sharing the same freedom level d. However, there are also known the solutions like the quantum version of k-nearest neighbours algorithm [17] where in one quantum register qubits and qudits are used.

Remark 1. It is a very important issue that not all quantum states can be expressed with use of a tensor product. If such a decomposition is not possible then the state is entangled. The quantum entanglement is an extremely important element of many algorithms and protocols, e.g. quantum teleportation [3], quantum communications [14].

The evolution of quantum register's state, in so-called quantum circuit model, is performed with use of unitary operations and the operation of measurement.

Remark 2. To solve problems presented in this chapter the operation of measurement is not required, so we omit the details concerning this issue, but the further information can be found in quantum computing textbooks, e.g. [15].

The unitary operations are realized with use of quantum gates which may be classified as 1-qubit/1-qudit gates and n-qubit/n-qudit gates where $(n > 1)$. The 1-qubit quantum gates X, Y and Z are termed as Pauli gates. The controlled negation gate $(CNOT)$ is an example of 2-qubit gate.

An operation performed by a 1-qubit gate X (called also a NOT gate) can be expressed as:

$$X|\psi\rangle = X(\alpha|0\rangle + \beta|1\rangle) = \beta|0\rangle + \alpha|1\rangle = |\psi'\rangle,$$
$$X(\psi') = X(\beta|0\rangle + \alpha|1\rangle) = \alpha|0\rangle + \beta|1\rangle = |\psi\rangle. \tag{5}$$

The X gate swaps the values of α and β, so the double application of this gate allows to return to the initial state of the qubit $|\psi\rangle$. For qudits, when the basic states are defined with value d, the X gate performs:

$$X|j\rangle = |j \overset{d}{\oplus} 1\rangle, \tag{6}$$

where $j \overset{d}{\oplus} 1$ stands for the operation $(j + 1) \mod d$.

The circuit performing the operation of quantum switch needs the $CNOT$ gate. Formally the operation realized by the $CNOT$ gate is expressed as:

$$|ab\rangle = |a, a \oplus b\rangle, \tag{7}$$

where $\oplus$ stands for addition modulo 2. This operation for qudits is the addition modulo d.

Remark 3. The generalized $CNOT$ gate is not self-adjoint: $CNOT \cdot CNOT^\dagger \neq I$ where † denotes the Hermitian adjoint and I is an identity matrix. It means that the operation of $SWAP$ cannot be realized for qudits with use of three $CNOT$ gates (see Fig. 1).

Remark 3 implies that it is necessary to present a new $CNOT$-like gate. This gate should allow to build the $SWAP$ gate in a similar way as we can do it for qubits. In [7] a gate $C\tilde{X}$ is described. This gate is self-adjoint and the $SWAP$ gate can be built with use of the $C\tilde{X}$ gate like in Fig. 1.

An operation performed by the $C\tilde{X}$ gate on qudits $|a\rangle, |b\rangle$ expressed in the standard computation base is

$$C\tilde{X}|ab\rangle = |a, -a \overset{d}{\ominus} b\rangle,\tag{8}$$

where $-a \overset{d}{\ominus} b = (-a - b) \mod d$.

Remark 4. For $d = 2$ the gate $C\tilde{X}$ is an equivalent of $CNOT$ gate because $(-a - b) \mod 2 = a + b \mod 2$.

(a)   (b)   (c)

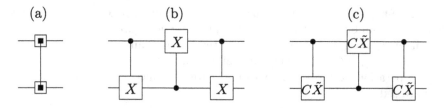

Fig. 1. The $SWAP$ gate and the circuits (see also paper [2]) implementing it for qubits and qudits. The graphical representation of the $SWAP$ gate used in this chapter is presented as a subfigure (a), its implementation is given in (b) and the $SWAP$ gate for qudits with the $C\tilde{X}$ gate is presented as a subfigure (c)

The construction of quantum switch needs also a gate with two control qubits. The mentioned gate is called Toffoli gate. For qubit states Toffoli gate performs the following operation:

$$T|abc\rangle = |ab(ab \oplus c)\rangle.\tag{9}$$

Generalization of this gate needs of course using an addition modulo d:

$$T^Q|abc\rangle = |ab((ab + c) \mod d)\rangle.\tag{10}$$

However, in the definition of quantum switch an additional gate $CC\tilde{X}$ is used. The $CC\tilde{X}$ gate needs two control signals $|a\rangle$ and $|b\rangle$:

$$CC\tilde{X}|abc\rangle = |ab((-a \cdot -b - c) \mod d)\rangle.\tag{11}$$

Remark 5. The $CC\tilde{X}$ gate is not an equivalent to T^Q gate. There are some configurations of control lines (signals) where these two gates work in the same way, e.g. for control signals with the first qutrit equal to $|0\rangle$. In the other cases, when the first qutrit is in the state $|1\rangle$ or $|2\rangle$, mentioned gates operate differently.

3 Quantum Switch for Qutrits

In [16] a quantum switch is defined as a 3-qubit, so-called, Controlled-$SWAP$ gate which swaps the states of first two qubits according to state of the third qubit. It means that we can divide the operations performed by the quantum switch into two cases.

Let us denote the state of the first qubit as $|A\rangle$ and the state of the second qubit as $|B\rangle$. In the first case the state of the third qubit is $|0\rangle$:

$$|A\rangle|B\rangle|0\rangle \Rightarrow |A\rangle|B\rangle|0\rangle. \tag{12}$$

As we can see in this case the quantum switch does not affect the quantum state. In the second case the state of the third qubit is $|1\rangle$:

$$|A\rangle|B\rangle|1\rangle \Rightarrow |B\rangle|A\rangle|1\rangle. \tag{13}$$

Now the states of first two qubits are swapped and this action is the aim of quantum switch's work. Because of the presence of the third qubit, taking state zero or one, it may be stated that the quantum switch is controlled with classical Boolean values.

Utilizing the $CNOT$ and Toffoli gate, defined in Sect. 2, we are able to specify the circuits realizing the quantum switch. The examples of mentioned circuits, described in [16], are shown at Fig. 2.

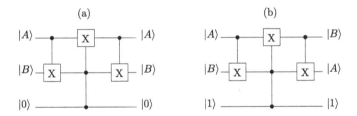

Fig. 2. The circuits realizing the operation of quantum switch for qubits. If the control state is $|0\rangle$ (case (a)) the switch does not change the order of first two input states. When the control state is expressed as $|1\rangle$ (case (b)) the quantum switch swaps the input states

The quantum switch for qudits may be defined similarly like the switch for qubits. Let us assume that $|A\rangle$ and $|B\rangle$ represent qudit states

$$|A\rangle|B\rangle|n\rangle \Rightarrow |A\rangle|B\rangle|n\rangle, \tag{14}$$

and $n = 0, 1, 2, 3, \ldots, d-2$.

The states $|A\rangle$ and $|B\rangle$ will be swapped when the state of the third qudit equals $|d-1\rangle$:

$$|A\rangle|B\rangle|d-1\rangle \Rightarrow |B\rangle|A\rangle|d-1\rangle. \tag{15}$$

Unfortunately, utilizing the gates $C\tilde{X}$ and $CC\tilde{X}$ is not sufficient to obtain a correctly working switch. If the state of control qudit is $|d-1\rangle$ then the switch swaps the states of first two qudits. In the other cases an error in the probability amplitudes distribution will be introduced into the system because of the construction of $C\tilde{X}$ and $CC\tilde{X}$ gates.

To avoid the mentioned above situation the error correction is necessary when the state of control qudit is $|n\rangle$. The appropriate gate have to be added to the circuit realizing quantum switch. Figure 3 depicts the circuit for the generalized switch. The gate marked as CB performs the correction only if the state of the first qudit is not $|d-1\rangle$.

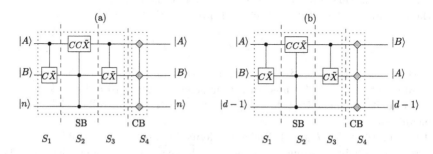

Fig. 3. The circuits realizing the operation of quantum switch for qudits. If the control state is $|n\rangle$ where $n = 0, 1, 2, 3, \ldots, d-2$ the switch does not change the order of first two input states. When the control state is expressed as $|d-1\rangle$ the quantum switch swaps the input states. The block SB denotes the swap operations and CB represents the gate for rearranging amplitudes in subspaces where the state of the third qudit is different than $|d-1\rangle$

3.1   Correctness of Qutrit Switch Gate

In this section a proof of algebraic correctness is presented. The proof concerns the circuit realizing a quantum switch for qutrits.

It is assumed that there are three qutrits. The states of $|A\rangle$ and $|B\rangle$ are unknown:

$$|A\rangle = \alpha_0|0\rangle + \beta_0|1\rangle + \gamma_0|2\rangle, \quad |B\rangle = \alpha_1|0\rangle + \beta_1|1\rangle + \gamma_1|2\rangle. \tag{16}$$

The third qutrit works as a control state and the only values that may be assigned to it are: $|0\rangle$, $|1\rangle$ or $|2\rangle$.

The first case to analyze is when the control qutrit is in the state $|2\rangle$. The initial state then may be expressed as:

$$\begin{aligned}
|\psi_0^2\rangle = (|A\rangle \otimes |B\rangle \otimes |2\rangle)) = {} & \alpha_0\alpha_1|002\rangle + \alpha_0\beta_1|012\rangle + \alpha_0\gamma_1|022\rangle \\
& + \alpha_1\beta_0|102\rangle + \beta_0\beta_1|112\rangle + \beta_0\gamma_1|122\rangle \\
& + \alpha_1\gamma_0|202\rangle + \beta_1\gamma_0|212\rangle + \gamma_0\gamma_1|222\rangle,
\end{aligned} \tag{17}$$

in the denotation of state $|\psi_0^2\rangle$ the superscript refers to the control state while the subscript describes the number of the computational step S_i (zero stands for the initial state).

Performing the computational step S_1, that is the $C\tilde{X}$ operation, when the second qutrit is the control signal, the following state will be obtained:

$$
\begin{aligned}
|\psi_1^2\rangle = {} & \alpha_0\alpha_1|002\rangle + \alpha_0\gamma_1|012\rangle + \alpha_0\beta_1|022\rangle \\
& + \beta_0\gamma_1|102\rangle + \beta_0\beta_1|112\rangle + \alpha_1\beta_0|122\rangle \\
& + \beta_1\gamma_0|202\rangle + \alpha_1\gamma_0|212\rangle + \gamma_0\gamma_1|222\rangle.
\end{aligned}
\tag{18}
$$

The step S_2 causes the reorganization of the probability amplitudes distribution:

$$
\begin{aligned}
|\psi_2^2\rangle = {} & \alpha_0\alpha_1|002\rangle + \alpha_0\gamma_0|012\rangle + \alpha_0\beta_0|022\rangle \\
& + \beta_1\gamma_0|102\rangle + \beta_0\beta_1|112\rangle + \alpha_0\beta_1|122\rangle \\
& + \beta_0\gamma_1|202\rangle + \alpha_0\gamma_1|212\rangle + \gamma_0\gamma_1|222\rangle.
\end{aligned}
\tag{19}
$$

After the step S_3 we obtain the correct final state for this case:

$$
\begin{aligned}
|\psi_3^2\rangle = {} & \alpha_0\alpha_1|002\rangle + \alpha_1\beta_0|012\rangle + \alpha_1\gamma_0|022\rangle \\
& + \alpha_0\beta_1|102\rangle + \beta_0\beta_1|112\rangle + \beta_1\gamma_0|122\rangle \\
& + \alpha_0\gamma_1|202\rangle + \beta_0\gamma_1|212\rangle + \gamma_0\gamma_1|222\rangle = |B\rangle \otimes |A\rangle \otimes |2\rangle.
\end{aligned}
\tag{20}
$$

The step S_4 does not change the values of amplitudes. The error correction operation is needed when the control qutrit equals $|0\rangle$ or $|1\rangle$ (in these two cases we expect the final state to be, respectively, $|AB0\rangle$ or $|AB1\rangle$).

When the control qutrit is $|0\rangle$ then the state of the system after the step S_3 is:

$$
\begin{aligned}
|\psi_3^0\rangle = {} & \alpha_0\alpha_1|000\rangle + \alpha_0\beta_1|010\rangle + \alpha_0\gamma_1|020\rangle \\
& + \gamma_0\gamma_1|100\rangle + \alpha_1\gamma_0|110\rangle + \beta_1\gamma_0|120\rangle \\
& + \beta_0\beta_1|200\rangle + \beta_0\gamma_1|210\rangle + \alpha_1\beta_0|220\rangle
\end{aligned}
\tag{21}
$$

The correct final state in this case we will obtain after the change of amplitudes' distribution in the computational step S_4:

$$
\begin{aligned}
|\psi_4^0\rangle = {} & \alpha_0\alpha_1|000\rangle + \alpha_0\beta_1|010\rangle + \alpha_0\gamma_1|020\rangle \\
& + \gamma_0\gamma_1|100\rangle + \alpha_1\gamma_0|110\rangle + \beta_1\gamma_0|120\rangle \\
& + \beta_0\beta_1|200\rangle + \beta_0\gamma_1|210\rangle + \alpha_1\beta_0|220\rangle
\end{aligned}
\tag{22}
$$

For the control qutrit $|1\rangle$ the state of the system after the step S_3 is:

$$
\begin{aligned}
|\psi_3^1\rangle = {} & \alpha_0\alpha_1|001\rangle + \gamma_0\gamma_1|011\rangle + \beta_0\beta_1|021\rangle \\
& + \alpha_1\beta_0|101\rangle + \alpha_0\gamma_1|111\rangle + \beta_1\gamma_0|121\rangle \\
& + \alpha_1\gamma_0|201\rangle + \beta_0\gamma_1|211\rangle + \alpha_0\beta_1|221\rangle
\end{aligned}
\tag{23}
$$

Just like before the error correction is needed in the computational step S_4 to obtain the final state:

$$\begin{aligned}
|\psi_4^1\rangle = {} & \alpha_0\alpha_1|001\rangle + \alpha_0\beta_1|011\rangle + \alpha_0\gamma_1|021\rangle \\
& + \alpha_1\beta_0|101\rangle + \beta_0\beta_1|111\rangle + \beta_0\gamma_1|121\rangle \\
& + \alpha_1\gamma_0|201\rangle + \beta_1\gamma_0|211\rangle + \gamma_0\gamma_1|221\rangle.
\end{aligned} \tag{24}$$

Remark 6. The control qutrit (generally the control qudit) accepts the classical states, so the switch may be called a quantum switch with the classical control.

4   The Level of Entanglement During Information Switching

The correct operating of quantum switch can be evaluated by the level of entanglement. In the computational steps S_1 and S_2 the $CNOT$ gate was used, so it is expected to cause the entanglement in the system. The level of this entanglement should be decreased after the step S_3 where the initial states are to be swapped. If the swapping is completed, the level of entanglement is equal to zero. Performing the step S_4 does not affect the level of entanglement.

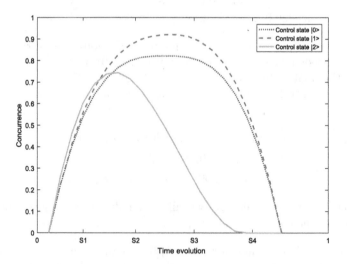

Fig. 4. The values of entanglement level during the work of quantum switch for qutrits. The results were obtained with use of Concurrence measure after every computational step: S_1, S_2, S_3, S_4 (the other values are interpolated). Three cases were analyzed respecting the value of control qutrit: $|2\rangle$, $|1\rangle$, $|0\rangle$

If the signals were not swapped then the level of entanglement does not decrease after the step S_3. In this case we will not obtain the state $|AB0\rangle$ neither

$|AB1\rangle$. However, after the error correction in step S_4 the level of entanglement should be equal to zero.

To evaluate the level of entanglement the Concurrence measure is used:

$$C(|AB\rangle) = \sqrt{2\left(1 - \mathrm{Tr}\left(\sigma_A^2\right)\right)}, \tag{25}$$

where σ_A^2 is a square of the state matrix after the operation of partial trace and after the rejection of $|B\rangle$ state and the state of control qutrit.

Figure 4 presents the values of entanglement level calculated with use of the Concurrence measure. The measure was utilized only for states obtained after the steps S_1, S_2, S_3 and S_4. The other values are interpolated. Despite the simplification, introduced by the interpolation, it may be noticed that the level of entanglement changes with the following computational steps. If the states are swapped the entanglement decreases to zero after the step S_3.

The chart refers to the switch performing in two states:

$$|A\rangle = \frac{2}{3}|0\rangle + \frac{1}{3}|2\rangle, \quad |B\rangle = \frac{3}{5}|0\rangle + \frac{2}{5}|1\rangle. \tag{26}$$

Naturally, before the first computational step S_1 the value of entanglement is zero because the input state for the switch is a fully separable state.

5   Conclusions

In this chapter we presented the quantum switch for qutrits. The switch is controlled by the state of the third qutrit. The first two states are swapped when the state of the third qudit is $|2\rangle$. The proposed solution is similar to the switch for qubits. Our aim was to define a switch which imitates the behavior of previously known solution, but works for qutrits. The presented circuit and its correctness (see: Sect. 3.1) allow to conclude that it is possible to build such a switch. However, the construction of qubit switch using $CNOT$ and Toffoli gates is not open for the simple generalization for qudits. As it is presented at Fig. 3, the circuit realizing the quantum switch for qutrits needs the $CC\tilde{X}$ gate which is not equivalent to Toffoli gate according to Remark 5.

Acknowledgements. We would like to thank for useful discussions with the Q-INFO group at the Institute of Control and Computation Engineering (ISSI) of the University of Zielona Góra, Poland. We would like also to thank to anonymous referees for useful comments on the preliminary version of this paper. The numerical results were done using the hardware and software available at the "GPU μ-Lab" located at the Institute of Control and Computation Engineering of the University of Zielona Góra, Poland.

References

1. Babiarz, A., Czornik, A., Klamka, J., Niezabitowski, M.: The selected problems of controllability of discrete-time switched linear systems with constrained switching rule. Bull. Pol. Acad. Sci. Tech. Sci. **63**(3), 657–666 (2015). doi:10.1515/bpasts-2015-0077

2. Balakrishnan, S.: Various constructions of Qudit SWAP Gate. Phys. Res. Int. **2014**, Article ID 479320 (2014)
3. Bennett, C.H., Brassard, G., Crepeau, C., Jozsa, R., Peres, A., Wootters, W.: Teleporting an unknown quantum state via dual classical and EPR channels. Phys. Rev. Lett. **70**, 1895–1899 (1993)
4. Brüning, E., Petruccione, F.: Theoretical Foundations of Quantum Information Processing and Communication. Springer, Berlin (2010)
5. Daboul, J., Wang, X., Sanders, B.C.: Quantum gates on hybrid qudits. J. Phys. A: Math. Gen. **36**(10), 2525–2536 (2003)
6. Di, Y.M., Wei, H.R.: Synthesis of multivalued quantum logic circuits by elementary gates. Phys. Rev. A **87**, 012325 (2013)
7. Garcia-Escartin, J.C., Chamorro-Posada, P.: A SWAP gate for qudits. Quant. Inf. Process. **12**(12), 3625–3631 (2013)
8. Hayashi, M., Ishizaka, S., Kawachi, A., Kimura, G., Ogawa, T.: Introduction to Quantum Information Science. Springer, Berlin (2015)
9. Klamka, J., Gawron, P., Miszczak, J., Winiarczyk, R.: Structural programming in quantum octave. Bull. Pol. Acad. Sci. Tech. Sci. **58**(1), 77–88 (2010)
10. Kuppusamy, L., Mahendran, A.: Modelling DNA and RNA secondary structures using matrix insertion-deletion systems. Int. J. Appl. Math. Comput. Sci. **26**(1), 245–258 (2016)
11. Landau, A., Aharonov, Y., Cohen, E.: Realisation of Qudits in coupled potential wells. Int. J. Quant. Inf. **14**(5), 1650029 (2016)
12. Metodi, T.S., Thaker, D.D., Cross, A.W., Chong, F.T., Chuang, I.L.: A quantum logic array microarchitecture: scalable quantum data movement and computation. In: Proceedings of 38th Annual IEEE/ACM International Symposium on Microarchitecture, MICRO-38, p. 12 (2005)
13. Mohammadi, M.: Radix-independent, efficient arrays for multi-level n-qudit quantum and reversible computation. Quant. Inf. Process. **14**, 2819–2832 (2015)
14. Muralidharan, S., Li, L., Kim, J., Lütkenhaus, N., Lukin, M.D., Jiang, L.: Optimal architectures for long distance quantum communication. Sci. Rep. **6**, Article no. 20463 (2016)
15. Nielsen, M.A., Chuang, I.L.: Quantum Computation and Quantum Information. Cambridge University Press, New York (2000)
16. Ratan, R., Shukla, M.K., Oruc, A.Y.: Quantum switching networks with classical routing. In: 2007 41st Annual Conference on Information Sciences and Systems, Baltimore, MD, pp. 789–793 (2007)
17. Schuld, M., Sinayskiy, I., Petruccione, F.: Quantum computing for pattern classification. In: Pham, D.-N., Park, S.-B. (eds.) PRICAI 2014. LNCS (LNAI), vol. 8862, pp. 208–220. Springer, Cham (2014). doi:10.1007/978-3-319-13560-1_17

SLA Life Cycle Automation and Management for Cloud Services

Waheed Aslam Ghumman and Alexander Schill[(✉)]

Technische Universität Dresden, Dresden, Germany
{Waheed-Aslam.Ghumman,Alexander.Schill}@tu-dresden.de

Abstract. Cloud service providers mostly offer service level agreements (SLAs) in descriptive format which is not directly consumable by a machine or a system. Manual management of SLAs with growing usage of cloud services can be a challenging, erroneous and tedious task especially for the cloud service users (CSUs) acquiring multiple cloud services. The necessity of automating the complete SLA life cycle (which includes SLA description in machine readable format, negotiation, monitoring and management) becomes imminent due to complex requirements for the precise measurement of quality of service (QoS) parameters. In this work, the complete SLA life cycle management is presented using an extended SLA specification to support multiple CSU locations. A time efficient SLA negotiation technique is integrated with the extended SLA specification for concurrently negotiating with multiple cloud service providers (CSPs). After a successful negotiation process, the next major task in the SLA life cycle is to monitor the cloud services for ensuring the quality of service according to the agreed SLA. A distributed monitoring approach for the cloud SLAs is elaborated, in this work, which is suitable for services being used at single or multiple locations. The discussed monitoring approach reduces the number of communications of SLA violations to a monitoring coordinator by eliminating the unnecessary communications. The presented work on the complete SLA life cycle automation is evaluated and validated with the help of experiments and simulations.

1 Introduction

The quality of a cloud service is parameterized by a service level agreement (SLA) between a cloud service user (CSU) and a cloud service provider (CSP). An agreement (SLA) between a CSU and a CSP is finalized by following different steps of the SLA life cycle, i.e. definition of business objectives, transforming the business objectives to service definitions, discovering the appropriate service providers and negotiating with them over the quality of service (QoS) parameters. Subsequently, monitoring and management of the final SLA are important phases of the SLA life cycle. The decision of choosing between a traditional solution and acquisition of a cloud service as a better replacement, is influenced by budget constraints, financial benefits, performance expectations, expeditious availability and rapid elasticity of resources. These utilities of a cloud service over the

© Springer International Publishing AG 2017
P. Gaj et al. (Eds.): CN 2017, CCIS 718, pp. 305–318, 2017.
DOI: 10.1007/978-3-319-59767-6_25

traditional solution may diminish due to inadequacy of the automation processes
in different phases of the SLA life cycle (i.e., definition, negotiation, monitoring
and management). For instance, manually negotiating with multiple cloud service providers may consume a significant amount of time and may compel costly
delays. Cloud service providers generally offer SLAs in descriptive/natural language format which is not directly consumable by a machine/system. A cloud
service user is conventionally responsible itself to monitor and enforce a natural language based SLA by first manually transforming the SLA details into
a suitable machine readable format. In this paper, the complete SLA life cycle
management is presented which is based on different automation components
that are joined collectively. The rest of the paper is organized as given in the
following. Section 2 describes a specification for the cloud SLAs as a basic structure to describe the SLAs into a machine readable format. Specification for the
distributed SLAs (deployed on more then one locations) is also presented in
Sect. 2 which extends the SLA specification presented in [5]. A time-efficient and
automated negotiation technique for multiple providers, along with a suitable
negotiation protocol, is elaborated in Sect. 3. Section 4 describes a distributed
monitoring approach for cloud SLAs. Section 5 describes the overview of the
complete framework. Results and analysis of different experiments are explained
in Sect. 6. A comparison with related work is described in Sect. 7. In the end,
conclusions and future directions are summarized in Sect. 8.

2   SLA Specification

In this section, a brief description of the "Structural Specification for the SLAs
in Cloud Computing (S3LACC)" [5] is given and also extended (as S3LACC+)
to support distributed SLAs. The S3LACC is a basic structure to define cloud
SLAs in a machine readable format.

An SLA contains one or more service level objectives (SLOs), e.g. service
availability, throughput or response time. Each SLO is measured by one or more
metrics, e.g. service availability SLO may contain metrics such as availability
percentage, maximum number of outages per month and maximum duration
per outage. S3LACC combines the SLA template parameters, negotiation, monitoring and management parameters in a single structure. The basic structure
of S3LACC is based on the cloud SLAs specific standards and guidelines.[1,2]
S3LACC classifies the fundamental components (service level objectives and
metrics) of the cloud SLAs in a hierarchical way which makes the extension
of its basic structure an uncomplicated task. A priority level can be assigned
to an SLO by changing its Weight. A metric can be a qualitative (e.g. service
reliability) or a quantitative (e.g. service availability) metric and both types of

---

[1] NIST Special publication 500-307, available online at: http://www.nist.gov/itl/
cloud/upload/RATAX-CloudServiceMetricsDescription-DRAFT-20141111.pdf.

[2] Cloud service level agreement standardization guidelines by European Commission (2014). Available online at: https://ec.europa.eu/digital-single-market/news/
cloud-service-level-agreement-standardisation-guidelines.

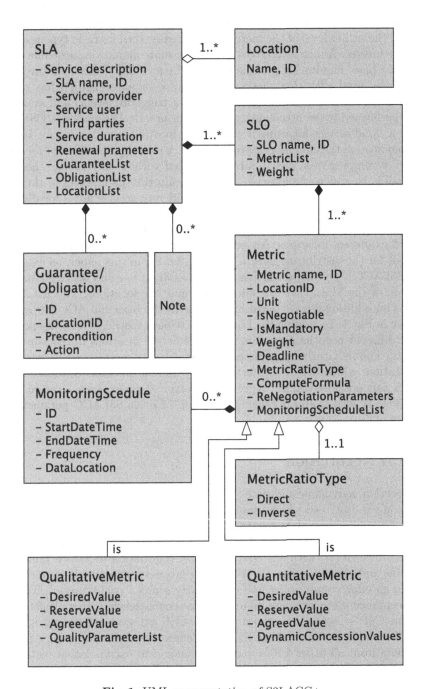

Fig. 1. UML representation of S3LACC+

metrics contain different types of values (descriptive and numeric, respectively). A metric can have a direct or inverse ratio type, i.e. if by increasing the value of a metric, the utility level of the metric also increases then its ratio type is direct, otherwise inverse. A metric also contains one or more monitoring schedule which defines the basic monitoring parameters, e.g. the location of the monitored data that is associated with that metric. A guarantee or an obligation has similar structure, i.e. a precondition (which works as a triggering event) and an action that is performed if the precondition is true. In practical scenarios, a CSU may acquire a cloud service for using it at multiple locations simultaneously. In such cases, monitoring the cloud service to detect SLA violations at different locations becomes a complex task, e.g. if S3LACC is used without any change then it is not possible to assign the location specific monitoring parameters in the same SLA. Moreover, one cloud service may offer different service levels for different global locations (a feature that may become part of the future cloud services, if not available in the present services), which leads to different negotiation parameters for different locations. An extension to the S3LACC is required to fulfill the needs for the distributed nature of the SLAs. So, in this paper, an extension to the S3LACC is presented which is termed as S3LACC+ and which enables the definition of the SLAs that are suitable for multiple locations. The S3LACC+ contains an additional class Location along with the basic S3LACC parameters as shown in Fig. 1. By adding the LocationID to each metric, it becomes possible to have different negotiation parameters for different locations in the same SLA template. The monitoring parameters of only relevant metrics can be passed to each location, which helps to enforce privacy among different locations. A single SLA can possibly contain segregated metrics data for each location using S3LACC+. A formal and detailed description of basic S3LACC parameters is available in [5].

3   SLA Negotiation

In this section, a dynamic and time-efficient SLA negotiation strategy (termed as flip-flop [4]) is described which uses S3LACC basically but the same negotiation strategy is applied to S3LACC+ to support multiple locations in an SLA. The negotiation protocol used in flip-flop strategy is based on Rubinstein's alternating offer protocol [17] with few alterations, i.e. negotiation process is limited by a deadline and offer acceptance is a two step process (one party sends an acceptance to an offer/counter-offer and other party sends a confirmation). The two step acceptance protocol is helpful to negotiate concurrently with multiple CSPs, i.e. if one CSP sends an acceptance then the CSU can evaluate the negotiations with other CSPs before making the final decision by estimating the expected final offers from all other CSUs using the concession extrapolation method as part of the flip-flop negotiation process. The flip-flop negotiation strategy uses time duration as a deadline rather than fixed number of negotiation rounds. Cloud services generally require quick provisioning and having fixed number of negotiation rounds as a deadline may delay the negotiation time duration due to network delays. The flip-flop negotiation strategy works on a principle that opponent's expected final offer at the end of the negotiation process is predicted after each negotiation round (before preparing an offer) and the same expected final offer from the opponent is tried to be reached earlier in time using the flip-flop strategy. This strategy aims to reduce the negotiation process time which can be very advantageous in time-critical cloud services. The flip-flop negotiation strategy operates as given in the following:

- A CSU prepares its initial offers (until third negotiation round) according to the negotiation parameters as described in the SLA template using the S3LACC+. Initial offer contains the most suitable values for the CSU.
- From the fourth offer, the previous three counter offers (along with the time at which they were received) are used as input to derive a function $(\alpha_j(T_u)$ for each metric M_j at time T_u [4]) using the polynomial interpolation method, where:

$$\alpha_j(T_u) = \sum_{\substack{\iota=v \\ 0 \le v \le q}}^{\iota+2} \left(\prod_{\substack{\kappa=v+1 \\ \kappa \ne \iota}}^{v+3} \frac{T_u - T(CO_\kappa^{b\longrightarrow a})}{T(CO_\iota^{b\longrightarrow a}) - T(CO_\kappa^{b\longrightarrow a})} \right) \zeta_{\iota,j}^{b\longrightarrow a} \qquad (1)$$

such that $T(CO_\kappa^{b\longrightarrow a})$ represents the point in time during the negotiation process when a κ-th counter offer from the CSP b is received to the CSU a and $\zeta_{\iota,j}^{b\longrightarrow a}$ is the ι-th concession that the opponent b has offered between two counter offers.

- The CSP's final offer is predicted with the help of function $\alpha_j(T_u)$ (using the polynomial extrapolation method).
- The CSU computes its concession to generate the next offer and adjusts the new offer with an extra amount of concession. The extra amount of concession is calculated using the decremented (by one) number of negotiation of rounds, i.e. if with CSU's normal concession value was reaching an agreement in n number of rounds then the extra concession value reaches the agreement in $n - 1$ number of negotiation rounds. This step of increasing the CSU's concession is termed as *flip*. In formal terms, CSU's concession is set by using a partial function $\gamma_j(T_u)$ as given in the following (Eq. 2):

$$\gamma_j(T_u) = \begin{cases} \dfrac{V_{j,q}^{b\longrightarrow a} - V_{j,k}^{a\longrightarrow b}}{NR_{rem}} & if\ V_{j,w} > V_{j,q}^{b\longrightarrow a} \\[3mm] \dfrac{V_{j,w} - V_{j,k}^{a\longrightarrow b}}{NR_{rem}} & if\ V_{j,w} \le V_{j,q}^{b\longrightarrow a} \end{cases} \qquad (2)$$

where $V_{j,q}^{b\longrightarrow a}$ is the expected final offer value from the CSP b to the CSU a for the metric M_j, $V_{j,w}$ is the worst-possible/reserve value, $V_{j,k}^{a\longrightarrow b}$ is the offer value from the CSU a to the provider b with normal concession and NR_{rem} is the expected number of remaining negotiation rounds (considering the time consumed per negotiation round and the total amount of negotiation time).

- If the CSP responds with an equivalent or more percentage increase in its concession then the CSU continues the flip step. However, in case of a negative response (due to a greedy strategy by the CSP), the CSU adjusts its regular concession and decreases it by the same amount that it was increased in the previous flip step. This process of decreasing the CSU's concession is termed as flop step.
- This flip-flop process continues in every negotiation round until an agreement is reached or the negotiation process times-out.

In context of S3LACC+, a custom negotiation strategy (e.g. linear, conceder or Boulware) can be dynamically integrated with the existing SLA structure by updating the DynamicConcessionValues parameter in the QuantitativeMetric class. The flip-flop negotiation strategy is evaluated in Sect. 6, outlining its potential benefits in detail.

4   SLA Monitoring and Management

After the successful negotiation process with one of the CSPs, a final/agreed SLA is stored at a central location by a monitoring coordinator (MC). The monitoring coordinator distributes the monitoring parameters to each location that intends to use the related cloud service. A distributed monitoring strategy is needed to efficiently report SLA violations at each location. The number of communications from each location to the MC is important due the fact that an excessive number of communications may consume the network bandwidth and a lesser number of communications may result in missing an important event at a location. So, a precise method of distributed monitoring can decrease the unnecessary communications by ensuring the fact that important events are reported to the MC. The monitoring parameters defined in an SLA (using the S3LACC+ implementation) work as a basis of the SLA monitoring process. The distributed SLA monitoring strategy using partial violations [6] is combined with S3LACC+ implementation to perform the SLA monitoring for multiple locations as described in the following:

- Each SLO is assigned the minimum number of violations V_{min} that must occur at a location before reporting to the MC.
- A partial violation value v (from the interval $[0, 1]$) is assigned to each metric at design time. This partial violation value is based on the type of violation, e.g. if violation is minor then a smaller v is assigned to that violation and vice versa.
- When a violation occurs, its corresponding v value is calculated and added to the existing violation value total S. Whenever $S \geq 1$, then one violation is added to the the existing number of violations ($V_{current}$) for the SLO and if $V_{current} \geq V_{min}$ then all collected violations are reported to the MC along with the necessary information. After reporting, counters are set to zero for the next monitoring round. In Sect. 6, the benefits of this threshold-based reporting approach are validated and further illustrated.

As a location reports SLA violation(s) to the MC, the guarantees and obligations parameters in the SLA are checked for the related location and if a precondition in a guarantee or obligation class is fulfilled then the corresponding action is taken. An action in a guarantee/obligation may include different SLA management tasks, e.g. a claim to be sent to the CSP for service credits along with the SLA violation data that is received from a location, sending a message to a financial system for deductions from the monthly payments to the CSP, renegotiating with the CSP depending on the *ReNegotiationParameters* in the SLA or adjusting the service usage if an obligation (on the CSU side) requires so. A custom management task can be embedded in the action function of a guarantee or obligation which makes this SLA specification very useful in context of the cloud services.

5   S3LACC+ Framework Overview

In this section, a collective functioning of the different phases of the SLA life cycle is described using S3LACC+. Generally, an SLA template and a final/agreed SLA are two different documents, whereas S3LACC+ combines them in a single document and each CSU and a CSP maintain their personal copy of the SLA (which includes template parameters, negotiation, monitoring and management parameters). It is assumed that all CSPs share the same SLA structure using the S3LACC+ implementation. The complete SLA life cycle for the S3LACC+ framework is described in the following (see also Fig. 2):

- A CSU requests the SLA templates from all of the CSPs. These SLA templates contain the basic information, e.g. CSP name, maximum negotiation time or names of possible negotiable SLOs and their respective metrics. Each CSP may set its own deadline (different from other CSPs) for the negotiation process in the SLA template. A CSU may also set the deadline for the negotiation process with few constraints, i.e. the CSU can not set the deadline that is greater than the minimum of all CSPs' deadlines and the minimum deadline among all the CSPs and the CSU is shared with all CSPs included in the concurrent negotiation process.
- The CSU prepares its SLA template (based on the SLA templates acquired from the CSPs) and adds the negotiation parameters according to its business objectives.
- The concurrent negotiation service starts the negotiation process with the each CSP according to the SLA negotiation described in the Sect. 3.
- After the successful negotiation process, with one of the CSPs, the CSU adds the monitoring parameters for each location and communicates them to the respective locations.
- Each location starts using the cloud service and marked SLA parameters are monitored according to the monitoring approach described in the Sect. 4.
- SLA violations are reported to the MC according to the rules defined in the monitoring parameters. The MC sends the SLA violations to the SLA management service which consults the guarantees and obligations parts of the SLAs to take the appropriate action against each SLA violation.

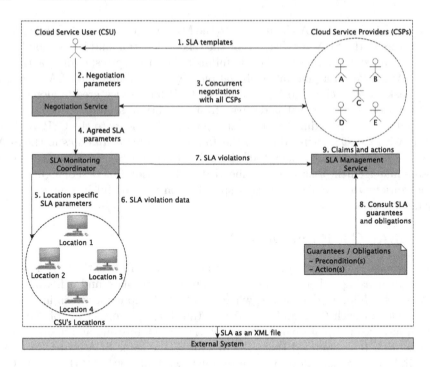

Fig. 2. An overview of the complete SLA life cycle management using S3LACC+

– As the SLA built using the S3LACC+ is based on an object oriented approach, so it can be serialized to an XML file for interaction(s) with external
system(s).

The S3LACC+ framework is useful for continuous change management and for
repeating the whole SLA life cycle by muting the already negotiated parameters
(that require no further change) and by only marking a few metrics as negotiable
based on the changes in business objectives.

6   Experiments and Validation

The S3LACC+ framework is implemented in Java and multiple experiments are
performed to evaluate the complete SLA life cycle. The sample experiments are
conducted with different number of SLOs and metrics including quantitative and
qualitative metrics. The negotiation process is tested by comparing the result
(overall agreement utility) achieved with and without using the flip-flop negotiation strategy in a concurrent setup. A CSP is allowed to adopt a greedy strategy
with 33% chances for all of the experiments. For instance, Table 1 shows the
experiment results for the negotiation process which is concurrently completed
with 10 CSPs. It can be noted that, in most of the cases, the flip-flop negotiation strategy performs better than a normal concession strategy. This automated

negotiation strategy eliminates the human-intervention during the complex negotiation scenarios and modifies its concession amount depending on the response from the opponent, i.e. if an opponent responds positively then this strategy seeks to reach the same agreement (that is expected at the end of negotiation process) in a lesser amount of time. If an opponent is responding with a greedy approach to benefit from the increased concession (during the flip step) from the CSU, then the flop step aims to recover the loss made during the previous step by reducing the CSU's concession. A graphical representation of the experiment results of Table 1 are shown in Fig. 3.

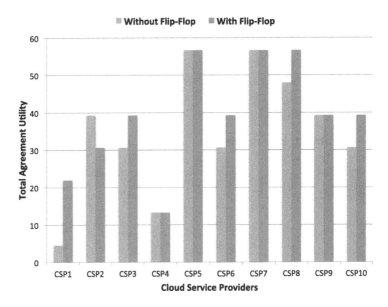

Fig. 3. Comparison of overall agreement utility achieved with and without flip-flop negotiation strategy

A simulation service (also used in the previous related work [6]) is implemented which induces the SLA violations for different numbers of SLOs to evaluate the effect on the total number of communications made to the monitoring coordinator. Table 2 shows results of one experimental simulation for 10 SLOs. Figure 4 gives the graphical representation of monitoring simulation for 4 SLOs.

The monitoring simulation data for an experiment (shown in Table 2) includes the increasing number of induced partial violations in its second column. The columns number 3 to number 7 (in Table 2) classify the number of values that fall under the interval (mentioned in the second row of respective columns). The last column (in Table 2) represents the resulted number of communications to the monitoring coordinator. The partial violation limits for each metric of an SLO is set randomly. Multiple experiments with different numbers of SLOs show the similar behavior in total number of communications which validates the consistency of the monitoring approach used for the S3LACC+ framework.

Table 1. Experimental results for overall agreement utility achieved with and without using the flip-flop negotiation strategy

	Agreement utility	
	Without flip-flop	With flip-flop
CSP1	4.55	21.91
CSP2	39.27	30.59
CSP3	30.59	39.27
CSP4	13.23	13.23
CSP5	56.63	56.63
CSP6	30.59	39.27
CSP7	56.63	56.63
CSP8	47.95	56.63
CSP9	39.27	39.27
CSP10	30.59	39.27

Table 2. Experiment data and results with 10 SLOs

SLOs	Total partial violations	Number of partial violations per interval					Number of communications
		[0,.2[	[.2,.4[	[.4,.6[	[.6,.8[	[.8,1]	
10	20	8	0	1	7	4	0
10	40	5	9	8	7	11	2
10	60	11	15	10	7	17	4
10	80	13	17	18	15	17	8
10	100	18	16	20	26	20	7
10	120	19	22	26	25	28	12
10	140	32	33	20	28	27	12
10	160	34	33	36	27	30	17
10	180	46	32	32	39	31	14
10	200	35	37	47	37	44	18

7 Related Work and Analysis

In this section, a comparison of existing approaches for SLA specification, monitoring and management is described. First, an overall analysis is given that compares different SLA specification languages with S3LACC+ and its features with respect to capabilities of SLA negotiation and monitoring/management. Qualitative metrics play an important role in cloud SLAs specifically, e.g. reliability is a major factor while selecting a cloud service which is a qualitative metric in general, requiring a different method of specification and negotiation than quantitative metrics. Most of the specification languages compared in Table 3 contain

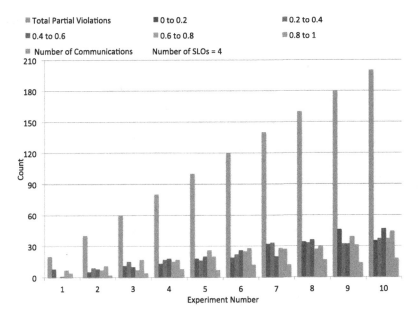

Fig. 4. Experiment results using the monitoring simulation for 4 SLOs

no method for processing the qualitative metrics whereas S3LACC+ provides a comprehensive support for the qualitative metrics. Table 3 shows a brief feature based comparison of WSLA [9], WS-Agreement [1], SLAng [12], SLA* [8], SLALOM [2], Stamou *et al.* [18], Joshi and Pearce [7], CSLA [11], Kotsokalis *et al.* [10] and S3LACC+.

In the second column of the Table 3, target domain (original domain for which the specification was given) is mentioned, next columns show if the SLA negotiation, monitoring and management are supported by the specification or not. The word Partial in the negotiation column represents that either negotiation parameters are partially definable or negotiation strategy is not integrated within the specification. S3LACC+ enables complete integration of the static and dynamic negotiation parameters. Also, S3LACC+ enables a user to include any custom negotiation strategy within the SLA template. Another feature of S3LACC+ adds the capability of merging the SLA template and the final SLA as a single document. Partial in monitoring/management column represents that either full SLA monitoring is not supported by the specification or a customizable SLA monitoring technique is not possible to integrate using the specification.

SLA management of cloud services includes tasks such as preparing claims in case of service violations, updating SLA parameters if requirements change or performing an action triggered due to a monitoring event. Zhang [19] present an approach for life cycle based SLA management for web services. An SLA management platform is presented in [19] to define SLAs for web services, registration of SLAs, monitoring and mapping of provider supplied parameters to service user's

Table 3. Comparative analysis of S3LACC+ framework with other approaches

Source	Original domain	Negotiation	Monitoring/management
WSLA	Web services	Yes (static)	Yes
WS-Agreement	Web services	Yes (static)	Partial
SLAng	Internet/web services	Yes (static)	Partial
SLA*	Domain independent	Partial	No
SLALOM	IT services	No	Partial
Stamou et al.	Cloud data services	No	Partial
Joshi et al.	Cloud services	Partial	Yes
CSLA	Cloud services	No	Yes
SLAC	Cloud services	Partial	Yes
Kotsokalis et al.	IT services	Yes	Yes
S3LACC	Cloud services	Yes	Yes

QoS parameters. In a most recent survey, Faniyi et al. [3] present an overview of SLA management for cloud services in which it is argued that cloud SLAs have still not standardized enough to be automatically deployed. It is also concluded in [3] (based on detailed analysis) that majority of approaches related to SLAs have considered between one to three SLA parameters. Rak et al. [15] base their work for SLA monitoring on the mOSAIC API [14] (which offers development of inter-operable, portable and provider independent cloud applications). In [16], mOSAIC API is used as basis for user-centric SLA management. Maarouf et al. [13] present a model for the SLA life cycle management in a more recent paper where different phases of the SLA life cycle are discussed and modelled using UML (unified modeling language) diagrams. However, this work does not includes any SLA specification itself.

8   Conclusions and Future Work

In this paper, automation and management of the complete SLA life cycle is presented which extends an existing SLA specification (S3LACC) for multiple locations. S3LACC consists a core structure (with favorable relationship among SLA elements), which is easily extensible to meet the customer specific requirements and it can also be easily modified for future changes, i.e. an extension (S3LACC+) is presented in this work to support SLAs for multiple locations based cloud services. The extended S3LACC+ specification targets the complete SLA life cycle whereas most of the existing specifications lack one or other critical phases of SLA life cycle. An SLA specification is not very beneficial if one of the SLA life cycle phases is not supported or complete features of the cloud service specific SLAs are not supported. The negotiation strategy used in this work (flip-flop negotiation) enables a CSU and a CSP to conclude the negotiation process in lesser time, hence efficient use of cloud resources is ensured which is an essence of cloud computing. The flip-flop negotiation strategy can be easily integrated in an SLA template using S3LACC+ and a CSU can make use of this efficient negotiation strategy without making any changes to the SLA template. Similarly, the monitoring strategy used in this work enables distributed and continuous monitoring which can be joined with the S3LACC+ easily as well. The used monitoring approach decreases the number of communications made from different service locations towards the monitoring coordinator. Also, this monitoring approach helps a monitoring coordinator to define different monitoring parameters for different locations rather than a global monitoring strategy. The future directions of this work include the extension of S3LACC+ for a CSP perspective and to design a negotiation strategy that enables offline negotiations to reduce the number of round trips between a CSU and a CSP during the negotiation process. These extensions require special considerations with respect to security and privacy issues as well.

References

1. Andrieux, A., Czajkowski, K., Dan, A., Keahey, K., Ludwig, H., Nakata, T., Pruyne, J., Rofrano, J., Tuecke, S., Xu, M.: Web services agreement specification (WS-Agreement). Open Grid Forum. **128**, 216 (2007)
2. Correia, A., Amaral, V., et al.: SLALOM: a language for SLA specification and monitoring (2011). arXiv preprint arXiv:1109.6740
3. Faniyi, F., Bahsoon, R.: A systematic review of service level management in the cloud. ACM Comput. Surv. **48**(3), 43:1–43:27 (2015)
4. Ghumman, W.A., Schill, A., Lässig, J.: The flip-flop SLA negotiation strategy using concession extrapolation and 3D utility function. In: IEEE 2nd International Conference on Collaboration and Internet Computing, pp. 159–168, November 2016
5. Ghumman, W.A., Schill, A.: Structural specification for the SLAs in cloud computing (S3LACC). In: 13th International Conference on the Economics of Grids, Clouds, Systems, and Services, September 2016
6. Ghumman, W.A., Schill, A.: Continuous and distributed monitoring of cloud SLAs using S3LACC. In: The 11th IEEE International Symposium on Service-Oriented System Engineering, April 2017
7. Joshi, K.P., Pearce, C.: Automating cloud service level agreements using semantic technologies. In: 2015 IEEE International Conference on Cloud Engineering (IC2E), pp. 416–421. IEEE (2015)
8. Kearney, K.T., Torelli, F., Kotsokalis, C.: SLA*: an abstract syntax for service level agreements. In: 11th IEEE/ACM International Conference on Grid Computing, pp. 217–224, October 2010
9. Keller, A., Ludwig, H.: The WSLA framework: specifying and monitoring service level agreements for web services. J. Netw. Syst. Manage. **11**(1), 57–81 (2003)
10. Kotsokalis, C., Yahyapour, R., Rojas Gonzalez, M.A.: Modeling service level agreements with binary decision diagrams. In: Baresi, L., Chi, C.-H., Suzuki, J. (eds.) ICSOC/ServiceWave-2009. LNCS, vol. 5900, pp. 190–204. Springer, Heidelberg (2009). doi:10.1007/978-3-642-10383-4_13
11. Kouki, Y., Ledoux, T.: CSLA: a language for improving cloud SLA management. In: International Conference on Cloud Computing and Services Science, CLOSER, vol. 2012, pp. 586–591 (2012)

12. Lamanna, D.D., Skene, J., Emmerich, W.: Specification language for service level agreements. EU IST 34069 (2003)
13. Maarouf, A., Marzouk, A., Haqiq, A.: Practical modeling of the SLA life cycle in cloud computing. In: 2015 15th International Conference on Intelligent Systems Design and Applications (ISDA), pp. 52–58, December 2015
14. Moscato, F., Aversa, R., Martino, B.D., Forti, T.F., Munteanu, V.: An analysis of mOSAIC ontology for cloud resources annotation. In: Federated Conference on Computer Science and Information Systems, pp. 973–980, September 2011
15. Rak, M., Venticinque, S., Máhr, T., Echevarria, G., Esnal, G.: Cloud application monitoring: the mOSAIC approach. In: IEEE Third International Conference on Cloud Computing Technology and Science (CloudCom), pp. 758–763, November 2011
16. Rak, M., Aversa, R., Venticinque, S., Martino, B.: User centric service level management in mOSAIC applications. In: Alexander, M., et al. (eds.) Euro-Par 2011. LNCS, vol. 7156, pp. 106–115. Springer, Heidelberg (2012). doi:10.1007/978-3-642-29740-3_13
17. Rubinstein, A.: Perfect equilibrium in a bargaining model. Econometrica **50**(1), 97–109 (1982)
18. Stamou, K., Kantere, V., Morin, J.H., Georgiou, M.: A SLA graph model for data services. In: Proceedings of the Fifth International Workshop on Cloud Data Management, pp. 27–34, October 2013
19. Zhang, S., Song, M.: An architecture design of life cycle based SLA management. In: Proceedings of the 12th International Conference on Advanced Communication Technology, pp. 1351–1355, February 2010

Queueing Theory

Performance Modeling
Using Queueing Petri Nets

Tomasz Rak[✉]

The Faculty of Electrical and Computer Engineering,
Rzeszow University of Technology, Rzeszów, Poland
trak@kia.prz.edu.pl
http://trak.kia.prz.edu.pl

Abstract. In this paper, a performance model is used for studying distributed web systems (an J2EE web application with Oracle backend-database). Performance evaluation is done by obtaining load test measurements. Queueing Petri Nets (QPN) formalism supports modeling and performance analysis of distributed World Wide Web environments. The proposed distributed web systems modeling and design methodology has been applied for evaluation of several system architectures under different external loads. Experimental analysis is based on benchmark with realistic workload. Furthermore, performance analysis is done to determine the system response time.

Keywords: Distributed web system models · Response time analysis · Queueing Petri Nets · Performance modeling

1 Introduction

Distributed Web systems development assumes that the systems consist of a set of distributed nodes. Groups of nodes (clusters) are organized in layers conducting predefined services. This approach makes possible easily to scale the system. An example of a web system is a stock trading system used by professional traders. In this system, it may be a requirement for certain positions to be bought or sold when market events occur.

Modeling and design stock trading systems as distributed web systems develope in two ways. On the one hand, formal models which can be used to analyze performance parameters are proposed. To describe systems such formal methods like Queueing Petri (QN) and Petri Nets (PN) [7,10,15,16] are used. For example [10], a closed queueing model of SPECjAppServer2002 benchmark comprising of client, application server cluster, database server and production line stations is described. Sometimes elements of the control theory are used to manage the movement of packages in web servers [19]. Experiments are the second way [26]. Applying experiments and models greatly influences validity of the systems being developed.

Our approach described in this paper may be treated as extension of selected solutions summed up in [13,14], where we propose a QPN [2] models for Internet

© Springer International Publishing AG 2017
P. Gaj et al. (Eds.): CN 2017, CCIS 718, pp. 321–335, 2017.
DOI: 10.1007/978-3-319-59767-6_26

system The final QPN based model can be executed and used for modeled system performance prediction. In our solution we propose QPN models for one kind of distributed web system with all types of quotes [14]. The models have been used as a background for developing a programming tool which is able to map timed behavior of QN by means of simulation. Subsequently we developed our individual method of modeling and analysis of distributed Web systems. The well known software toolkits as Queueing Petri net Modeling Environment (QPME) [9] can be naturally used for our models simulation and performance analysis.

The remaining work is organized as follows. Section 3 presents distributed Web system architecture and describe modeling approach. Section 4 presents performance analysis results. The final section contains concluding remarks.[1]

2   Related Work

In this section, we review some related work in the area of performance web systems modeling. Several approaches have been proposed for previous performance analysis. Existing modeling approaches are mostly based on stochastic models such as QN (classical product-form, extended or layered) or stochastic PN. Building such models requires experience in stochastic modeling and analysis. The research community has proposed high-level network modeling approaches that support automatic generation of low-level predictive models. Most existing model-based approaches are based either on black-box statistical models [3] or on highly detailed protocol-level simulation models [23]. Several recent surveys review performance modeling tools, with a focus on model-based prediction [1], evaluation of component-based systems [11], and analyzing software architectures. An outlook into the future and directions for future research in software performance engineering are given in [24]. The related work can be divided into publications based on analysis of QN and PN.

2.1   Queueing Nets Models

Layered Queueing Network (LQN) performance models have been used for studying of software and web systems [20]. Cao et al. [6] proposes queueing model of a web server. Kattepur and Nambiar [8] have proposed a theoretically model performance of multi-tiered applications and use queuing networks and Mean Value Analysis models. Similarly, Tiwari and Myanampati compare LQN with QPN using the LQNS and HiQPN tools to model the SPECjAppServer2001 benchmark application [22].

2.2   Petri Nets Models

PN is the first choice for system modeling because it is applicable for the evaluation and prediction of web systems both in terms of theoretical support and

---

[1] We assumed that the reader is familiar with Queueing Petri Net formalism [2] and performance analysis tool [9].

quantitative analysis. Many efficient simulation techniques for stochastic PN are available. Regarding the stochastic PN, such as QPNs, so far they have mainly been used to model specific components at a high level of detail [25]. Additionally, in other work [18], nodes appear as a part of a larger modeling landscape and are typically modeled as a black-box. Traditional methods based on mathematical derivation and theoretical explanation fail to solve the problem of the explosive growths of data quantity, data amount and calculation amount existing in the analytical process of the formalized model of web systems. Some authors use it for modeling:

- databases [17],
- web architectures [4],
- grid environments [12],
- cloud systems [5],
- virtualized environments [21].

Performance models are an abstraction of a combined hardware and software system describing its performance relevant structure and behavior. It substantiate this, here we demonstrate the usage of popular modeling technique QPN. They are very popular and helpful for the qualitative and quantitative analysis. QPN in addition to quantitatively performance modeling of sources, it also can depict the dependency relationship among multilayered systems considering the characteristics of web system involving diverse application service, complex service behavior and superimposed hierarchical structures. The modeling approach presented in this paper differs from that of previous work because we model different types of requests. Moreover, our QPN models have a more intuitive structure that maps the infrastructure to the QPN model. This makes the model easier to comprehend by system developers. In the current work we utilize the versatility of QPN to study a web system. System analysis is often needed with respect to both qualitative and quantitative aspects. From a practical point of view, QPN paradigm provides a number of benefit over conventional modeling paradigm. Using QPNs one can integrate hardware and software aspects of system behavior into the same model. QPN have greater expressive power than QN (quantitative analysis) and PN (qualitative analysis). QNs have a queue, scheduling discipline and are suitable for modeling competition of equipment (hardware contention). PNs have tokens representing the tasks and are suitable for modeling software. They easy lend themselves to modeling blocking and synchronization. QPNs have the advantages of QNs (e.g., evaluation of the system performance, the network efficiency) and PNs (e.g., logical assessment of the system correctness).

3   Distributed Web System

3.1   Distributed Web System Architecture

Distributed Internet system architecture is made up of several layers:

- The first one presents information (system offer) for clients in the web pages form and contains clusters.
- The second one manages transactions (clients requests) and provides the clustering functionality that allows load balancing.
- The third one controls the transactions, as a single element of this layer or multiple servers with database replication.
- the fourth one is a data storage system.

In our approach the presented architecture has been simplified to two layers:

- Front-end layer is based on the presentation and processing mechanisms.
- Back-end layer keeps the system data.

Architecture composed of these layers is used for e-business systems. The presented double-layer system architecture realizes distributed Web system functions. Access to the system is realized through transactions. Clustering mechanism was used in a fron-end layer. These simplifications have no influence on the modeling process, which has been shown repeatedly e.g. [10].

The characteristic feature of many distributed Web systems is a large number of clients using the Internet services (e.g. stock trading system) at the same time. The distributed Web system clients have different response time requirements. In the case of the described systems class, clients are often focused on an one event related to the same system offer (the same database resources). Based on these futures we used stock trading system as a benchmark with two-layered architecture (a cluster in a front-end layer and one database in a back-end layer).

3.2 Modeling of Distributed Web System

Typically, distributed Web systems are composed of layers where each layer consists of a set of servers - a server cluster. The layers are dedicated for proper tasks and exchanged requests between each other. To explain our approach to distributed Web system modeling a typical structure will be modeled and simulated. The first layer (front-end) is responsible for presentation and processing of client requests. Nodes of this layer are modeled by Processor Sharing queues. The next layer (back-end) implements system data handling. A node of this layer is modeled by using the First In First Out queue. Requests are sent to the system and then can be processed in both layers. The successfully processed requests are send back to the client.

Consequently, an executable (in a simulation sense) QPN model is obtained. Tokens generated by arrival process are transferred in sequence by models of a front-end layer and by a back-end layer. QPN extend coloured stochastic PN by incorporating queues and scheduling strategies into places forming queueing places. This very powerful modeling formalism has the synchronization capabilities of PN while also being capable of modeling queueing behaviors. The queue mean service time, the service time probability distribution function and the number of servicing units defined for each queueing system in the model are the main parameters of the modeled system. In the demonstrated model it has been assumed that queues belonging to a front-end layer have identical parameters. QPN consists of a set of connected queueing places. Each queueing place is described by arrival process, waiting room, service process and additionally depository. We used several queueing systems (e.g. $-$/M/PS/∞, $-$/M/FIFO/∞) most frequently used to represent properties of distributed Web system components. For example -/M/1/PS/∞ is: exponential service times, single server, Processor Sharing service discipline and unlimited number of arrivals in the system. In the QPME software tool, it is possible to construct QPN with queueing systems having Processor Sharing and First In First Out disciplines. As it was mentioned above, the main application of the software tool presented in the paper is modeling and evaluation of distributed Web systems. To effectively model the Internet requests from the clients, a separate token colour (type) has been used.

4 Response Time Analysis

In our solution we propose a very popular formal method - QPN [2]. This method is based on QN and PN. Queuing theory deals with modeling and optimizing different types of service units. QN usually consists of a set of connected queuing systems. The various queue systems represent computer components. QN are very popular for the quantitative analysis. To analyze any queue system it is necessary to determine: arrival process, service distribution, service discipline and waiting room (scheduling strategies). PN are used to specify and analyze the concurrency in systems. The system dynamics is described by rules of tokens flow. The net scheme can be subjected to a formal analysis in order to carry out a qualitative analysis, based on determining its logical validity. PN are referred as the connection between the engineering description and the theoretical approach. PN are well-known models used to describe and analyze the service units. PN cannot be used for a quantitative analysis due to lack of time aspects. The studies focus on incoming load measuring, e.g. measure of the response time or presentation of an overall modeling plan. Queuing Nets - quantitative analysis - have a queue and scheduling discipline and are suitable for modeling competition of equipment. PN - qualitative analysis - have tokens representing the tasks and are suitable for modeling a software. QPN have the advantages of QN and PN.

QPN formalism is a very popular formal method of functional and performance modeling (performance analysis). These nets provide sufficient power to express modeling and analyzing of complex on-line systems. The choice of QPN was caused by a possibility of obtaining the different character information. The main idea of QPN is to add queueing and timing aspects to the net places. QN consists of a collection of service stations and clients. The service stations (queue) represent system resources while the clients represent users or transactions. A service station is composed of one or more servers and a waiting area. Tokens enter the queueing place through the firing of input transitions, as in other PN. When a request arrives at a service station, it is immediately serviced if a free server is available. Otherwise, the request has to wait in the waiting area. Different scheduling strategies can be used to serve the requests waiting in the waiting area. Places (QPN) are of two types: ordinary and queued. A queued place (resource or state) is composed of a queue (service station) and a depository for tokens that completed their service at a queue. After being served by the service station (coloured) tokens are placed onto a depository. Input transitions are fired and then tokens are inserted into a queueing place according to the queue's scheduling strategy. Queueing places can have variable scheduling strategies and service distributions (timed queueing places). Tokens in the queue are not available for output transitions while tokens in a depository are available to all output transitions of the queued place. Immediate queueing places impose a scheduling discipline on arriving tokens without a delay [2].

QPN have been recently applied in the performance evaluation of component-based distributed systems, databases and grid environments because they are more expressive to represent simultaneous resource possession and blocking. Here, QPN models are used to predict the distributed Web system performance. The response time was chosen to analyze from many Performance Engineering parameters.

4.1 Experiments Parameters

First, we present results of our experimental analysis. The goal is to check - among others - the service demand parameter for front-end and back-end nodes.

The application servers, considered as a front-end tier/layer, are responsible for the method execution. All sensitive data is stored in a database system (back-end tier/layer). When a method has to retrieve or update data, the application server makes the corresponding calls to the database system. Deployment details are as follows: Gbit LAN network and six nodes (HP ProLiant DL180 G6). Software environment consists of: 64-bit Linux operating systems, workload generator, Apache Tomcat Connector (as a load balancer), GlassFish (as a application server - first is Domain Administration Server) and Oracle (as a database server). Distributed Web systems are usually built on middleware platforms such as J2EE. We use the DayTrader [27] performance benchmark which is available as open source application. Overall, the DayTrader application is primarily used for performance research on a wide range of software components and platforms. DayTrader is a suite of workloads that allows performance analysis of J2EE application server. DayTrader is benchmark application built around the paradigm of an online stock trading system. It drives a trade scenario that allows to monitor the stock portfolio, inquire about stock quotes, buy or sell stock. The load generator is implemented using multi-threaded Java application connected to DayTrader benchmark (Fig. 1). By client business transactions we mean the stock-broker operations: Buy Quote, Sell Quote, Update Profile, Show Quote, Get Home, Get Portfolio, Show Account and Login/Logout. Each business transaction emulates a specific class (type) of clients [14].

Some of the most important requests are Buy Quote and Sell Quote (Fig. 2), but now we used requests class (type) in our simulations.

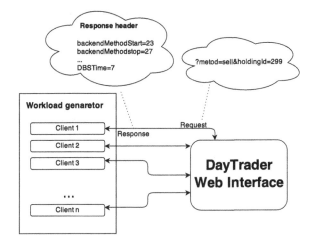

Fig. 1. Workload generator schema

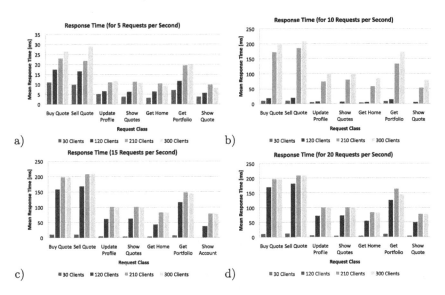

Fig. 2. Real response time for: (a) 5 requests per second workload, (b) 10 requests per second workload, (c) 15 requests per second workload, (a) 20 requests per second workload

Experiments - (Table 1) - (Test 1 - one node in a front-end (FE) and one node in a back-end (BE) layer) have shown that mean number of requests per second for a front-end layer is about 1300. Respectively the mean measured number of requests per second for a back-end layer is about 7500 requests per second. We can also see, that the delay in the requests processing is mainly caused by the waiting time for service in a BE node in all cases (Figs. 3(a) and 4(a)), but

Table 1. Parameters of cluster experiments

Server/Parameter	Test 1	Test 2	Test 3
Hardware			
Client	10.10.10.1	10.10.10.1	10.10.10.1
Load balancer	No	10.10.10.3	10.10.10.3
GlassFish application server nodes[a]	**10.10.10.3**	**10.10.10.4, 10.10.10.5**	**10.10.10.4, 10.10.10.5, 10.10.10.6**
Oracle database node	10.10.10.2	10.10.10.2	10.10.10.2
Software			
Application server threads pool	30	2×30	3×30
Database connections pool	40	2×40	3×40
Client workload			
Number of requests per second	15	15	15
Number of clients[b]	30, 120, 210, 300	30, 120, 210, 300	30, 120, 210, 300
Experiments			
Experiment time [s]	300	300	300

[a] With the domain administration server (10.10.10.3) as a specially designated Glass-Fish Server instance that hosts administrative applications.
[b] Four subtests in all cases.

the main problem is the performance of the system response time (System - one node in a FE layer and one node in a BE layer).

Starting the server cluster in front-end layer requires a mechanism that would allow for an equable distribution of load. It must also be a gateway that transfers requests and responses between a user and an application. In such scenario, only a gateway is visible from the outside and - on the basis of the request - it determines which part of the system (application server), and how, will be used to perform the request. Built-in load balancer is not available in the free version of the GlassFish server. Apache Tomcat Connector has been used as the load balancer. Also cluster - (Table 1) - (Test 2 - two or Test 3 - three nodes in a FE and one node in a BE layer) experiments have shown that the mean number of requests per second for a front-end layer is about 2400. The mean measured number of requests per second for a back-end layer is the same as earlier. We can also see, that the delay in the requests processing is mainly caused by the waiting time for service in the BE node in all cases (Figs. 3(b), (c) and 4(b), (c)), but the main problem is still the performance of the system response time (System - two or three nodes in a FE layer and one node in a BE layer). We use experimental results in our simulations.

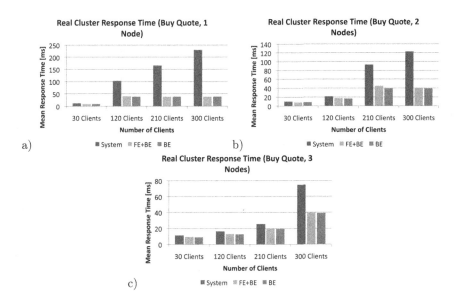

Fig. 3. Real response time for one class requests Buy Quote: (a) 1 node in a front-end layer, (b) 2 nodes in a front-end layer, (c) 3 nodes in a front-end layer

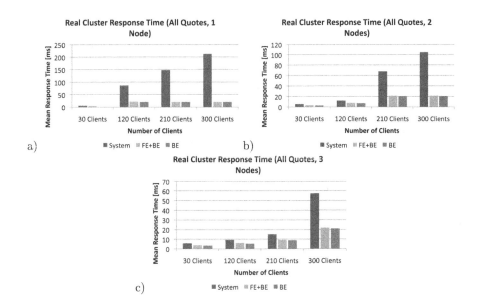

Fig. 4. Real response time for all classes requests: (a) 1 node in a front-end layer, (b) 2 nodes in a front-end layer, (c) 3 nodes in a front-end layer

4.2   Queueing Petri Net Models

Multiple front-end nodes and one back-end node are the main configuration scenario. QPN models (Fig. 5) are used to predict the system response time. We use QPME [9] tool. QPME is an open-source tool for stochastic modeling and analysis based on the QPN modeling formalism used in many works [9,10].

Software and client workload parameters are the same as in the experiment environment. Clients think time are modeled by Infinite Server scheduling strategy (*CLIENTS* place). Servers of a front-end layer are modeled using the Processor Sharing queuing systems (*FE_CPU* places). A back-end server is modeled by First In First Out queue (*BE_I/O* place). Places (*FE* and *BE*) used to stop incoming requests when they await application server threads and database server connections respectively. Application server threads and database server connections are modeled respectively by *THREADS* and *CONNECTIONS* places (Fig. 5). The process of requests arrival to the system is modeled by exponential distribution with the λ parameter (client think time) corresponds to the number of clients request per second. Service in all queueing places is modeled by an exponential distribution. Service demands in layers based on experimental results in Sect. 4.1:

- $d_{FE\_CPU} = 0,714$ [ms],
- $d_{BE\_I/O} = 0,133$ [ms].

Initial marking for places corresponds to input parameters of cluster experiment:

- number of clients (number of tokens in *CLIENTS* place),
- application server threads pool (number of tokens in *THREADS* place),
- database server connections pool (number of tokens in *CONNECTIONS* place).

In these models we have many types[2] of tokens:

- quotes (Buy Quote, Sell Quote, Update Profile, Show Quote, Get Home, Get Portfolio, Show Account and Login/Logout),
- application server threads,
- connections to the database server.

Total response time is a sum of all individual response times of queues and depositories in a simulation model without the client queue response time (client think time).

4.3   Simulation Results

Many simulations were performed for various input parameters:

- 30 threads for one front-end node, 60 threads for two front-end nodes, 90 threads for three front-end nodes

---

[2] A colour specifies a type of tokens that can be resided in the place.

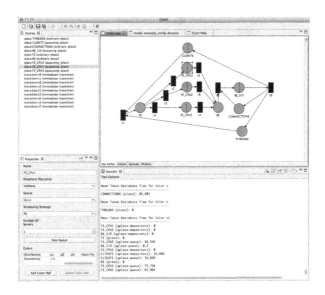

Fig. 5. Model of distributed web system with front-end cluster (selected example)

- 40 connections for one front-end node, 80 connections for two front-end nodes, 120 connections for three front-end nodes
- Client think time for all types of tokens equals 66,67 [ms]
- Simulation time 300 [s].

The number of clients was increasing in accordance with values (*CLIENTS* place). We used some scenarios in which we have a nine requests classes - transactions (Buy Quote, Sell Quote, Update Profile, Show Quote, Get Home, Get Portfolio, Show Account and Login/Logout). The scenarios involves the response time of the entire system (Sys), the response time of a front-end layer and a back-end layer (FE+BE) and the response time of a back-end layer (BE). The number of application server nodes is 1, 2 and 3. QPN model was used to predict the performance of the system for the scenarios mentioned above and it was developed using QPME 2.0.[3]

We investigate the behavior of the system as the workload intensivity increases. As a result, the response time of transactions is improved for cases with more number of front-end nodes. Increasing number of nodes is resulted in simultaneously increasing number of application server threads and connections to the database.

As we can see (Fig. 6) overall response time decreases while the number of nodes was increasing (15 requests per second). Response time of one front-end

---

[3] Queueing Petri net Modeling Environment 2.0 does not support timed transitions, so it was approximated by a serial network consisting of an immediate transition (Black rectangles represent immediate transitions.), a timed queueing place and a second immediate transition [9].

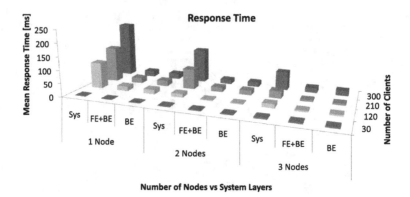

Fig. 6. Mean response time of simulation results (system, fron-end and back-end layer, back-end layer) for different number of nodes and clients (15 requests per second workload)

node architecture for all cases is the biggest. A difference of response time between 2 and 3 nodes is much smaller. When more nodes in front-end layer are added the analysis of their impact on other elements of the system should be preluded.

Overall system response time increases with increasing workload, even with a larger number of nodes.

The convergence of simulation results with the real systems results confirms a correctness of the modeling methods. The validation results show that the model is able to predict the performance with a relative error about 20% for different number of nodes in front-end layer (Table 2).

Table 2. Modeling response time error for scenario with 300 clients

Number of nodes	Model [ms]	Measured [ms]	Error [%]
1	241,22	211,92	13,8
2	127,76	105,19	21,4
3	69,22	57,50	20,3

5 Conclusions

We can not always add new devices to improve performance, because the initial cost and maintenance will become too large. Also not every system can or should be virtualized or put in the cloud computing. Because the overall system capacity is unknown we propose a combination of benchmarking and modeling solution.

It is still an open issue how to obtain an appropriate distributed Web system. Our earlier works propose Performance Engineering frameworks [15, 16] to evaluate performance during the different phases of their life cycle. The demonstrated research results are an attempt to apply QPN formalism to the development of a software tool that can support distributed Web system design. The idea of using QPN was proposed previously in [10,13,14]. In the presented approach an alternative implementation of QPN has been proposed. The rules of modeling and analysis of distributed Web systems applying described net structures was introduced. Our present approach predicts response time for distributed Web system and the benchmark used in our work has got realistic workload. The modeling approach presented in this paper differs from our previous works because all types of tokens (requests classes) were not used earlier. This paper deals with the problem of calculating performance values like the response time in distributed Web systems environment.

At present, we have demonstrated the potential in modeling distributed Web systems using QPN, understood as monolithic modeling technique used to assess the performance. We develop a framework that helps to identify performance requirements. Next we analyze the response time characteristics for several different workloads and configuration scenarios. The study demonstrates the modeling power and shows how discussed models can be used to represent system bahaviour.

References

1. Balsamo, S., Di Marco, A., Inverardi, P., Simeoni, M.: Model-based performance prediction in software development: a survey. IEEE Trans. Softw. Eng. **30**(5), 295–310 (2004)
2. Bause, F.: Queueing Petri Nets - A Formalism for the Combined Qualitative and Quantitative Analysis of Systems. IEEE Press, New York (1993)
3. Becker, S., Koziolek, H., Reussner, R.: The palladio component model for model-driven performance prediction. J. Syst. Softw. **82**(1), 3–22 (2009)
4. Brosig, F., Meier, P., Becker, S., Koziolek, A., Koziolek, H., Kounev, S.: Quantitative evaluation of model-driven performance analysis and simulation of component-based architectures. IEEE Trans. Softw. Eng. **41**(2), 157–175 (2015)
5. Cao, Y., Lu, H., Shi, X., Duan, P.: Evaluation model of the cloud systems based on queuing petri net. In: Wang, G., Zomaya, A., Perez, G.M., Li, K. (eds.) ICA3PP 2015. LNCS, vol. 9532, pp. 413–423. Springer, Cham (2015). doi:10.1007/978-3-319-27161-3_37
6. Cao, J., Andersson, M., Nyberg, C., Kihl, M.: Web server performance modeling using an M/G/1/K*PS queue. In: International Conference on Telecommunications, vol. 2 (2003)
7. Chen, X., Ho, C.P., Osman, R., Harrison, P.G., Knottenbelt, W.J.: Understanding, modelling and improving the performance of web applications in multi-core virtualised environments. In: ACM/SPEC International Conference on Performance Engineering, pp. 197–207 (2014)
8. Kattepur, A., Nambiar, M.: Performance modeling of multi-tiered web applications with varying service demands. In: IEEE International Parallel and Distributed Processing Symposium Workshop, pp. 415–424 (2015)

9. Kounev, S., Spinner, S., Meier, P.: QPME 2.0 - a tool for stochastic modeling and analysis using queueing petri nets. In: Sachs, K., Petrov, I., Guerrero, P. (eds.) From Active Data Management to Event-Based Systems and More. LNCS, vol. 6462, pp. 293–311. Springer, Heidelberg (2010). doi:10.1007/978-3-642-17226-7_18

10. Kounev, S., Rathfelder, C., Klatt, B.: Modeling of event-based communication in component-based architectures: state-of-the-art and future directions. J. Electr. Notes Theor. Comput. Sci. **295**, 3–9 (2013)

11. Koziolek, H.: Performance evaluation of component-based software systems: a survey. Perform. Eval. **67**(8), 634–658 (2010)

12. Nou, R., Kounev, S., Julia, F., Torres, J.: Autonomic QoS control in enterprise grid environments using online simulation. J. Syst. Softw. **82**(3), 486–502 (2009)

13. Rak, T.: Response time analysis of distributed web systems using QPNs. Math. Prob. Eng. **2015**, Article ID 490835, 1–10 (2015)

14. Rak, T.: Performance analysis of distributed internet system models using QPN simulation. IEEE Ann. Comput. Sci. Inf. Syst. **2**, 769–774 (2014)

15. Rak, T., Werewka, J.: Performance analysis of interactive internet systems for a class of systems with dynamically changing offers. In: Szmuc, T., Szpyrka, M., Zendulka, J. (eds.) CEE-SET 2009. LNCS, vol. 7054, pp. 109–123. Springer, Heidelberg (2012). doi:10.1007/978-3-642-28038-2_9

16. Rak, T., Samolej, S.: Distributed internet systems modeling using TCPNs. In: IEEE International Multiconference on Computer Science and Information Technology, vol. 1 and 2, pp. 559–566 (2008)

17. Rygielski, P., Kounev, S.: Data Center network throughput analysis using queueing petri nets. In: IEEE International Conference on Distributed Computing Systems Workshops, pp. 100–105 (2014)

18. Rygielski, P., Kounev, S., Zschaler, S.: Model-based throughput prediction in data center networks. In: IEEE International Workshop on Measurements and Networking, pp. 167–172 (2013)

19. Samolej, S., Szmuc, T.: HTCPNs–based modelling and evaluation of dynamic computer cluster reconfiguration. In: Szmuc, T., Szpyrka, M., Zendulka, J. (eds.) CEE-SET 2009. LNCS, vol. 7054, pp. 97–108. Springer, Heidelberg (2012). doi:10.1007/978-3-642-28038-2_8

20. Shoaib, Y., Das, O.: Web application performance modeling using layered queueing networks. Electr. Notes Theor. Comput. Sci. **275**, 123–142 (2011)

21. Spinner, S., Walter, J., Kounev, S.: A reference architecture for online performance model extraction in virtualized environments. In: International Conference on Performance Engineering, pp. 57–62 (2016)

22. Tiwari, N., Mynampati, P.: Experiences of using LQN and QPN tools for performance modeling of a J2EE application. Comput. Meas. Group Conf. **1**, 537–548 (2006)

23. de Wet, N., Kritzinger, P.: Using UML models for the performance analysis of network systems. Comput. Netw. **49**(5), 627–642 (2005)

24. Woodside, M., Franks, G., Petriu, C.D.: The future of software performance engineering. In: Future of Software Engineering, pp. 171–187 (2007)

25. Zaitsev, D.A., Shmeleva, T.R.: A parametric colored petri net model of a switched network. Netw. Syst. Sci. **4**, 65–76 (2011)

26. Zatwarnicki, K.: Operation of cluster-based web system guaranteeing web page response time. In: Bădică, C., Nguyen, N.T., Brezovan, M. (eds.) ICCCI 2013. LNCS, vol. 8083, pp. 477–486. Springer, Heidelberg (2013). doi:10.1007/978-3-642-40495-5_48
27. DayTrader. https://geronimo.apache.org/GMOxDOC22/daytrader-a-more-complex-application.html

Self-similarity Traffic and AQM Mechanism Based on Non-integer Order $PI^\alpha D^\beta$ Controller

Adam Domański[1(✉)], Joanna Domańska[2], Tadeusz Czachórski[2], and Jerzy Klamka[2]

[1] Institute of Informatics, Silesian Technical University,
Akademicka 16, 44-100 Gliwice, Poland
`adamd@polsl.pl`
[2] Institute of Theoretical and Applied Informatics,
Polish Academy of Sciences, Baltycka 5, 44-100 Gliwice, Poland
`{joanna,tadek,jerzy.klamka}@iitis.gliwice.pl`

Abstract. In this paper the performance of fractional order PID controller as AQM mechanism and impact of traffic self-similarity on network utilization are investigated with the use of discrete event simulation models. The researches show the influence of selection of PID parameters and degree of traffic self-similarity on queue behavior. During the tests we analyzed the length of the queue, the number of rejected packets and waiting times in queues. In particular, the paper uses fractional Gaussian noise as a self-similar traffic source. The quantitative analysis is based on simulation.

1 Introduction

Most AQM mechanism proposed by IETF to control the network congestions are based on preventive packed dropping. For the most known active mechanism the number of discarded packets grows with the increase in queue occupancy. The basic active queue management algorithm is Random Early Detection (RED) algorithm. It was primarily proposed in 1993 by Sally Floyd and Van Jacobson [1]. Since that time a number of studies how to improve the basic algorithm have been proposed. We have also proposed and evaluated a few variants, [2–7].

In 2001 the use of the PI controller as AQM mechanism was proposed by C.V. Hollot, V. Misra and D. Towsley [8]. Based on the first implementation, a number of PI controllers have been proposed later [9–11].

In recent years the fractional order calculus becomes very popular. The articles [12–14] show that non-integer order controllers may have better performance than classic integer order. The first application of the fractional order PI controller as a AQM policy in fluid flow model of a TCP connection was presented in [15]. The detailed influence of fractional order PI controller on queue behavior was presented in article [16].

Measurements and statistical analysis (performed already in the 90s) of packet network traffic show that this traffic displays a complex statistical nature.

P. Gaj et al. (Eds.): CN 2017, CCIS 718, pp. 336–350, 2017.
DOI: 10.1007/978-3-319-59767-6_27

It is related to such statistic phenomena as self-similarity, long-range dependence and burstiness [17–20].

Self-similarity of a process means that the change of time scales does not influence the statistical characteristics of the process. It results in long-distance autocorrelation and makes possible the occurrence of very long periods of high (or low) traffic intensity. These features have a great impact on a network performance [21]. They enlarge the mean queue lengths at buffers and increase the probability of packet losses, reducing this way the quality of services provided by a network [22].

As a consequence of this fact, it is needed to propose new or to adapt known types of stochastic processes when modeling these negative phenomena in network traffic. Several models have been introduced for the purposes of modeling self-similar processes in the network traffic area. These models of traffic use fractional Brownian Motion [23], chaotic maps [24], fractional Autoregressive Integrated Moving Average (fARIMA) [25], wavelets and multifractals and processes based on Markov chains: SSMP (Special Semi-Markov Process) [26], MMPP (Markov-Modulated Poisson Process) [27,28], HMM (Hidden Markov Model) [29].

The main purpose of the paper is to present simulation results for the AQM mechanism in which fractional discrete calculus is used. Section 2 presents theoretical bases for $PI^\alpha D^\beta$ controller next used in simulation. Section 3 briefly describes a self-similar traffic used in this article and presents the obtained results.

2 An AQM Mechanism Based on $PI^\alpha D^\beta$ Controller

A proportional-integral-derivative controller (PID controller) is a traditional mechanism used in feedback control systems. The article [12] indicates that the introduction non-integer controllers may improve closed loop control quality. Therefore here we propose to use the $PI^\alpha D^\beta$ (PID controller with non integer integral and derivative order) instead of the RED mechanism to determine the probability of packet drop. Equation (1) is based on our proposition discussed in [16] for PI^α controller and extended here to the case of $PI^\alpha D^\beta$.

This probability is calculated in the following way:

$$p = max\{0, -(K_P e_k + K_I \Delta^\alpha e_k + K_D \Delta^\beta e_k)\} \tag{1}$$

where K_P, K_I, K_D are tuning parameters, e_k is the error in current slot $e_k = q - q_d$, q - actual queue size, q_d - desired queue size and $\Delta^\alpha e_k$ is defined as follows:

$$\Delta^\alpha e_k = \sum_{j=0}^{k}(-1)^j \binom{\alpha}{j} e_{k-1} \tag{2}$$

where $\alpha \in R$ is generally a not-integer fractional order, e_k is a differentiated discrete function and generalized Newton symbol $\binom{\alpha}{j}$ is defined as follows:

$$\binom{\alpha}{j} = \begin{cases} 1 & \text{for } j = 0 \\ \dfrac{\alpha(\alpha - 1)(\alpha - 2)..(\alpha - j + 1)}{j!} & \text{for } j = 1, 2, \ldots \end{cases} \tag{3}$$

This definition unifies the definition of derivative and integral to one differintegral definition. We have the fractional integral of the considered function e_k for $\alpha < 0$. If the parameter α is positive, we obtain in the same way a fractional derivative and, to distinguish, we denote this parameter as β. If $\alpha = 0$ the operation (2) does not influence the function e_k.

Figure 1 presents a comparison of the increase of packet dropping probability in PI^α and PD^β controllers as a function of the queue length increased due to arrivals of packets. Naturally, the response depends on the choice of parameters. As can be seen, the integral order affects the time of controller reaction (below a certain threshold there is no packet dropping). The derivative order influences on increases packet dropping probability.

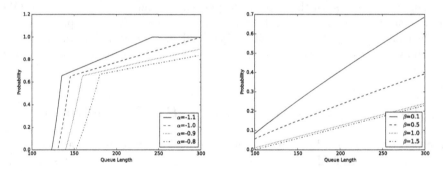

Fig. 1. Packet dropping probability in PI^α controller (the influence of the integral order α, $K_P = 0.00115$, $K_I = 0.0011$) (left), and in PD^β controller (the influence of the derivative order β, $K_P = 0.00115$, $K_D = 0.01$) (right)

3 $PI^\alpha D^\beta$ Controller Under Self-similar Traffic

In this article we use fractional Gaussian noise as an example of exactly self-similar traffic source. Fractional Gaussian noise (fGn) has been proposed as a model [30] for the long-range dependence postulated to occur in a variety of hydrological and geophysical time series. Nowadays, fGn is one of the most commonly used self-similar processes in network performance evaluation. The fGn process is the stationary Gaussian process that is exactly self-similar [31]. The Hurst parameter H characterizes a process in terms of the degree of self-similarity. The degree of self-similarity increases with the increase of H [32]. A Hurst value smaller or equal to 0.5 means the lack of long range dependence.

We use a fast algorithm for generating approximate sample paths for a fGn process, introduced in [33]. We have generated the sample traces with the Hurst parameter with the range of 0.5 to 0.90. After each trace generation, the Hurst parameter was estimated. The simulations were done using the Simpy Python simulation packet.

During the tests we analyzed the following parameters of the transmission with AQM: the length of the queue, queue waiting times and the number of rejected packets. The service time represented the time of a packet treatment and dispatching. Considered input traffic intensities were $\lambda = 0.5$, independently of Hurst parameter. The distribution of service time was also geometric. Its parameter changed during the test. The high traffic load was considered for parameter $\mu = 0.25$. The average traffic load we obtained for $\mu = 0.5$. A small network traffic was considered for parameter $\mu = 0.75$.

Table 1. FIFO queue

μ	Hurst parameter	Mean queue length	Mean waiting time	Rejected packets	
0.25	0.50	299.099	119.380	249520	49.90%
0.25	0.70	298.118	119.158	249879	49.97%
0.25	0.80	296.878	118.883	250354	50.07%
0.25	0.90	248.553	102.061	256587	51.32%
0.50	0.50	163.7547	32.7147	889	0.17%
0.50	0.70	145.8734	29.8820	13342	2.66%
0.50	0.80	141.3440	29.5832	23828	4.76%
0.50	0.90	133.8558	32.5300	89659	17.93%
0.75	0.50	1.2930	0.1586	0	
0.75	0.70	3.0506	0.5101	0	
0.75	0.80	6.9882	1.2976	0	
0.75	0.90	55.9574	11.5942	21484	4.29%

In order to better demonstrate the influence of degree of selfsimilarity on queue behavior first experiment focused on the FIFO queue. The Fig. 2 presents the distribution of the queue length. This figure clearly shows dependence of the queue occupancy on the degree of traffic selfsimilarity. The figure shows three situations: most overloaded network node ($\rho = \frac{\lambda}{\mu} = 2$), medium overloaded situation ($\rho = 1$) and almost empty buffer for ($\rho = \frac{2}{3}$). The detailed results obtained during the simulation present Table 1. For overloaded buffer ($\mu = 0.25$ and $\mu = 0.50$) the number of dropped packets increased with the traffic degree of selfsimilarity increasing. This effect becomes more evident with congestion decrease. In the case of an unloaded buffer ($\mu = 0.75$) packet loss occur only in the case of traffic with a high degree of selfsimilarity (Hurst parameter $H = 0.9$).

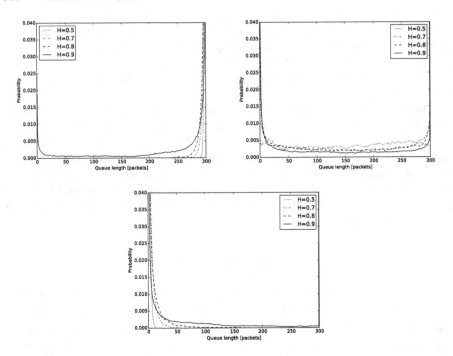

Fig. 2. The influence of degree of traffic selfsimilarity on queue distribution, FIFO queue, queue size = 300, $\lambda = 0.5$, $\mu = 0.25$ (left), $\mu = 0.5$ (right), $\mu = 0.75$ (bottom)

The presented results show how models that do not consider self-similar traffic may underestimate the queues occupancy and packet lost in routers.

In a first phase of the research we consider the influence of the PI^α controller on queue behavior. During the simulation the controller parameters were set as follows: $K_P = 0.00115$, $K_I = 0.0011$. The integral orders α changed and I received the following values: $-0.8, -1.0$ and -1.2. For the integral orders $\alpha = -1$ the controller becomes standard PI control loop feedback mechanism. The Tables 2, 3 and 4 present the obtained results. The queues distribution are presented in Figs. 3 and 4 (the queue distribution for controller with parameter $\alpha = 0.8$ is similar to distribution shown in Fig. 3). The controller desired point was set at 100 packet. It should be noted that regardless of the integral order the controller behaved properly.

These studies showed a very interesting controller behavior. In the case of overloaded FIFO queue for traffic of the high degree of self-similarity ($H = 0.9$) compared to less self-similar traffic the mean queue length decreases rapidly (see Table 1). This phenomenon also occurs in the case of standard AQM mechanisms [34]. In the case of PI^α occurrence of this phenomenon depends on the integral term and becomes less noticeable with the decrease in α. Comparing the mean queue length for $H = 0.9$ and $H = 0.8$ can be stated that for $\alpha = -1.2$ the mean queue length decreases by 19% for $\rho = 2$ and decreases by 3% for $\rho = 1$. For $\alpha = -1.0$ the mean queue length decreases by 8% for $\rho = 2$ and decreases by 2% for $\rho = 1$. Whereas for $\alpha = -0.8$ the mean queue length increases by 6% for $\rho = 2$ and increases by 2% for $\rho = 1$. On the other hand, the number of discarded packets analyze shows that for traffic with high degree of self-similarity ($H = 0.9$ and $H = 0.8$) with integral order growth decreases the number of dropped packets (for standard AQM queue, the situation is exactly opposite).

Interesting results were also obtained for the low traffic intensity. The mean queue length grows with integral order decreasing. The Fig. 1 explains these phenomena. The controller response to increasing queue in depends on the queue previous moments. The controller reaction is delayed with the integral order increasing.

Table 2. PI^{α} queue, $K_P = 0.00115$, $K_I = 0.0011$, $\alpha = -0.8$

μ	Hurst parameter	Mean queue length	Mean waiting time	Rejected packets	
0.25	050	105.2024	42.0998	250646	50.13%
0.25	070	112.1730	44.8029	250112	50.02%
0.25	080	118.2300	47.1973	249966	49.99%
0.25	090	126.4218	53.4780	263974	52.79%
0.50	050	53.8331	10.7526	3954	0.79%
0.50	070	55.8842	11.6289	23524	4.7%
0.50	080	52.0427	11.1830	38757	7.75%
0.50	090	54.3019	13.9019	112126	22.43%
0.75	050	1.2806	0.1561	0	
0.75	070	2.9819	0.4962	0	
0.75	080	6.6740	1.2359	440	0.08%
0.75	090	26.1258	5.4831	32047	6.40%

The second phase of the researches shows how derivative term changes the queue occupancy and packet waiting times. The Figs. 5, 6 and 7 present the queue distribution for PID^{β} controller ($\alpha = -1$). The results for PI controller were present in Fig. 3 and Table 3. Comparing the figures does not show a significant visual amendments. Differences in the controllers responses show Tables 5, 6 and 7.

The most interesting results were obtained for controller with derivative terms $\beta = 0.8$. For high traffic ($\mu = 0.25$ and $\mu = 0.5$) the controller reduces the mean queue length and at the same time reduces the number of packet losses. The further derivative order increasing (Tables 6 and 7) reduces the mean queue length and at the same time increases number of dropped packets. However, these differences are much smoother as in the case of integral order α decreasing (see Table 4).

The last phase of the simulation evaluates the impact of derivate term on PI^{α} controller. Controller with integral term $\alpha = -1.2$ is an example of strong mechanism. For this controller the lowest values of mean queue length and waiting

Table 3. PI^α queue, $K_P = 0.00115$, $K_I = 0.0011$, $\alpha = -1.0$

μ	Hurst parameter	Mean queue length	Mean waiting time	Rejected packets	
0.25	050	105.4769	42.0835	249914	49.98%
0.25	070	109.6279	43.6825	249538	49.90%
0.25	080	112.7636	45.0588	250246	50.04%
0.25	090	103.7300	43.9278	264321	52.86%
0.50	050	50.8356	10.1493	4024	0.80%
0.50	070	51.4054	10.6806	23202	4.64%
0.50	080	48.9152	10.5024	38644	7.72%
0.50	090	47.8680	12.2617	112735	22.54%
0.75	050	1.2892	0.1578	0	
0.75	070	3.0249	0.5047	0	
0.75	080	6.4268	1.1868	587	0.11%
0.75	090	25.3451	5.3177	32175	6.43%

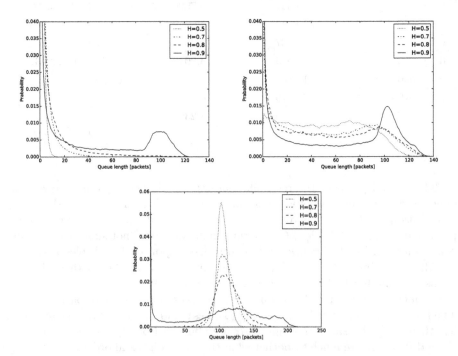

Fig. 3. The influence of degree of traffic selfsimilarity on queue distribution, PI^α queue, queue size = 300, $K_P = 0.00115$, $K_I = 0.0011$, $\alpha = -1.0$, $\lambda = 0.5$, $\mu = 0.75$ (left), $\mu = 0.5$ (right), $\mu = 0.25$ (bottom)

Table 4. PI^α queue, $K_P = 0.00115$, $K_I = 0.0011$, $\alpha = -1.2$

μ	Hurst parameter	Mean queue length	Mean waiting time	Rejected packets	
0.25	050	102.9224	41.0802	250026	50.0%
0.25	070	103.0052	41.1281	250100	50.01%
0.25	080	102.2977	40.8632	250214	50.04%
0.25	090	81.8945	34.7584	265013	53.0 %
0.50	0.50	49.525675	9.8806	3789	0.75%
0.50	0.70	50.429557	10.4798	23351	4.67%
0.50	0.80	47.869046	10.2771	38708	7.74%
0.50	0.90	46.207823	11.8211	112356	22.47%
0.75	0.50	1.283052	0.1566	0	
0.75	0.70	3.016572	0.5031	0	
0.75	0.80	6.400407	1.1817	655	0.13%
0.75	0.90	24.920464	5.2269	32173	6.43%

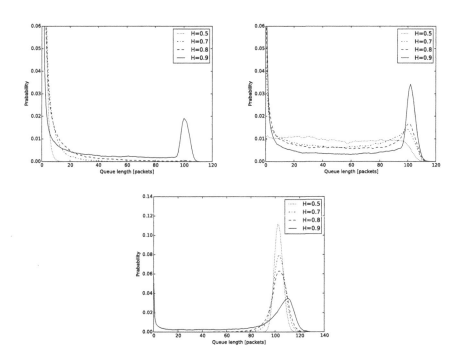

Fig. 4. The influence of degree of traffic selfsimilarity on queue distribution, PI^α queue, queue size $= 300$, $K_P = 0.00115$, $K_I = 0.0011$, $\alpha = -1.2$, $\lambda = 0.5$, $\mu = 0.75$ (left), $\mu = 0.5$ (right), $\mu = 0.25$ (bottom)

Table 5. PID queue, $K_P = 0.00115$, $K_I = 0.0011$, $\alpha = -1.0$, $K_D = 0.01$, $\beta = 0.8$

μ	Hurst parameter	Mean queue length	Mean waiting time	Rejected packets	
0.25	0.50	105.2382	41.9506	249687	49.93%
0.25	0.70	109.0697	43.5297	249941	49.98%
0.25	0.80	111.5187	44.5991	250457	50.0%
0.25	0.90	102.2642	43.2207	263868	52.77%
0.50	0.50	52.9857	10.5912	4385	0.87%
0.50	0.70	51.2877	10.6640	23537	4.70%
0.50	0.80	49.3514	10.6034	38920	7.78%
0.50	0.90	47.5944	12.1854	112550	22.50%
0.75	0.50	1.2878	0.1575	0	
0.75	0.70	3.0133	0.5025	0	
0.75	0.80	6.3137	1.1643	0	
0.75	0.90	25.2454	5.2913	31734	6.34%

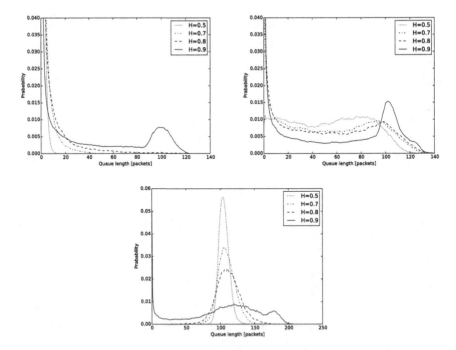

Fig. 5. The influence of degree of traffic selfsimilarity on queue distribution, $PI^{\alpha}D^{\beta}$ queue, queue size = 300, $K_P = 0.00115$, $K_I = 0.0011$, $\alpha = -1.0$, $K_D = 0.01$, $\beta = 0.8$, $\lambda = 0.5$, $\mu = 0.75$ (left), $\mu = 0.5$ (right), $\mu = 0.25$ (bottom)

Table 6. PID queue, $K_P = 0.00115$, $K_I = 0.0011$, $\alpha = -1.0$, $K_D = 0.01$, $\beta = 1.0$

μ	Hurst parameter	Mean queue length	Mean waiting time	Rejected packets	
0.25	0.50	105.1122	41.9762	250138	50.02%
0.25	0.70	109.4128	43.4958	248962	49.79%
0.25	0.80	112.3291	44.8387	249994	49.99%
0.25	0.90	104.6598	44.3687	264588	52.91%
0.50	0.50	52.6690	10.5310	4544	0.90%
0.50	0.70	51.1150	10.6226	23324	4.66%
0.50	0.80	48.9330	10.5084	38738	7.74%
0.50	0.90	47.6431	12.1726	111753	22.35%
0.75	0.50	1.2823	0.1564	0	
0.75	0.70	2.9884	0.4975	0	
0.75	0.80	6.4988	1.2012	581	0.11%
0.75	0.90	25.2075	5.2839	31798	6.35%

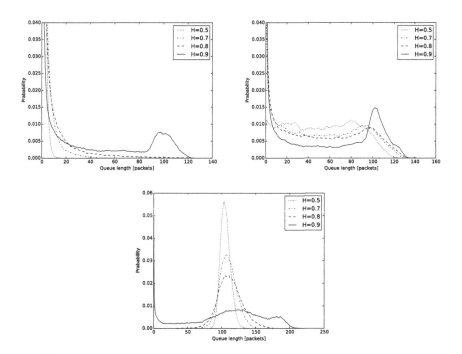

Fig. 6. The influence of degree of traffic selfsimilarity on queue distribution, $PI^{\alpha}D^{\beta}$ queue, queue size $= 300$, $K_P = 0.00115$, $K_I = 0.0011$, $\alpha = -1.0$, $K_D = 0.01$, $\beta = 1.0$, $\lambda = 0.5$, $\mu = 0.75$ (left), $\mu = 0.5$ (right), $\mu = 0.25$ (bottom)

Table 7. PID queue, $K_P = 0.00115$, $K_I = 0.0011$, $\alpha = -1.0$, $K_D = 0.01$, $\beta = 1.2$

μ	Hurst parameter	Mean queue length	Mean waiting time	Rejected packets	
0.25	0.50	105.3953	42.0064	249638	49.92%
0.25	0.70	109.8195	43.8410	250000	50.00%
0.25	0.80	112.8646	44.9803	249592	49.91%
0.25	0.90	105.0905	44.5343	264473	52.89%
0.50	0.50	50.5028	10.0713	3482	0.69%
0.50	0.70	50.6084	10.5039	22747	4.54%
0.50	0.80	48.2802	10.3441	37731	7.54%
0.50	0.90	47.9325	12.2640	112284	22.45%
0.75	0.50	1.2882	0.1576	0	
0.75	0.70	3.0119	0.5022	0	
0.75	0.80	6.3877	1.1789	561	0.11%
0.75	0.90	25.1610	5.2733	31731	6.34%

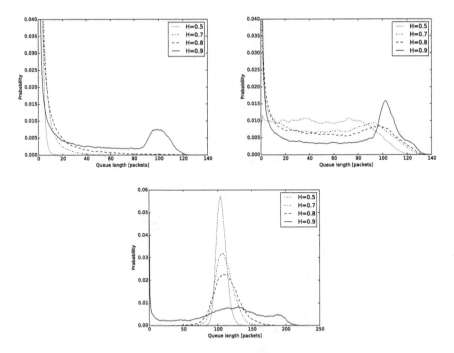

Fig. 7. The influence of degree of traffic selfsimilarity on queue distribution, $PI^\alpha D^\beta$ queue, queue size $= 300$, $K_P = 0.00115$, $K_I = 0.0011$, $\alpha = -1.0$, $K_D = 0.01$, $\beta = 1.2$, $\lambda = 0.5$, $\mu = 0.75$ (left), $\mu = 0.5$ (right), $\mu = 0.25$ (bottom)

times were obtained. At the same time increase the number of dropped packets is insignificant. The results of the controller with derivative term and derivative order $\beta = 0.8$ are shown in Fig. 8. Table 8 presents the detailed results. In this case, the controller with derivative term response is softer.

Table 8. PID queue, $K_P = 0.00115$, $K_I = 0.0011$, $\alpha = -1.2$, $K_D = 0.01$, $\beta = 0.8$

μ	Hurst parameter	Mean queue length	Mean waiting time	Rejected packets	
0.25	0.50	102.8881	41.0135	249696	49.93%
0.25	0.70	102.9803	41.1081	250038	50.00%
0.25	0.80	102.2453	40.8194	250084	50.01%
0.25	0.90	81.9062	34.7162	264699	52.93%
0.50	0.50	52.6182	10.5227	4659	0.93%
0.50	0.70	49.8837	10.3693	23532	4.70%
0.50	0.80	48.0090	10.3122	38918	7.78%
0.50	0.90	46.0250	11.7805	112565	22.51%
0.75	0.50	1.2859	0.1571	0	
0.75	0.70	3.0204	0.5039	0	
0.75	0.80	6.5480	1.2112	610	0.12%
0.75	0.90	24.9491	5.2310	31991	6.39%

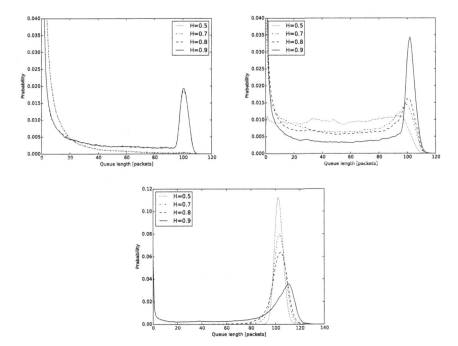

Fig. 8. The influence of degree of traffic selfsimilarity on queue distribution, $PI^\alpha D^\beta$ queue, queue size $= 300$, $K_P = 0.00115$, $K_I = 0.0011$, $\alpha = -1.2$, $K_D = 0.01$, $\beta = 0.8$, $\lambda = 0.5$, $\mu = 0.75$ (left), $\mu = 0.5$ (right), $\mu = 0.25$ (bottom)

4   Conclusions

Our article presents the impact of the degree of self-similarity (ex- pressed in Hurst parameter) on the length of the queue, queue waiting times and the number of rejected packets. Obtained results are closely related to the degree of self-similarity. The experiments are carried out for the four types of traffic ($H = 0.5, 0.7, 0.8, 0.9$). During the test we also changed the parameter of distribution of service time. This change allowed us to consider the different queues loading.

The article presents an evaluation of the fractional order $PI^\alpha D^\beta$ controller used as an active queue management mechanism. The effectiveness of the controller as an AQM mechanism depends on proper parameters of the PID selection. In the case of fractional order controller we need to consider two additional parameters: fractional derivative (β) and integral (α) orders. The controllers behavior was also compared to FIFO queue.

The results showed the usefulness of the $PI^\alpha D^\beta$ controller as AQM mechanism. The proper selection of the controller parameters is important in adaptation to various types of traffic (degree of self-similarity or various intensity).

References

1. Floyd, S., Jacobson, V.: Random early detection gateways for congestion avoidance. IEEE/ACM Trans. Netw. **1**(4), 397–413 (1993)
2. Domańska, J., Domański, A., Czachórski, T., Klamka, J.: Fluid flow approximation of time-limited TCP/UDP/XCP streams. Bull. Pol. Acad. Sci. Tech. Sci. **62**(2), 217–225 (2014)
3. Domański, A., Domańska, J., Czachórski, T.: Comparison of AQM control systems with the use of fluid flow approximation. In: Kwiecień, A., Gaj, P., Stera, P. (eds.) CN 2012. CCIS, vol. 291, pp. 82–90. Springer, Heidelberg (2012). doi:10.1007/978-3-642-31217-5_9
4. Domańska, J., Augustyn, D., Domański, A.: The choice of optimal 3rd order polynomial packet dropping function for NLRED in the presence of self-similar traffic. Bull. Pol. Acad. Sci. Tech. Sci. **60**(4), 779–786 (2012)
5. Augustyn, D.R., Domański, A., Domańska, J.: A choice of optimal packet dropping function for active queue management. In: Kwiecień, A., Gaj, P., Stera, P. (eds.) CN 2010. CCIS, vol. 79, pp. 199–206. Springer, Heidelberg (2010). doi:10.1007/978-3-642-13861-4_20
6. Domańska, J., Domański, A., Augustyn, D., Klamka, J.: A RED modified weighted moving average for soft real-time application. Int. J. Appl. Math. Comput. Sci. **24**(3), 697–707 (2014)
7. Domańska, J., Domański, A., Czachórski, T.: The drop-from-front strategy in AQM. In: Koucheryavy, Y., Harju, J., Sayenko, A. (eds.) NEW2AN 2007. LNCS, vol. 4712, pp. 61–72. Springer, Heidelberg (2007). doi:10.1007/978-3-540-74833-5_6
8. Hollot, C., Misra, V., Towsley, D., Gong, W.: On designing improved controllers for AQM routers supporting TCP flows. In: IEEE/INFOCOM 2001, pp. 1726–1734 (2001)
9. Michiels, W., Melchor-Aquilar, D., Niculescu, S.: Stability analysis of some classes of TCP/AQM networks. Int. J. Control **79**, 1136–1144 (2006)

10. Melchor-Aquilar, D., Castillo-Tores, V.: Stability analysis of proportional-integral AQM controllers supporting TCP flows. Computacion y Sistemas **10**, 401–414 (2007)
11. Ustebay, D., Ozbay, H.: Switching resilient pi controllers for active queue management of TCP flows. In: Proceedings of the 2007 IEEE International Conference on Networking, Sensing and Control, pp. 574–578 (2007)
12. Podlubny, I.: Fractional order systems and $PI^\lambda D^\mu$ controllers. IEEE Trans. Autom. Control **44**(1), 208–214 (1999)
13. Chen, Y., Petras, I., Xue, D.: Fractional order control - a tutorial. In: American Control Coference, pp. 1397–1411 (2009)
14. Babiarz, A., Czornik, A., Klamka, J., Niezabitowski, M.: Theory and Applications of Non-integer Order Systems. Lecture Notes in Electrical Engineering, vol. 407. Springer, Heidelberg (2017)
15. Krajewski, W., Viaro, U.: On robust fractional order PI controller for TCP packet flow. In: BOS Coference: Systems and Operational Research, Warsaw, Poland, September 2014
16. Domanski, A., Domanska, J., Czachorski, T., Klamka, J.: Use of a non integer order PI controller with an active queue management mechanism. Int. J. Appl. Math. Comput. Sci. **26**, 777–789 (2016)
17. Crovella, M., Bestavros, A.: Self-similarity in world wide web traffic: evidence and possible causes. IEEE/ACM Trans. Netw. **5**, 835–846 (1997)
18. Domański, A., Domańska, J., Czachórski, T.: The impact of self-similarity on traffic shaping in wireless LAN. In: Balandin, S., Moltchanov, D., Koucheryavy, Y. (eds.) NEW2AN 2008. LNCS, vol. 5174, pp. 156–168. Springer, Heidelberg (2008). doi:10.1007/978-3-540-85500-2_14
19. Domańska, J., Domańska, A., Czachórski, T.: A few investigations of long-range dependence in network traffic. In: Czachórski, T., Gelenbe, E., Lent, R. (eds.) Information Sciences and Systems 2014, pp. 137–144. Springer, Cham (2014). doi:10.1007/978-3-319-09465-6_15
20. Domańska, J., Domański, A., Czachórski, T.: Estimating the intensity of long-range dependence in real and synthetic traffic traces. In: Gaj, P., Kwiecień, A., Stera, P. (eds.) CN 2015. CCIS, vol. 522, pp. 11–22. Springer, Cham (2015). doi:10.1007/978-3-319-19419-6_2
21. Domańska, J., Domański, A.: The influence of traffic self-similarity on QoS mechanism. In: Proceedings of the International Symposium on Applications and the Internet, SAINT, Trento, Italy, pp. 300–303 (2005)
22. Stallings, W.: High-Speed Networks: TCP/IP and ATM Design Principles. Prentice-Hall, New York (1998)
23. Norros, I.: On the use of fractional brownian motion in the theory of connectionless networks. IEEE J. Sel. Areas Commun. **13**(6), 953–962 (1995)
24. Erramilli, A., Singh, R., Pruthi, P.: An application of deterministic chaotic maps to model packet traffic. Queueing Syst. **20**(1–2), 171–206 (1995)
25. Harmantzis, F., Hatzinakos, D.: Heavy network traffic modeling and simulation using stable farima processes. In: 19th International Teletraffic Congress, Beijing, China, pp. 300–303 (2005)
26. Robert, S., Boudec, J.: New models for pseudo self-similar traffic. Perform. Eval. **30**(1–2), 57–68 (1997)
27. Andersen, A.T., Nielsen, B.F.: A Markovian approach for modeling packet traffic with long-range dependence. IEEE J. Sel. Areas Commun. **16**(5), 719–732 (1998)

28. Domańska, J., Domański, A., Czachórski, T.: Modeling packet traffic with the use of superpositions of two-state MMPPs. In: Kwiecień, A., Gaj, P., Stera, P. (eds.) CN 2014. CCIS, vol. 431, pp. 24–36. Springer, Cham (2014). doi:10.1007/978-3-319-07941-7_3

29. Domańska, J., Domański, A., Czachórski, T.: Internet traffic source based on hidden Markov model. In: Balandin, S., Koucheryavy, Y., Hu, H. (eds.) NEW2AN/ruSMART -2011. LNCS, vol. 6869, pp. 395–404. Springer, Heidelberg (2011). doi:10.1007/978-3-642-22875-9_36

30. Mandelbrot, B., Ness, J.: Fractional brownian motions, fractional noises and applications. SIAM Rev. **10**, 422–437 (1968)

31. Samorodnitsky, G., Taqqu, M.: Stable Non-Gaussian Random Processes: Stochastic Models with Infinite Variance. Chapman and Hall, New York (1994)

32. Rutka, G.: Neural network models for internet traffic prediction. Electron. Electr. Eng. **4**(68), 55–58 (2006)

33. Paxson, V.: Fast, approximate synthesis of fractional Gaussian noise for generating self-similar network traffic. ACM SIGCOMM Comput. Commun. Rev. **27**(5), 5–18 (1997)

34. Domański, A., Domańska, J., Czachórski, T.: The impact of the degree of self-similarity on the NLREDwM mechanism with drop from front strategy. In: Gaj, P., Kwiecień, A., Stera, P. (eds.) CN 2016. CCIS, vol. 608, pp. 192–203. Springer, Cham (2016). doi:10.1007/978-3-319-39207-3_17

Stability Analysis of a Basic Collaboration System via Fluid Limits

Rosario Delgado[1(✉)] and Evsey Morozov[2]

[1] Departament de Matemàtiques, Universitat Autònoma de Barcelona,
Edifici C- Campus de la UAB., Av. de l'Eix Central s/n.,
08193 Bellaterra (Cerdanyola del Vallès), Barcelona, Spain
delgado@mat.uab.cat
[2] Institute of Applied Mathematical Research, Russian Academy of Sciences,
Petrozavodsk State University, Petrozavodsk, Russia
emorozov@karelia.ru

Abstract. In this work, the fluid limit approach methodology is applied to find a sufficient and necessary stability condition for the Basic Collaboration (BC) system with feedback allowed, which is a generalization of the so-called W-model. In this queueing system, some customer classes need cooperation of a subset of (non-overlapping) servers. We assume that each customer class arrives to the system following a renewal input with general i.i.d. inter-arrival times, and general i.i.d. service times are also assumed. Priority is given to customer classes that can not be served by a single server but need a cooperation.

Keywords: Stability · Fluid limit approach · Skorokhod problem · Workload · BC system · W-model

1 Introduction

In this paper, we study a generalization of the so-called queueing W-model which, in the simplest setting, consists of two single-server stations, $1, 2$, and three infinite-capacity buffers, 1, 2, 3, with independent renewal inputs of class-k customers, respectively, $k = 1, 2, 3$. Server i processes class-i costumers, $i = 1, 2$, but both servers are required to process class-3 customers which have preemptive-resume priority. (For more detailed description see [9,15].)

We generalize the W-model, which is in turn a particular case of the so-called sparsely connected model [15]. More exactly, we consider a Basic Collaboration (BC) system with J infinite buffer servers and $K \geq J$ customer classes. Each customer class needs cooperation of a subset of (non-overlapping) servers (it is called concurrent service). At the same time, there may be customer classes that

R. Delgado—Supported by Ministerio de Economía y Competitividad, Gobierno de España, project ref. MTM2015 67802-P (MINECO/FEDER, UE).

E. Morozov—Supported by Russian Foundation for Basic Research, projects 15-07-02341, 15-07-02354, 15-07-02360.

P. Gaj et al. (Eds.): CN 2017, CCIS 718, pp. 351–365, 2017.
DOI: 10.1007/978-3-319-59767-6_28

only need one server to be served. Concurrent customer classes on a server can only occur between a class that needs its cooperation with another server(s), and a class that only needs it to be served, without cooperation. In this setting, to keep work-conserving service discipline, we assume the mentioned priority of the customer requiring cooperation. We assume i.i.d. general inter-arrival and i.i.d. service times. Such a system is also called joint service model [10], or concurrent server release [2]. Queueing systems with concurrent service have been considered in a number of works, and pioneering ones are [2,8,13,18]. For the buffer-less (loss) concurrent service systems, the performance analysis has been developed in a number of works [1,11,12,16–18]. However, analysis of the buffered concurrent service system is much more challenging.

A comprehensive study of the concurrent service system has been developed in [13], where the author used the matrix analytic method to deduce a stability condition. However, this condition requires solving a matrix equation of a large dimension, and moreover, the corresponding matrices are not explicitly defined. The authors of [14] study a multi-server system in which each customer requires a random number of servers simultaneously and a random but identical service time at all occupied servers, which describes the dynamics of modern high performance clusters. They assume exponential distributions and an arbitrary number of servers. In [14], a modification of the matrix-analytic method is developed to obtain stability criterion of the simultaneous service model in an explicit form. (Also see [14] for a broad bibliographic review on the subject including previous references.) Note that the paper [15] considers various sparsely connected models assuming saturated regime, while, in the present research, we are seeking for the stability conditions.

In the model presented in this work, a restricted type of feedback is allowed. Indeed, feedback from each customer class to itself is permitted, as well as feedback from each customer class needing cooperation of some (more than one) servers for processing, to any of the concurrent customer classes at any of these servers. No other is allowed. From now on, we name "non-crossing" this type of feedback.

Motivation. BC systems model real situations in which different agents are able to work together to solve complex problems. Consider the following scenario introduced in [19]. A user wishes to determine the best package price for a ski trip given the following criteria: a resort in the Alps, for a week in February, with slope-side lodging, and the lowest price for all expenses. To solve this problem, an agent obtains a list of appropriate ski resorts from a database before spawning other agents to query travel databases, possibly in different formats, for package prices at those resorts in February. Agents can perform this task more efficiently when they can correlate their results and adjust their computations based on the outcome of a collaboration. Suppose the agents visit local travel agencies and then share their intermediate results and collaborate before migrating to another travel agency. If an agent determines that a particular resort does not have any available lodging meeting the user's criteria, the agents may determine to drop queries about trips to that destination. As more information is gathered, agents may also make other decisions. As this example demonstrates, agents can perform complex distributed computations more effectively if they based on the combined results. To do it, they can divide a complex task into smaller pieces and delegate them to agents that migrate throughout the network to accomplish them. These agents perform computations, synchronously share results, and collaboratively determine any changes to future actions, giving service to the user.

Another example are medical centers and hospitals, in which different types of patients have different requirements concerning technical equipment, facilities, doctors and nurses, which can be considered as the servers.

We give a brief summary of the research. The main contribution of this work is that, in contrast to previous works on W-models and concurrent service systems in general, we obtain stability condition, following fluid stability analysis developed in [3]. Indeed, our model is more general than the Generalized Jackson network in [3] (Sect. 5), in which there is only one class of customers served at each single-server station. Instead, in our model each server can serve more than one class: at most one customer class requires cooperation with other servers (multiserver customers), but no limit on the number of customer classes that do not need cooperation (single-server customers). Note that multi-class customers but in a single-station network have been considered in [3] (Sect. 6), as well. In Theorem 1 we first establish the stability of the fluid limit model associated to the BC system under sufficient condition. The fluid limit model, which allows to transform the initial stochastic problem into a (related) deterministic one, is introduced in Proposition 1. The stability of the fluid limit model means that the fluid limit of the queue-size process reaches zero in a finite time interval and stays there. Then, using stability of the fluid limit model and Theorem 4.2 [3], we deduce positive Harris recurrence of the basic Markov process describing the network. Similarly to [3], functional laws of large numbers for the renewal processes or, in other words, the hydrodynamic scaling by the increasing value of the initial state, are used to obtain the stability of the fluid limit model via the solution of a Skorokhod problem. At that, the choice of an appropriate Lyapunov function is the key point of analysis. By the same approach, we show that if the necessary condition is violated, then the fluid limit model is weakly unstable. It means that, if the process starts at zero, then there exists a time at which the fluid limit of the queue-size process becomes positive. As a result, by Theorem 3.2 [4], the queueing network is unstable: the queue size grows infinitely with probability (w.p.) 1 as time increases.

In the paper [7], the fluid approach methodology [3] has been applied to stability analysis of a cascade network. In this network, known as N-model, awaiting customers from the preceding queue jumps to the following server, when it is free, to be served there immediately. That is, in this model each free server helps the previous one serving some of its customers. It is shown that sufficient and necessary conditions match if the network is composed by two stations, but not in general. In the current work, we use the same methodology to develop stability analysis of a completely different queueing model, which is a generalization of the known W-model [15, 19]. We deal with a BC system in which some customer classes need cooperation of a subset of the servers working together. Moreover, we allow some kind of the feedback. Although the methodology is the same, the stability conditions of these two different models turn out to be different as well. In particular, in the current work we find stability criterion. To the best of our knowledge, it is the first time the fluid approach have been used for this type of the collaboration models. In addition, the way this methodology is applied presents some interesting differences. The main difference is that in the present work we deal with a modification of the fluid limit of the workload process that takes into account feedback. Instead, the fluid limit of the queue length process has been used in [7]. In particular, a key point to prove sufficiency is that we now show that this process is part of a solution of the continuous dynamic complementary problem (DCP), also known as deterministic Skorokhod problem.

The paper is organized as follows. In Sect. 2, we give notation and describe the BC in more detail, introducing the associated queueing network equations. Section 3 contains fluid stability analysis, at that, in Sect. 3.1, the fluid limit model is constructed, and the proof of stability condition is given in Sect. 3.2 (Theorem 1).

2   Notation and Description of the BC with "Non-crossing" Feedback

We first give basic notation. Vector are column vectors and (in)equalities are interpreted component-wise. v^T denotes the transpose of a vector (or a matrix). For any integer $d \geq 1$, let $\mathbb{R}^d_+ = \{ v \in \mathbb{R}^d : v \geq 0 \}$, $\mathbb{Z}^d_+ = \{ v = (v_1, \ldots, v_d)^{\mathbf{T}} \in \mathbb{R}^d : v_i \in \mathbb{Z}_+ \}$. For a vector $v = (v_1, \ldots, v_d)^{\mathbf{T}} \in \mathbb{R}^d$, let $|v| = \sum_{i=1}^d |v_i|$. We denote $diag(v)$ the diagonal matrix with diagonal entries being the components of vector v, and I is the d-dimensional identity matrix. We say that a sequence of vectors $\{v^n\}_{n\geq 1}$ converges to a vector v as $n \to \infty$ if $|v^n - v| \to 0$, and denote it as $\lim_{n\to\infty} v^n = v$. (This convergence is equivalent to the component-wise convergence.) For $n \geq 1$, let $\phi^n \colon [0, \infty) \to \mathbb{R}^d$ be right continuous functions having limits on the left on $(0, \infty)$, and let function $\phi \colon [0, \infty) \to \mathbb{R}^d$ be continuous. We say that ϕ^n converges to ϕ as $n \to \infty$ uniformly on compacts (u.o.c.) if for any $T \geq 0$,

$$\|\phi^n - \phi\|_T := \sup_{t\in[0,T]} |\phi^n(t) - \phi(t)| \to 0 \quad \text{as } n \to \infty, \tag{1}$$

and write it as $\lim_{n\to\infty} \phi^n = \phi$. If function ϕ is differentiable at a point $s \in (0, \infty)$ then s is a *regular* point of ϕ, and we denote the derivative by $\dot{\phi}(s)$.

Recall that we consider a BC system with J infinite buffer servers and $K \geq J$ customer classes. In what follows, we use index k to denote the quantities related to class-k customers, $k \in \{1, 2, \ldots, K\}$. Let $s(k) \subset \{1, \ldots, J\}$ be the set of servers that need to work together to service a class-k customer. Note that the capacity $\#s(k) \geq 1$ and that, if $\#s(k) = 1$, then server collaboration is not required.

Evidently, $\cup_{k=1}^{K} s(k) = \{1, \ldots, J\}$, and we assume non-overlapping property: for each two classes $k \neq k'$,

$$s(k) \cap s(k') = \emptyset \quad \text{if} \quad \min\{s(k), s(k')\} > 1. \tag{2}$$

Define the customer classes $C(j) = \{k = 1, \ldots, K : j \in s(k)\}$ served by server $j \in \{1, \ldots, J\}$, and assume that, for each j, the capacity

$$\#\{k \in C(j) : \#s(k) > 1\} \le 1. \tag{3}$$

In other words, at most one class may capture a given server for cooperation. To ensure work-conserving (or non-idling) discipline, in addition to the non-overlapping property, we also assume that multiserver customers have preemptive-resume priority.

Let $\xi_k(i)$, $i \ge 2$, be the independent identically distributed (i.i.d.) inter-arrival times of the ith class-k customers arriving from outside the system after instant 0, and let $\eta_k(i)$, $i \ge 2$, be the i.i.d. service times of the ith class-k customers finishing service after instant 0 (this is time required by any server in the set $s(k)$). All sequences are assumed to be mutually independent. We denote the generic elements of these sequences by ξ_k and η_k, respectively. The residual arrival time $\xi_k(1)$ of the first class-k customer entering the network after instant 0 is independent of $\{\xi_k(i), i \ge 2\}$. Also the residual service time $\eta_k(1)$ of a class-k customer initially being served, if any, is independent of $\{\eta_k(i), i \ge 2\}$, and $\eta_k(1) =_{st} \eta_k$ if class k is initially empty.

For each $k = 1, \ldots, K$, we impose the following standard conditions on inter-arrival and service times, both with general distribution (see [3]):

$$\mathsf{E}\,\eta_k < \infty, \tag{4}$$
$$\mathsf{E}\,\xi_k < \infty, \tag{5}$$
$$\mathsf{P}(\xi_k \ge x) > 0, \quad \text{for any } x \in [0, \infty). \tag{6}$$

Then, in particular, the arrival rate $\alpha_k := 1/\mathsf{E}\xi_k \in (0, \infty)$ and the service rate $\mu_k := 1/\mathsf{E}\eta_k > 0$, and we denote $\alpha = (\alpha_1, \ldots, \alpha_K)^T$ and $\mu = (\mu_1, \ldots, \mu_K)^T$. Also we assume that the inter-arrival times are spread out, that is, for some integer $r > 1$ and functions $f_k \ge 0$ with $\int_0^\infty f_k(y)\,dy > 0$,

$$\mathsf{P}\left(a \le \sum_{i=2}^{r} \xi_k(i) \le b\right) \ge \int_a^b f_k(y)\,dy, \quad \text{for any } 0 \le a < b. \tag{7}$$

A restricted type of feedback (we name it "non-crossing") is allowed in our model: a class-k customer, when finishes service, re-enters the system and becomes class-k customer with a probability $P_{kk} \in [0, 1)$, and a class-k customer needing cooperation of servers $s(k)$ with $\#s(k) > 1$, when finishes service, becomes a class-ℓ customer if $\#s(\ell) = 1$ and $s(\ell) \in s(k)$, that is, if ℓ and k are concurrent customer classes at server $s(\ell)$, with probability $P_{k\ell} \in [0, 1)$. Then, with probability $1 - \sum_{\ell=1}^{K} P_{k\ell} \ge 0$, a class-$k$ customer leaves the system upon service. Thus, $P := (P_{k\ell})_{k,\ell=1}^{K}$ is the (sub-stochastic) routing (or flow) matrix of the network. It is assumed that spectral radius of P is strictly less than 1, and hence, the inverse matrix $Q = (I - P^{T})^{-1}$ is well defined.

Define vector $\lambda = (\lambda_1, \ldots, \lambda_K)^{T}$ as (the unique) solution to the traffic equation

$$\lambda = \alpha + P^{T}\lambda, \quad \text{equivalently,} \quad \lambda = Q\alpha, \tag{8}$$

where λ_k can be interpreted as the potential long run arrival rate of class-k customers into the system. Let $\rho_j = \sum_{k \in C(j)} \lambda_k/\mu_k$ be the traffic intensity for server j, and $\rho := (\rho_1, \ldots, \rho_J)^{T}$.

Now we introduce the following primitive processes describing the dynamics of the queueing network: the exogenous arrival process $E = \{E(t) := (E_1(t), \ldots, E_K(t))^{T}, t \geq 0\}$, where

$$E_k(t) = \max \left\{ n \geq 1 : \sum_{i=1}^{n} \xi_k(i) \leq t \right\} \tag{9}$$

is the total number of class-k arrivals from outside to the system in interval $[0, t]$. We also introduce the process $S = \{S(t) := (S_1(t), \ldots, S_K(t)), t \geq 0\}$, where the renewal process

$$S_k(t) = \max \left\{ n \geq 1 : \sum_{i=1}^{n} \eta_k(i) \leq t \right\} \tag{10}$$

is the total number of class-k customers that would be served in interval $[0, t]$, provided all servers from $s(k)$ devote all time to class-k customers. (By definition, $E(0) = S(0) = 0$.) The routing process $\Phi = \{\Phi(n)\}_{n \in \mathbb{N}}$ is defined as follows:

$$\Phi_k(n) = \sum_{i=1}^{n} \phi^{k}(i), \tag{11}$$

where, for each $i \in \mathbb{N}$, K-dimensional vectors $\phi^{k}(i) = \{\phi_{\ell}^{k}(i), \ell = 1, \ldots, K\}$ are i.i.d. (independent of the inter-arrival and service time processes), with at most one component equals 1, and the rest components being equal 0. If $\phi_{j}^{k}(i) = 1$ then the ith class-k customer becomes class-j, while $\phi^{k}(i) = 0$ means the departure from the network.

Now we introduce the descriptive processes to measure the performance of the network. For any $t \geq 0$ and k, let $A_k(t)$ be the number of class-k arrivals (from outside and by feedback) by time t, $D_k(t)$ be the number of class-k departures (to other classes or outside the system), and let $Z_k(t)$ be the number of class-k customers being served at time t, so $Z_k(t) \in \{0, 1\}$. Also let $T_k(t)$ be the total service time devoted to class-k customers in interval $[0, t]$. Denote $Y_j(t)$ the idle time of server j in $[0, t]$, and let $Q_j(t)$ be the number of customers in the buffer of station j at time t, $j \in \{1, \ldots, J\}$. In an evident notation, processes D, T and Y are non-decreasing and satisfy initial conditions $D(0) = T(0) = Y(0) = 0$. We note that $A(0) = 0$, and assume that $Z(0)$ and $Q(0)$ are mutually independent and independent of all above given quantities.

For each t and k, we define the remaining time $U_k(t)$ until the next exogenous class-k arrival, and the remaining service time $V_k(t)$ of class-k customer being served at time t, if any. We introduce (in an evident notation) processes U and V, assume that they are right-continuous, and define $V_k(t) = 0$ if $Z_k(t) = 0$. Note that $U_k(0) = \xi_k(1)$, while $V_k(0) = \eta_k(1)$ if $Z_k(0) = 1$. Now we define the process $X = \{X(t), t \geq 0\}$ describing the dynamics of the network, where $X(t) := (Q(t), Z(t), U(t), V(t))^{\mathbf{T}}$, with the state space $\mathbb{X} = \mathbb{Z}_+^K \times \{0,1\}^K \times \mathbb{R}_+^K \times \mathbb{R}_+^K$. The process X is a piecewise-deterministic Markov process which satisfies Assumption 3.1 [5], and is a strong Markov process (p. 58, [3]).

We define the workload process $W = \{W(t) := (W_1(t), \ldots, W_J(t))^T, t \geq 0\}$, where $W_j(t)$ is the (workload) time needed to complete service of all class-k customers present in the system at time t, for any $k \in C(j)$. We introduce the cumulative service time process

$$\Upsilon = \{\Upsilon(n) := (\Upsilon_1(n_1), \ldots, \Upsilon_K(n_K))^T, \ n = (n_1, \ldots, n_K) \in \mathbb{N}^K\}, \tag{12}$$

where $\Upsilon_k(n_k)$ is the total amount of service time of the first n_k class-k customers (including the remaining service time at time 0 for the first one), by any of the servers in the set $s(k)$. Note that this time is the same for each server from $s(k)$, and that $\Upsilon_k(0) = 0$.

The following queueing network equations, which are easy to verify, hold for all $t \geq 0$, $k = 1, \ldots, K$ and $j = 1, \ldots, J$:

$$A(t) = E(t) + \sum_{k=1}^{K} \Phi_k(D_k(t)), \tag{13}$$

$$D_k(t) = S_k\left(T_k(t)\right), \tag{14}$$

$$Q_k(t) = Q_k(0) + A_k(t) - \left(D_k(t) + Z_k(t)\right), \tag{15}$$

$$\sum_{k \in C(j)} T_k(t) + Y_j(t) = t, \tag{16}$$

$$\int_0^\infty W_j(t)\, dY_j(t) = 0, \tag{17}$$

$$W(t) = C\left(\Upsilon(Q(0) + A(t)) - T(t)\right), \tag{18}$$

where $e = (1, \ldots, 1)^T \in \mathbb{R}^d$ and C is the $J \times K$ matrix defined by:

$$C_{jk} = \begin{cases} 1, & \text{if } j \in s(k), \quad \text{equivalently, if and only if } k \in C(j), \\ 0, & \text{otherwise.} \end{cases} \tag{19}$$

Note that Eq. (17) reflects the work-conserving property introduced above. Also we note that Eq. (16) can be written as $CT(t) + Y(t) = t\,e$.

We assume that the service discipline is head-of-the-line (HL): only the oldest customer of each class can receive service. It gives the additional equation:

$$\Upsilon(D(t)) \leq T(t) < \Upsilon(D(t) + e). \tag{20}$$

3 Stability Analysis of the BC System

By definition, a queueing network is stable if its associated underlying Markov process X is positive Harris recurrent, that is, it has a unique invariant probability measure. To prove stability of the network it is enough to establish stability of the associated fluid limit model [3].

3.1 The Fluid Limit Model

Now we present, without proof, an analogue of Theorem 4.1 [3] (see also Proposition 1 [7]). If $X(0) = (Q(0), Z(0), U(0), V(0))^{\mathbf{T}} = x$, then we denote X as X^x (and analogously, for the processes E, S, D, T, Y, W).

Proposition 1. *Consider the BC system. Then, for almost all sample paths and any sequence of initial states $\{x_n\}_{n\geq 1} \subset \mathbb{X}$ with $\lim_{n\to\infty} |x_n| = \infty$, there exists a subsequence $\{x_{n_r}\}_{r\geq 1} \subseteq \{x_n\}_{n\geq 1}$ with $\lim_{r\to\infty} |x_{n_r}| = \infty$ such that the following limit*

$$\lim_{r\to\infty} \frac{1}{|x_{n_r}|} X^{x_{n_r}}(0) := \bar{X}(0), \tag{21}$$

exists, and moreover the following u.o.c. limit exists for each $t \geq 0$,

$$\lim_{r\to\infty} \frac{1}{|x_{n_r}|} \left(X^{x_{n_r}}(|x_{n_r}|t), D^{x_{n_r}}(|x_{n_r}|t), T^{x_{n_r}}(|x_{n_r}|t), Y^{x_{n_r}}(|x_{n_r}|t), W^{x_{n_r}}(|x_{n_r}|t) \right)$$
$$:= \left(\bar{X}(t), \bar{D}(t), \bar{T}(t), \bar{Y}(t), \bar{W}(t) \right), \tag{22}$$

where (in evident notation)

$$\bar{X}(t) := \left(\bar{Q}(t), \bar{Z}(t), \bar{U}(t), \bar{V}(t) \right)^{\mathbf{T}}, \tag{23}$$

and the components of vectors $\bar{U}(t)$, $\bar{V}(t)$ have, respectively, the form

$$\bar{U}_k(t) = (\bar{U}_k(0) - t)^+, \quad \bar{V}_k(t) = (\bar{V}_k(0) - t)^+, \quad k = 1,\ldots,K. \tag{24}$$

Furthermore, the following equations are satisfied for any $t \geq 0$, $k = 1, \ldots, K$
and $j = 1, \ldots, J$:

$$\bar{A}(t) = t\,\alpha + P^T \bar{D}(t), \tag{25}$$

$$\bar{D}(t) = M^{-1} \bar{T}(t), \tag{26}$$

$$\bar{Z}_k(t) = 0, \tag{27}$$

$$\bar{Q}(t) = \bar{Q}(0) + \bar{A}(t) - \bar{D}(t) = \bar{Q}(0) + t\,\alpha - (I - P^T)\,\bar{D}(t), \tag{28}$$

$$C\bar{T}(t) + \bar{Y}(t) = t\,e, \tag{29}$$

$$\int_0^\infty \bar{W}_j(t)\, d\bar{Y}_j(t) = 0, \tag{30}$$

$$\bar{W}(t) = C\left(M\left(\bar{Q}(0) + \bar{A}(t)\right) - \bar{T}(t)\right) = C\,M\,\bar{Q}(t), \tag{31}$$

where diagonal matrix M *is defined as*

$$M = diag\left(\left(\frac{1}{\mu_1}, \ldots, \frac{1}{\mu_K}\right)^T\right). \tag{32}$$

We note that

$$\rho = C\,M\,\lambda. \tag{33}$$

Any limit $(\bar{X}, \bar{D}, \bar{T}, \bar{Y}, \bar{W})$ in (21), (22) is called a fluid limit associated with the BC system, [3]. Thus, Proposition 1 states that any fluid limit associated with the BC system satisfies the fluid model Eqs. (25)–(31).

Remark 1. By Lemma 5.3 in [3], hereinafter we will assume without loss of generality that $\bar{U}(0) = \bar{V}(0) = 0$, which, by (24), implies $\bar{U}(t) = \bar{V}(t) = 0$ for all $t > 0$. We denote it $\bar{U} = \bar{V} = 0$ and identify $\bar{X}$ with $\bar{Q}$.

Definition 1. *The fluid limit* $(\bar{Q}, \bar{D}, \bar{T}, \bar{Y}, \bar{W})$ *associated with a queueing network is stable, if there exists* $t_1 \geq 0$ *(depending on the input and service rates only) such that if* $|\bar{Q}(0)| = 1$, *then*

$$\bar{Q}(t) = 0 \ \text{for all}\ t \geq t_1. \tag{34}$$

3.2   The Stability Criterion

Now we are ready to introduce and prove the stability criterion of the BC system with "non-crossing" feedback, following Theorem 5.1 [3]. We prove this result under a technical Assumption (A) concerning the routing matrix P, which is equivalent to the "non-crossing" feedback assumption. We first introduce a process $\widetilde{W}$ by

$$\widetilde{W}(t) = C\,M\,Q\,\bar{Q}(t), \quad t \geq 0. \tag{35}$$

Remark 2. Process $\widetilde{W}$ has a key role in the proof of Theorem 1 and deserves a few words. Intuitively, from (31) the fluid limit of the workload process $\bar{W}$ is related with the fluid limit of the queue length process $\bar{Q}$ by $\bar{W}(t) = C\,M\,\bar{Q}(t)$. Therefore, $\widetilde{W}$ defined in this way is a correction of $\bar{W}$ that takes into account feedback. Note that $\widetilde{W} = \bar{W}$ in case of no feedback, since in this case $P = 0$ and then $Q = I$. Consider a non-trivial example of W-model with $J = 2$ and $K = 3$ (see Introduction) in which only feedback from class-3 to other classes is allowed. Then,

$$P = \begin{pmatrix} 0 & 0 & 0 \\ 0 & 0 & 0 \\ p_{31} & p_{32} & 0 \end{pmatrix}. \tag{36}$$

Since $C = \begin{pmatrix} 1 & 0 & 1 \\ 0 & 1 & 1 \end{pmatrix}$ for this model, we obtain

$$\widetilde{W}_1(t) = \frac{1}{\mu_1}\,\bar{Q}_1(t) + \left(\frac{p_{31}}{\mu_1} + \frac{1}{\mu_3}\right)\bar{Q}_3(t), \tag{37}$$

$$\widetilde{W}_2(t) = \frac{1}{\mu_2}\,\bar{Q}_2(t) + \left(\frac{p_{32}}{\mu_2} + \frac{1}{\mu_3}\right)\bar{Q}_3(t). \tag{38}$$

Because

$$\bar{W}_1(t) = \frac{1}{\mu_1}\,\bar{Q}_1(t) + \frac{1}{\mu_3}\,\bar{Q}_3(t), \quad \bar{W}_2(t) = \frac{1}{\mu_2}\,\bar{Q}_2(t) + \frac{1}{\mu_3}\,\bar{Q}_3(t), \tag{39}$$

we have then that

$$\widetilde{W}_1(t) = \bar{W}_1(t) + \frac{p_{31}}{\mu_1}\,\bar{Q}_3(t), \quad \widetilde{W}_2(t) = \bar{W}_2(t) + \frac{p_{32}}{\mu_2}\,\bar{Q}_3(t), \tag{40}$$

expressions from which it is evident that differences between $\widetilde{W}$ and $\bar{W}$ are due to feedback from class-3 customers to the other classes.

Assumption (A). ("non-crossing" feedback):
The matrix P is such that for any $t \geq 0$ and $j = 1, \ldots, J$,

$$\bar{W}_j(t) = 0 \text{ if and only if } \widetilde{W}_j(t) = 0. \tag{41}$$

Remark 3. Assumption (A) has a key role in proving property (c) in the proof of sufficiency in Theorem 1. It is trivially accomplished if no feedback is allowed since in this case $\widetilde{W} = \bar{W}$. We can illustrate this assumption by introducing a simple example. Consider example introduced in Remark 2 but with (possible) feedback from any customer class to itself (that is, in the more general "non-crossing" feedback setting). Then,

$$P = \begin{pmatrix} p_{11} & 0 & 0 \\ 0 & p_{22} & 0 \\ p_{31} & p_{32} & p_{33} \end{pmatrix} \tag{42}$$

and

$$\widetilde{W}_1(t) = \frac{1}{\mu_1\,(1-p_{11})}\,\bar{Q}_1(t) + \Big(\frac{p_{31}}{\mu_1\,(1-p_{11})\,(1-p_{33})} + \frac{1}{\mu_3\,(1-p_{33})}\Big)\,\bar{Q}_3(t),$$
$$(43)$$

$$\widetilde{W}_2(t) = \frac{1}{\mu_2\,(1-p_{22})}\,\bar{Q}_2(t) + \Big(\frac{p_{32}}{\mu_2\,(1-p_{22})\,(1-p_{33})} + \frac{1}{\mu_3\,(1-p_{33})}\Big)\,\bar{Q}_3(t).$$
$$(44)$$

Because

$$\bar{W}_1(t) = \frac{1}{\mu_1}\,\bar{Q}_1(t) + \frac{1}{\mu_3}\,\bar{Q}_3(t), \quad \bar{W}_2(t) = \frac{1}{\mu_2}\,\bar{Q}_2(t) + \frac{1}{\mu_3}\,\bar{Q}_3(t), \qquad (45)$$

and all coefficients are positive, it is immediate to check that Assumption (A) holds.

Analogously to [6], the crucial fact in the proof of the stability criterion (Theorem 1) is that the fluid limit $\widetilde{W}$ turns out to be a part of a solution of a linear Skorokhod problem, while the fluid limit process $\bar{Q}$ is instead used in [3]. We note that in some settings, the workload is better adapted to the use of the methodology of the Skorokhod problems than the queue-size process. On the other hand, a key point in the proof is the adequate choice of the Lyapunov function.

Theorem 1. *If the non-overlapping BC system with "non-crossing" feedback given by fluid model Eqs. (25)–(31) and Assumption (A), satisfy conditions (4)–(7), then a sufficient stability condition is*

$$\max_{j=1,\ldots,J} \rho_j < 1, \qquad (46)$$

while $\max_{j=1,\ldots,J} \rho_j \le 1$ *is the necessary stability condition.*

Proof. Sufficiency: By the Eqs. (25)–(31),

$$\bar{A}(t) = t\,\alpha + P^T\big(\bar{Q}(0) + \bar{A}(t) - \bar{Q}(t)\big), \qquad (47)$$

implying

$$\big(I - P^T\big)\,\bar{A}(t) = t\,\alpha + P^T\big(\bar{Q}(0) - \bar{Q}(t)\big). \qquad (48)$$

It in turn implies

$$\bar{A}(t) = t\,\lambda + Q\,P^T\big(\bar{Q}(0) - \bar{Q}(t)\big). \qquad (49)$$

By (49) we obtain

$$\begin{aligned}
\bar{W}(t) &= C\,M\,\bar{Q}(0) + C\,M\,\bar{A}(t) - C\,\bar{T}(t) \\
&= C\,M\,\bar{Q}(0) + C\,M\,\Big(t\,\lambda + Q\,P^T\big(\bar{Q}(0) - \bar{Q}(t)\big)\Big) - t\,e + \bar{Y}(t) \\
&= C\,M\,\bar{Q}(0) + (\rho - e)\,t + C\,M\,Q\,P^T\big(\bar{Q}(0) - \bar{Q}(t)\big) + \bar{Y}(t) \\
&= C\,M\,Q\,\bar{Q}(0) + (\rho - e)\,t - C\,M\,Q\,P^T\,\bar{Q}(t) + \bar{Y}(t). \qquad (50)
\end{aligned}$$

Since $\bar{W}(t) = C M \bar{Q}(t)$, it then follows from (50) that

$$C M Q \bar{Q}(t) = C M Q \bar{Q}(0) + (\rho - e)\, t + \bar{Y}(t), \qquad (51)$$

or, denoting $\tilde{X}(t) = C M Q \bar{Q}(0) + (\rho - e)\, t$,

$$\widetilde{W}(t) = \tilde{X}(t) + \bar{Y}(t). \qquad (52)$$

It is easy to check that the following properties hold:

(a) $\tilde{X}(\cdot)$ has continuous paths with $\tilde{X}(0) \geq 0$,
(b) $\widetilde{W}(t) \geq 0$ for all $t \geq 0$,
(c) $\bar{Y}(\cdot)$ has nondecreasing paths, $\bar{Y}(0) = 0$, and $\bar{Y}_j(\cdot)$ increases only at times t such that $\bar{W}_j(t) = 0$ for $j = 1, \ldots, J$ (see (30)). By Assumption (A), $\bar{Y}_j(\cdot)$ increases only when $\widetilde{W}_j(t) = 0$, $j = 1, \ldots, J$.

It follows that the paths of processes $(\widetilde{W}, \bar{Y})$ are solutions of the continuous dynamic complementarity problem (DCP) for $\tilde{X}$ (see Definition 5.1 [3]), also known as the deterministic Skorokhod problem. Moreover, it is easy to check that condition (5.1) in [3],

$$\widetilde{W}(s) + \tilde{X}(t + s) - \tilde{X}(s) \geq \theta t \quad \forall t,\, s \geq 0, \qquad (53)$$

is satisfied with $\theta := \rho - e$. Therefore, by Lemma 5.1 [3],

$$\dot{\bar{Y}}(s) \leq (e - \rho), \quad \text{if } s \geq 0 \text{ is a regular point of } \bar{Y}(\cdot). \qquad (54)$$

Define function f as

$$f(t) = |\widetilde{W}(t)| = e^T \widetilde{W}(t). \qquad (55)$$

It follows that

$$f(t) = e^T \left(\tilde{X}(t) + \bar{Y}(t) \right) = f(0) + e^T \left((\rho - e)\, t + \bar{Y}(t) \right)$$
$$= f(0) + \sum_{j=1}^{J} \left((\rho_j - 1)\, t + \bar{Y}_j(t) \right). \qquad (56)$$

Assume that $t > 0$ is a regular point for $\widetilde{W}$ (equivalently, for $\bar{Y}$).

If $f(t) > 0$, then there exists $j_0 \in \{1, \ldots, J\}$ such that $\widetilde{W}_{j_0}(t) > 0$. By definition of process $\widetilde{W}$, there exists some k_0 with $\bar{Q}_{k_0}(t) > 0$, which implies that for any $j_1 \in s(k_0)$, $\widetilde{W}_{j_1}(t) > 0$, which in turn implies that $\dot{\bar{Y}}_{j_1}(t) = 0$. Fix a $j_1 \in s(k_0)$. Hence, by (56) and (54),

$$\dot{f}(t) = \sum_{j=1}^{J} \left((\rho_j - 1) + \dot{\bar{Y}}_j(t) \right) = (\rho_{j_1} - 1) + \sum_{j \neq j_1} \left((\rho_j - 1) + \dot{\bar{Y}}_j(t) \right)$$
$$\leq \rho_{j_1} - 1 \leq \max_{j=1,\ldots,J} \rho_j - 1 = -\kappa, \qquad (57)$$

where $\kappa = 1 - \max_{j=1,\dots,J} \rho_j > 0$ by assumption.

As f is a nonnegative function that is absolutely continuous and, for almost surely all regular points t, $\dot{f}(t) \leq -\kappa$ whenever $f(t) > 0$, then, by Lemma 5.2 [3], f is non increasing and $f(t) = 0$ for $t \geq f(0)/\kappa$. That is,

$$\widetilde{W}(t) = 0, \ t \geq \delta := \frac{|\bar{W}(0)|}{1 - \max_{j=1,\dots,J} \rho_j}. \tag{58}$$

Finally, by definition of process $\widetilde{W}$, $\widetilde{W} = 0$ if and only if $\bar{Q} = 0$. Moreover, since

$$|\widetilde{W}(t)| = |C\,M\,Q\,\bar{Q}(t)| = \sum_{j=1}^{J}\Big(\sum_{k \in C(j)} a_{kj}\bar{Q}_k(t)\Big), \tag{59}$$

where a_{kj} depends on μ and matrix $Q = (q_{k\ell})_{k,\ell=1,\dots,K}$. More exactly, $0 \leq a_{kj} \leq M_K$, for any $k = 1,\dots,K$ and $j = 1,\dots,J$, where

$$M_K = \Big(\max_{k=1,\dots,K} \max_{\ell=1,\dots,K} q_{k\ell}\Big)\Big(\max_{k=1,\dots,K} \frac{1}{\mu_k}\Big)K > 0. \tag{60}$$

Then,

$$|\widetilde{W}(t)| \leq J\,M_K\,|\bar{Q}(t)|, \tag{61}$$

and we obtain

$$\bar{Q}(t) = 0, \quad t \geq \frac{J\,M_K\,|\bar{Q}(0)|}{1 - \max_{j=1\dots,J}\rho_j} \geq 0. \tag{62}$$

It means that the fluid model is stable (by Definition 1), and Theorem 4.2 [3] ensures the stability of the queueing network.

Necessity: To prove the necessity of condition $\max_{j=1,\dots,J}\rho_j \leq 1$, we assume $\rho_{j_0} > 1$ for some $j_0 \in \{1,\dots,J\}$. Consider the non-negative function

$$g(t) = \widetilde{W}_{j_0}(t) = g(0) + (\rho_{j_0} - 1)\,t + \bar{Y}_{j_0}(t) \geq (\rho_{j_0} - 1)\,t > 0, \quad t > 0. \tag{63}$$

Then $\widetilde{W}_{j_0}(t) > 0$, which implies $\bar{Q}(t) \neq 0$ since $\widetilde{W}_{j_0}(t) = \sum_{k \in C(j_0)} a_{kj_0}\bar{Q}_k(t)$ with $a_{kj_0} \geq 0$, finishing the proof. $\qquad\Box$

Remark 4. We note that in practice, condition (46) can be treated as stability criterion which, for the W-model in Remark 3, becomes

$$\max\{\rho_1, \rho_2\} < 1, \tag{64}$$

where

$$\rho_1 = \alpha_1 \frac{1}{\mu_1\,(1 - p_{11})} + \alpha_3\Big(\frac{p_{31}}{\mu_1\,(1 - p_{11})\,(1 - p_{33})} + \frac{1}{\mu_3\,(1 - p_{33})}\Big) \tag{65}$$

$$\rho_2 = \alpha_2 \frac{1}{\mu_2\,(1 - p_{22})} + \alpha_3\Big(\frac{p_{32}}{\mu_2\,(1 - p_{22})\,(1 - p_{33})} + \frac{1}{\mu_3\,(1 - p_{33})}\Big). \tag{66}$$

This can be easily seen since $\rho = C M Q \alpha$,

$$CM = \begin{pmatrix} \frac{1}{\mu_1} & 0 & \frac{1}{\mu_3} \\ 0 & \frac{1}{\mu_2} & \frac{1}{\mu_3} \end{pmatrix} \tag{67}$$

and

$$Q = (I - P^T)^{-1} = \begin{pmatrix} \frac{1}{1-p_{11}} & 0 & \frac{p_{31}}{(1-p_{11})(1-p_{33})} \\ 0 & \frac{1}{1-p_{22}} & \frac{p_{32}}{(1-p_{22})(1-p_{33})} \\ 0 & 0 & \frac{1}{1-p_{33}} \end{pmatrix}. \tag{68}$$

If the model does not allow feedback, then $p_{ij} = 0$ for all $i, j = 1, 2, 3$, and

$$\rho_1 = \frac{\alpha_1}{\mu_1} + \frac{\alpha_3}{\mu_3}, \quad \rho_2 = \frac{\alpha_2}{\mu_2} + \frac{\alpha_3}{\mu_3}. \tag{69}$$

4   Conclusion

We consider a non-overlapping Basic Collaboration queueing system, which is a multi-class queueing system with "non-crossing" feedback, that generalizes the so-called W-model [15]. In the system, some customer classes cooperate to be processed by a subset of non-overlapping servers, and feedback is allowed from each customer class to itself, and also from each customer class needing cooperation of some servers to any of the concurrent customer classes at any of these servers. We apply the fluid limit approach methodology [3] to find the stability condition of the system.

Acknowledgement. The authors wish to thank the anonymous referees for careful reading and helpful comments that resulted in an overall improvement of the paper.

References

1. Arthurs, E., Kaufman, J.S.: Sizing a message store subject to blocking criteria. In: Proceedings of the Third International Symposium on Modelling and Performance Evaluation of Computer Systems: Performance of Computer Systems, pp. 547–564. North-Holland Publishing Co., Amsterdam (1979)
2. Brill, P., Green, L.: Queues in which customers receive simultaneous service from a random number of servers: a system point approach. Manage. Sci. **30**(1), 51–68 (1984)
3. Dai, J.G.: On positive Harris recurrence of multiclass queueing networks: a unified approach via fluid limit models. Ann. Appl. Prob. **5**(1), 49–77 (1995)
4. Dai, J.G.: A fluid limit model criterion for unstability of multiclass queueing networks. Ann. Appl. Prob. **6**(3), 751–757 (1996)
5. Davis, M.H.A.: Piecewise deterministic Markov processes: a general class of non-diffusion stochastic models. J. Roy. Statist. Soc. Ser. B. **46**, 353–388 (1984)

6. Delgado, R.: State space collapse and stability of queueing networks. Math. Meth. Oper. Res. **72**, 477–499 (2010)
7. Delgado, R., Morozov, E.: Stability analysis of cascade networks via fluid models. Perform. Eval. **82**, 39–54 (2014)
8. Fletcher, G.Y., Perros, H., Stewart, W.: A queueing system where customers require a random number of servers simultaneously. Eur. J. Oper. Res. **23**, 331–342 (1986)
9. Garnet, O., Mandelbaum, A.: An introduction to Skills-Based Routing and its operational complexities. http://iew3.technion.ac.il/serveng/Lectures/SBR.pdf
10. Green, L.: Comparing operating characteristics of queues in which customers require a random number of servers. Manage. Sci. **27**(1), 65–74 (1980)
11. Kaufman, J.: Blocking in a shared resource environment. IEEE Trans. Commun. **29**(10), 1474–1481 (1981)
12. Kelly, F.P.: Loss networks. Ann. Appl. Prob. **1**(3), 319–378 (1991)
13. Kim, S.: M/M/s queueing system where customers demand multiple server use. Ph.D. thesis, Southern Methodist University (1979)
14. Rumyantsev, A., Morozov, E.: Stability criterion of a multiserver model with simultaneous service. Ann. Oper. Res. **252**(1), 29–39 (2015). doi:10.1007/s10479-015-1917-2
15. Talreja, R., Whitt, W.: Fluid models for overloaded multiclass many-server queueing systems with first-come, first-served routing. Manage. Sci. **54**, 1513–1527 (2008)
16. Tikhonenko, O.: Generalized erlang problem for service systems with finite total capacity. Probl. Inf. Transm. **41**(3), 243–253 (2005)
17. Van Dijk, N.M.: Blocking of finite source inputs which require simultaneous servers with general think and holding times. Oper. Res. Lett. **8**(1), 45–52 (1989)
18. Whitt, W.: Blocking when service is required from several facilities simultaneously. AT&T Tech. J. **64**(8), 1807–1856 (1985)
19. Wong, D., Paciorek, N., Walsh, T., DiCelie, J., Young, M., Peet, B.: *Concordia*: an infrastructure for collaborating mobile agents. In: Rothermel, K., Popescu-Zeletin, R. (eds.) MA 1997. LNCS, vol. 1219, pp. 86–97. Springer, Heidelberg (1997). doi:10.1007/3-540-62803-7_26

Erlang Service System with Limited Memory Space Under Control of AQM Mechanizm

Oleg Tikhonenko[1(✉)] and Wojciech M. Kempa[2]

[1] Faculty of Mathematics and Natural Sciences, College of Sciences,
Cardinal Stefan Wyszynski University in Warsaw,
Ul. Woycickiego 1/3, 01-938 Warsaw, Poland
o.tikhonenko@uksw.edu.pl
[2] Institute of Mathematics, Silesian University of Technology,
Ul. Kaszubska 23, 44-100 Gliwice, Poland
wojciech.kempa@polsl.pl

Abstract. We investigate the $M/G/n \leq \infty/(0, V)$-type Erlang loss service system with $n \leq \infty$ independent service stations and Poisson arrival stream in which volumes of entering demands and their processing times are generally distributed and, in general, are dependent random variables. Moreover, the total volume of all demands present simultaneously in the system is bounded by a non-random value V (system memory capacity). The enqueueing process is controlled by an AQM-type non-increasing accepting function. Two different acceptance rules are considered in which the probability of acceptance is dependent or independent on the volume of the arriving demand. Stationary queue-size distribution and the loss probability are found for both scenarios of the system behavior. Besides, some special cases are discussed. Numerical examples are attached as well.

Keywords: Active Queue Management (AQM) · Erlang service system · Loss probability · Queue-size distribution · Supplementary variables' technique

1 Introduction

Queueing systems with finite buffer capacities are commonly used, especially in the design and performance evaluation of telecommunication and computer networks. They are good models for describing different-type phenomena appearing in network nodes (e.g., IP routers). In packet-oriented networks problems of buffer overflows and packet losses are typical ones. The classical Tail Drop (TD) mechanism rejects the entering packet only when the buffer is completely saturated and hence, according to frequent complex nature of the traffic (e.g., in Internet), it can give some negative consequences, like, e.g., losing packets in series, too long delays in queueing or TCP synchronization.

The idea of the Active Queue Management (AQM) is in implementing the mechanism of the entering packets rejection that is possible even when the buffer

© Springer International Publishing AG 2017
P. Gaj et al. (Eds.): CN 2017, CCIS 718, pp. 366–379, 2017.
DOI: 10.1007/978-3-319-59767-6_29

is not saturated. Hence, the using of AQM can physically decrease the queue length at the node and allows for avoiding the hazard of buffer overflow by reducing the intensity of arrivals (as a consequence of packets dropping). The first AQM scheme, called RED (Random Early Detection), was proposed in [1], where a linearly dropping function was presented. Some modifications of the original RED mechanism can be found e.g. in [3,10], where an exponential and quadratic dropping functions were applied, respectively. An analysis of stationary characteristics of the $M/M/1/N$ queue with packet dropping was carried out e.g. in [2].

In the paper we study the queue-size distribution in the $M/G/n \leq \infty/(0, V)$-type Erlang loss service system with $n \leq \infty$ independent service stations and Poisson arrival stream, in which volumes of entering demands and their processing times are generally distributed and, in general, are dependent random variables. Moreover, the total volume of all demands present simultaneously in the system is bounded by a non-random value V (system memory capacity). We define an AQM-type accepting function which qualifies the arriving packet for service with probability depending on its volume and the total capacity of packets present in the system at the pre-arrival epoch. The basics of the theory of queueing models with randomly distributed packet sizes and bounded system capacity can be found in [4,5]. The representations for the steady-state queue-size distributions in such systems with Poisson arrivals were obtained in [7,8], where the exponentially and generally distributed service times were considered, respectively. Similar results for the multi-server model was derived in [9].

The paper is organized as follows. Section 2 contains a description of the queueing model and necessary notations. In Sect. 3 we define a Markovian process describing the evolution of the system. In Sect. 4 we build a system of partial differential equations for the transient system characteristics. Section 5 contains the formulae for the stationary queue-size distribution and loss probability and in Sect. 6 we analyze some special cases of the original model. Section 7 contains examples of numerical results illustrating theoretical formulae and the last Sect. 8 presents conclusions and final remarks.

2   The Model and Notation

We consider the Erlang $M/G/n \leq \infty/(0, V)$ service system and denote by a the parameter (intensity) of demands entrance flow. Let ζ and ξ be a demand volume and its service time, respectively. $F(x, t) = \mathsf{P}\{\zeta < x, \xi < t\}$ be the joint distribution function of the non-negative random variables ζ and ξ. Then, $L(x) = F(x, \infty)$ and $B(t) = F(\infty, t)$ are the distribution functions of the random variables ζ and ξ, respectively. Denote by $\beta_1 = \mathsf{E}\xi = \int_0^\infty t \, dB(t)$ the mean service time. Let $\sigma(t)$ be the total volume of demands present in the system at time instant t. The values of the process $\sigma(t)$ are bounded by a constant positive value V (system memory capacity). Let $\eta(t)$ be the number of demands present in the system at time instant t.

Consider the right continuous non-increasing (accepting) function $r(x)$ defined on the segment $[0; V]$ such that $r(V) \geq 0$, $r(0) \leq 1$.

Assume that at the epoch t a demand of volume x arrives to the system when the total volume of the other demands present in it is equal to y. In what follows, we analyze two scenarios of system behavior: (1) the arriving demand is accepted to service with probability $r(x + y)$ and removed from the system with probability $1 - r(x + y)$; (2) the arriving demand is accepted to service with probability $r(y)$ and removed from the system with probability $1 - r(y)$. In both cases, the arriving demand is also removed from the system if $x + y > V$ or $\eta(t^-) = n$ (if $n < \infty$).

If the customer of volume x arriving at the epoch t is not accepted to the system, then $\eta(t) = \eta(t^-)$ and $\sigma(t) = \sigma(t^-)$. If it is accepted to the system, then $\eta(t) = \eta(t^-) + 1$ and $\sigma(t) = \sigma(t^-) + x$. If servicing of the demand of volume x is completed at the epoch τ, then we get $\eta(\tau) = \eta(\tau^-) - 1$ and $\sigma(\tau) = \sigma(\tau^-) - x$.

Denote by $\sigma_j(t)$ the volume of the jth demand presenting in the system at the time instant t ($j = 1, \ldots, k$ if $\eta(t) = k$, $1 \le k \le n$). We agree to assume that the customers present in the system at an arbitrary time instant are numerated randomly, that is, if at the time instant t there are k demands in the system, then any of the possible $k!$ numerations can be used with the same probability $1/k!$. Denote by $\xi_j^*(t)$ the remaining service time of jth demand presenting in the system at time instant t.

Later on, the following notation for vectors will be used to reduce the relations:

$$Y_k = (y_1, \ldots, y_k), \ Y_k^j = (y_1, \ldots, y_{j-1}, y_{j+1} \ldots, y_k), (Y_k, u) = (y_1, \ldots, y_k, u). \tag{1}$$

3 Random Process and Functions Describing System Behavior

Behavior of the system under consideration we shall describe by the following Markovian process:

$$\left(\eta(t); \sigma_j(t), \xi_j^*(t), j = 1, \ldots, \eta(t) \right), \tag{2}$$

where the components $\sigma_j(t)$ and $\xi_j^*(t)$ are absent, if $\eta(t) = 0$. In this case, obviously, $\sigma(t) = 0$. Otherwise, we have $\sigma(t) = \sum_{j=1}^{\eta(t)} \sigma_j(t)$.

Process (2) we shall characterize by functions having probability sense as follows:

$$P_0(t) = \mathsf{P}\{\eta(t) = 0\}; \tag{3}$$

$$G_k(x, Y_k, t)\mathrm{d}x = \mathsf{P}\{\eta(t) = k, \sigma(t) \in [x; x + \mathrm{d}x), \xi_j^*(t) < y_j, j = 1, \ldots, k\}, \tag{4}$$

$$\Theta_k(Y_k, t) = \mathsf{P}\{\eta(t) = k, \xi_j^*(t) < y_j, j = 1, \ldots, k\} = \int_0^V G_k(x, Y_k, t)\mathrm{d}x, \tag{5}$$

$$P_k(t) = \mathsf{P}\{\eta(t) = k\} = \int_0^V G_k(x, \infty_k, t)\mathrm{d}x = \Theta_k(\infty_k, t), \ k = 1, \ldots, n, \tag{6}$$

where $\infty_k = (\underbrace{\infty, \ldots, \infty}_{k})$.

It follows from the random demands numeration that the functions $G_k(x, Y_k, t)$ and $\Theta_k(Y_k, t)$ are symmetrical with respect to permutations of components of the vector Y_k.

4   Equations for Introduced Functions and Their Solution

Consider a system whose behavior corresponds to the first scenario describing in Sect. 2.

It can be shown by supplementary variables' technique [6] and taking in consideration the aforementioned symmetry of the functions $G_k(x, Y_k, t)$ and $\Theta_k(Y_k, t)$ that the partial differential equations for the functions (3)–(5) have the form:

$$P_0'(t) = -aP_0(t) \int_0^V r(v)\mathrm{d}L(v) + \left.\frac{\partial\Theta_1(y, t)}{\partial y}\right|_{y=0}; \tag{7}$$

$$\frac{\partial\Theta_1(y, t)}{\partial t} - \frac{\partial\Theta_1(y, t)}{\partial y} + \left.\frac{\partial\Theta_1(y, t)}{\partial y}\right|_{y=0} = aP_0(t)\int_0^V r(v)\mathrm{d}_v F(v, y)$$
$$- a\int_0^V G_1(x, y, t)\left[\int_0^{V-x} r(x+v)\mathrm{d}L(v)\right]\mathrm{d}x + 2\left.\frac{\partial\Theta_2(y, u, t)}{\partial u}\right|_{u=0}; \tag{8}$$

$$\frac{\partial\Theta_k(Y_k, t)}{\partial t} - \sum_{j=1}^{k}\left[\frac{\partial\Theta_k(Y_k, t)}{\partial y_j} - \left.\frac{\partial\Theta_k(Y_k, t)}{\partial y_j}\right|_{y_j=0}\right]$$
$$= \frac{a}{k}\sum_{j=1}^{k}\int_0^V G_{k-1}(x, Y_k^j, t)\left[\int_0^{V-x} r(x+v)\mathrm{d}_v F(v, y_j)\right]\mathrm{d}x$$
$$- a\int_0^V G_k(x, Y_k, t)\left[\int_0^{V-x} r(x+v)\mathrm{d}L(v)\right]\mathrm{d}x + (k+1)\left.\frac{\partial\Theta_{k+1}(Y_k, u, t)}{\partial u}\right|_{u=0},$$
$$k = 2, \ldots, n-1; \tag{9}$$

$$\frac{\partial\Theta_n(Y_n, t)}{\partial t} - \sum_{j=1}^{n}\left[\frac{\partial\Theta_n(Y_n, t)}{\partial y_j} - \left.\frac{\partial\Theta_n(Y_n, t)}{\partial y_j}\right|_{y_j=0}\right]$$
$$= \frac{a}{n}\sum_{j=1}^{n}\int_0^V G_{n-1}(x, Y_n^j, t)\left[\int_0^{V-x} r(x+v)\mathrm{d}_v F(v, y_j)\right]\mathrm{d}x. \tag{10}$$

We say that the system is empty at time t if there are no demands in it at this time.

To explain (7)–(10), we first clear the probability sense of some integrals in these equations:

$\int_0^V r(v)\mathrm{d}L(v)$ is the probability that an arbitrary demand will be accepted to the system if it arrives at the moment when the system is empty;

$\int_0^V r(v)\mathrm{d}_vF(v,y)$ is the probability of the same event, but for a demand with service time less than y; $\int_0^V G_k(x,Y_k,t)\left[\int_0^{V-x} r(x+v)\mathrm{d}L(v)\right]\mathrm{d}x$ is the probability that a demand arriving at time t will be accepted if there are k other demands in the system at this moment and their remaining service times are less than $y_1,\ldots,y_k$, respectively;

$\int_0^V G_k(x,Y_{k+1}^j,t)\left[\int_0^{V-x} r(x+v)\mathrm{d}_vF(v,y_j)\right]\mathrm{d}x$ is the probability of the same event for the arriving demand with service time less than y_j when remaining service times of other demands present in the system are less than $y_1,\ldots,y_{j-1},$ $y_{j+1},\ldots,y_k$, respectively.

Suppose for simplicity that $n=2$. Then, the analysis of the Markovian process (2) provides to the following difference equations for the functions (3)–(5):

$$P_0(t+\Delta t)=P_0(t)\left[1-a\Delta t\int_0^V r(v)\mathrm{d}L(v)\right]+\Theta(\Delta t,t)+o(\Delta t);$$

$$\Theta_1(y,t+\Delta t)=a\Delta tP_0(t)\int_0^V r(v)\mathrm{d}_vF(v,y+\Delta t)+\Theta_1(y+\Delta t,t)-\Theta_1(\Delta t,t)$$

$$-a\Delta t\int_0^V G_1(x,y+\Delta t,t)\left[\int_0^{V-x} r(x+v)\mathrm{d}L(v)\right]\mathrm{d}x$$

$$+\Theta_2(\Delta t,y+\Delta t,t)+\Theta_2(y+\Delta t,\Delta t,t)+o(\Delta t);$$

$$\Theta_2(y_1,y_2,t+\Delta t)=\frac{a\Delta t}{2}\Theta_2(y_1+\Delta t,y_2+\Delta t,t)$$

$$+\frac{a\Delta t}{2}\int_0^V G_1(x,y_2+\Delta t,t)\left[\int_0^{V-x} r(x+v)\mathrm{d}_vF(v,y_1+\Delta t)\right]\mathrm{d}x$$

$$+\Theta_2(y_1+\Delta t,y_2+\Delta t,t)-\Theta_2(y_1+\Delta t,\Delta t,t)-\Theta_2(\Delta t,y_2+\Delta t,t)$$

$$-a\Delta t\int_0^V G_2(x,y_1+\Delta t,y_2+\Delta t,t)\left[\int_0^{V-x} r(x+v)\mathrm{d}L(v)\right]\mathrm{d}x$$

$$+\Theta_3(\Delta t,y_1+\Delta t,y_2+\Delta t,t)+\Theta_3(y_1+\Delta t,\Delta t,y_2+\Delta t,t)$$

$$+\Theta_3(y_1+\Delta t,y_2+\Delta t,\Delta t,t)+o(\Delta t).$$

$$(11)$$

From these difference equations, using standard technique and taking in consideration the symmetry of the functions $G_2(x,y_1,y_2,t)$, $\Theta_2(y_1,y_2,t)$ and $\Theta_3(y_1,y_2,y_3,t)$, we obtain the following partial differential equations:

$$P_0'(t) = -aP_0(t) \int_0^V r(v) dL(v) + \left. \frac{\partial \Theta_1(y,t)}{\partial y} \right|_{y=0} ;$$

$$\frac{\partial \Theta_1(y,t)}{\partial t} - \frac{\partial \Theta_1(y,t)}{\partial y} + \left. \frac{\partial \Theta_1(y,t)}{\partial y} \right|_{y=0} = aP_0(t) \int_0^V r(v) d_v F(v,y)$$

$$- a \int_0^V G_1(x,y,t) \left[\int_0^{V-x} r(x+v) dL(v) \right] dx + 2 \left. \frac{\partial \Theta_2(y,u,t)}{\partial u} \right|_{u=0} ;$$

$$\frac{\partial \Theta_2(y_1,y_2,t)}{\partial t} - \frac{\partial \Theta_2(y_1,y_2,t)}{\partial y_1} + \left. \frac{\partial \Theta_2(y_1,y_2,t)}{\partial y_1} \right|_{y_1=0}$$

$$- \frac{\partial \Theta_2(y_1,y_2,t)}{\partial y_2} + \left. \frac{\partial \Theta_2(y_1,y_2,t)}{\partial y_2} \right|_{y_2=0}$$

$$= \frac{a}{2} \int_0^V G_1(x,y_1,t) \left[\int_0^{V-x} r(x+v) d_v F(v,y_2) \right] dx$$

$$+ \frac{a}{2} \int_0^V G_1(x,y_2,t) \left[\int_0^{V-x} r(x+v) d_v F(v,y_1) \right] dx$$

$$- a \int_0^V G_2(x,y_1,y_2,t) \left[\int_0^{V-x} r(x+v) dL(v) \right] dx + 3 \left. \frac{\partial \Theta_3(y_1,y_2,u,t)}{\partial u} \right|_{u=0} .$$
(12)

For arbitrary k, this technique gives us Eqs. (7)–(10).
Let us introduce the notation

$$R(z,y) = \int_0^z r(V - z + v) d_v F(v,y).$$
(13)

Then Eqs. (7)–(10) can be rewritten as

$$P_0'(t) = -aP_0(t)R(V,\infty) + \left. \frac{\partial \Theta_1(y,t)}{\partial y} \right|_{y=0} ;$$
(14)

$$\frac{\partial \Theta_1(y,t)}{\partial t} - \frac{\partial \Theta_1(y,t)}{\partial y} + \left. \frac{\partial \Theta_1(y,t)}{\partial y} \right|_{y=0} = aP_0(t)R(V,y)$$

$$- a \int_0^V G_1(x,y,t)R(V-x,\infty)dx + 2 \left. \frac{\partial \Theta_2(y,u,t)}{\partial u} \right|_{u=0} ;$$
(15)

$$\frac{\partial \Theta_k(Y_k,t)}{\partial t} - \sum_{j=1}^k \left[\frac{\partial \Theta_k(Y_k,t)}{\partial y_j} - \left. \frac{\partial \Theta_k(Y_k,t)}{\partial y_j} \right|_{y_j=0} \right]$$

$$= \frac{a}{k} \sum_{j=1}^k \int_0^V G_{k-1}(x,Y_k^j,t)R(V-x,y_j)dx - a \int_0^V G_k(x,Y_k,t)R(V-x,\infty)dx$$

$$+ (k+1) \left. \frac{\partial \Theta_{k+1}(Y_k,u,t)}{\partial u} \right|_{u=0} , \quad k = 2, \ldots, n-1.$$
(16)

$$\frac{\partial \Theta_n(Y_n, t)}{\partial t} - \sum_{j=1}^{n} \left[\frac{\partial \Theta_n(Y_n, t)}{\partial y_j} - \frac{\partial \Theta_n(Y_n, t)}{\partial y_j} \bigg|_{y_j=0} \right]$$

$$= \frac{a}{n} \sum_{j=1}^{n} \int_0^V G_{n-1}(x, Y_n^j, t) R(V - x, y_j) dx. \tag{17}$$

If the inequality $\rho = a\beta_1 < \infty$ takes place, the steady state exists for the system under consideration, and the following limits exist in the sense of a weak convergence: $\eta(t) \Rightarrow \eta$; $\sigma(t) \Rightarrow \sigma$; $\xi_j^*(t) \Rightarrow \xi_j^*$, $j = 1, \dots, \eta$, where η, σ, ξ_j^* are the appropriate steady-state characteristics. Then, the following finite limits exist:

$$p_0 = \lim_{t \to \infty} P_0(t) = \mathsf{P}\{\eta = 0\}; \tag{18}$$

$$g_k(x, Y_k) = \lim_{t \to \infty} G_k(x, Y_k, t) \tag{19}$$

and

$$g_k(x, Y_k)dx = \mathsf{P}\{\eta = k; \sigma \in [x, x + dx); \xi_j^* < y_j, j = 1, \dots, k\};$$

$$\theta_k(Y_k) = \lim_{t \to \infty} \Theta_k(Y_k, t) = \mathsf{P}\{\eta = k; \xi_j^* < y_j, j = 1, \dots, k\} = \int_0^V g_k(x, Y_k)dx; \tag{20}$$

$$p_k = \lim_{t \to \infty} P_k(t) = \mathsf{P}\{\eta = k\} = \int_0^V g_k(x, \infty_k)dx = \theta_k(\infty_k) \tag{21}$$

where $k = 1, \dots, n$.

It is clear that the steady-state functions $g_k(x, Y_k)$ and $\theta_k(Y_k)$ are also symmetrical with respect to permutations of components of the vector Y_k.

In steady state, for the value (18) and the functions (19), (20), we obtain the following equations that follow from Eqs. (14)–(17):

$$0 = -ap_0 R(V, \infty) + \frac{\partial \theta_1(y)}{\partial y} \bigg|_{y=0}; \tag{22}$$

$$-\frac{\partial \theta_1(y)}{\partial y} + \frac{\partial \theta_1(y)}{\partial y} \bigg|_{y=0} = ap_0 R(V, y) - a \int_0^V g_1(x, y) R(V - x, \infty) dx$$

$$+ 2 \frac{\partial \theta_2(y, u)}{\partial u} \bigg|_{u=0}; \tag{23}$$

$$-\sum_{j=1}^{k} \left[\frac{\partial \theta_k(Y_k)}{\partial y_j} - \frac{\partial \theta_k(Y_k)}{\partial y_j} \bigg|_{y_j=0} \right]$$

$$= \frac{a}{k} \sum_{j=1}^{k} \int_0^V g_{k-1}(x, Y_k^j) R(V - x, y_j) dx - a \int_0^V g_k(x, Y_k) R(V - x, \infty) dx \tag{24}$$

$$+ (k+1) \frac{\partial \theta_{k+1}(Y_k, u)}{\partial u} \bigg|_{u=0}, \quad k = 2, \dots, n - 1.$$

$$-\sum_{j=1}^{n}\left[\frac{\partial\theta_n(Y_n)}{\partial y_j}-\frac{\partial\theta_n(Y_n)}{\partial y_j}\bigg|_{y_j=0}\right]=\frac{a}{n}\sum_{j=1}^{n}\int_0^V g_{n-1}(x,Y_n^j)R(V-x,y_j)\mathrm{d}x.$$

(25)

In steady state, the following boundary conditions (or equilibrium equations) hold:

$$a\int_0^V g_k(x,Y_k)R(V-x,\infty)\mathrm{d}x=(k+1)\frac{\partial\theta_{k+1}(Y_k,u)}{\partial u}\bigg|_{u=0},\quad k=1,\ldots,n-1.\quad(26)$$

It is clear that $R(z,y)$ is the probability that an arriving demand with service time less than y will be accepted to service, if immediately before its arrival the system had z units of free memory space. Then obviously, the function $R(z,\infty)$ represents the probability that an arbitrary demand is accepted to service, under condition that immediately before its arrival there were z units of free memory space in the system.

Consider the function $H(z,y)=R(z,\infty)-R(z,y)$ which defines the probability that a demand with service time greater than or equal to y is accepted to service under the same condition. Let us present the distribution function $F(x,t)$ as $F(x,t)=L(x)B(t|\zeta<x)$, where $B(t|\zeta<x)=\mathsf{P}\{\xi<t|\zeta<x\}$ is the conditional distribution function of the random variable ξ under condition $\zeta<x$. Then, we have

$$R(z,y)=\int_0^z r(V-z+v)B(y|\zeta=v)\mathrm{d}L(v)$$

(27)

and, consequently,

$$H(z,y)=\int_0^z r(V-z+v)[1-B(y|\zeta=v)]\mathrm{d}L(v).$$

(28)

Let us introduce the function

$$\Phi_y(z)=\int_0^y H(z,u)\mathrm{d}u=\int_0^z r(V-z+v)\left[\int_0^y(1-B(u|\zeta=v))\,\mathrm{d}u\right]\mathrm{d}L(v)\quad(29)$$

and notation

$$S(z)=R(z,\infty)=\frac{\partial\Phi_u(z)}{\partial u}\bigg|_{u=0}=\int_0^z r(V-z+v)\mathrm{d}L(v)$$

(30)

assuming that $B(0|\zeta=v)=0$. We also introduce the following notation for the Stieltjes convolution:

$$F_1*\cdots*F_k(x)=\mathop{*}_{j=1}^{k}F_j(x).$$

(31)

The kth order Stieltjes convolution of the function $D(x)$ we shall denote by $D_*^{(k)}(x)$.

Taking in consideration the aforementioned symmetry of the functions $\theta_k(Y_k)$ and the boundary conditions (26), one can easily show by direct substitution that the solution of the equation system (22)–(25) has the form:

$$g_k(x, Y_k)dx = C\frac{a^k}{k!}\, d_x \left(\mathop{*}_{j=1}^{k} \Phi_{y_j}(x) \right), \; k = 1, \ldots, n, \tag{32}$$

where C is a constant value to be defined later from the normalization condition. It follows from the last relation and (20) that

$$\theta(Y_k) = C\frac{a^k}{k!} \mathop{*}_{j=1}^{k} \Phi_{y_j}(V), \; k = 1, \ldots, n. \tag{33}$$

5   Steady-State Demands Number Distribution and Loss Probability

Let us introduce the function

$$A(z) = \lim_{y \to \infty} \Phi_y(z) = \int_0^\infty H(z, u)du$$
$$= \int_0^z r(V - z + v) \left[\int_0^\infty (1 - B(u|\zeta = v))\, du \right] dL(v). \tag{34}$$

Here the integral $\int_0^\infty (1 - B(u|\zeta = v))\, du = \mathsf{E}(\xi|\zeta = v)$ is the conditional expectation of the demand service time under condition $\zeta = v$.

We establish from (21) that

$$p_k = C\frac{a^k}{k!}A_*^{(k)}(V), \; k = 1, \ldots, n, \tag{35}$$

where the constant C is determined from the normalization condition as

$$C = p_0 = \left[1 + \sum_{k=1}^{n} \frac{a^k}{k!}A_*^{(k)}(V) \right]^{-1}. \tag{36}$$

The relation for the loss probability P_{loss} follows from the equilibrium condition

$$a(1 - P_{loss}) = \sum_{k=1}^{n} k\frac{\partial\theta_k(\infty_{k-1}, u)}{\partial u} \bigg|_{u=0}, \tag{37}$$

according to which the mean number of customers accepted to the system during a time unit in steady state is equal to the mean number of customers completed their service during the same time. After simple computations we get

$$P_{loss} = 1 - p_0 \left[S(V) + \sum_{k=1}^{n-1} \frac{a^k}{k!}S * A_*^{(k)}(V) \right]. \tag{38}$$

We recall that all above relations refer to the first scenario of system behavior (see Sect. 2). To study the second scenario, one needs to define the function $R(z, y)$ by the following relation differing from (13): $R(z, y) = r(V - z)F(z, y)$. It can be easily shown that Eqs. (22)–(26) stay the same for this function and, in this case, the functions $\Phi_y(z)$, $A(z)$ and $S(z)$ take the following forms:

$$\Phi_y(z) = r(V - z)L(z) \int_0^y [1 - B(u|\zeta < z)] \, du,$$

$$A(z) = r(V - z)L(z) \int_0^\infty [1 - B(u|\zeta < z)] \, du = r(V - z) \int_0^z \mathsf{E}(\xi|\zeta = v) dL(v),$$

$$S(z) = r(V - z)L(z). \tag{39}$$

The probability sense of the function $S(z)$ remains the same for both scenarios of system behavior.

6   Analysis of Some Special Cases

6.1   Demand Service Time Doesn't Depend on Its Volume

In this case we have $F(x, t) = L(x)B(t)$. Then, for both scenarios of system behavior, we obtain that $A(s) = \beta_1 S(z)$. Therefore, the demands number distribution and loss probability doesn't depend on the form of distribution function $B(t)$. These characteristics depend on the first moment of service time β_1 only. The steady-state demands number distribution in this case has the form:

$$p_0 = \left[1 + \sum_{k=1}^n \frac{\rho^k}{k!} S_*^{(k)}(V) \right]^{-1}, \tag{40}$$

$$p_k = p_0 \frac{\rho^k}{k!} S_*^{(k)}(V), \; k = 1, \ldots, n, \tag{41}$$

where $\rho = a\beta_1$, $S(z)$ is defined by (30), for the first scenario, and by (39), for the second one. The relation for the loss probability, as it follows from (38), takes the form:

$$P_{loss} = 1 - p_0 \left[S(V) + \sum_{k=1}^{n-1} \frac{\rho^k}{k!} S_*^{(k+1)}(V) \right]. \tag{42}$$

6.2   Classical Erlang System with AQM

Let us consider the system under the second scenario with demands of volume equals to 1. It is clear that, in this case, we can assume that service time doesn't depend on demand volume and demands number characteristics (including loss probability) depend on the first moment of service time, only. In this case, we denote the memory capacity of the system by N, $N \le n$.

Denote by $r_i = r(i)$ the probability that the arriving demand will be accepted to the system if immediately before its arrival there were i other demands in it, $i = 0, \ldots, N$. Obviously, $r_i = 1 - d_i$, where d_i is the classical dropping function [2]. Then, we obtain from (39)–(42) that

$$p_0 = \left(1 + \sum_{j=1}^{N} \frac{\rho^j}{j!} \prod_{j=0}^{j-1} r_i \right)^{-1} ; \quad p_k = p_0 \frac{\rho^k}{k!} \prod_{i=0}^{k-1} r_i, \ k = 1, \ldots, N;$$

$$P_{loss} = 1 - p_0 \sum_{k=0}^{N-1} \frac{\rho^k}{k!} \prod_{i=0}^{k} r_i. \tag{43}$$

6.3 Service Time Is Proportional to Demand Volume

Let us assume that $\xi = c\zeta$, $c > 0$. Then, we obviously have:

$$\mathsf{E}(\xi|\zeta = v) = \int_0^\infty [1 - B(y|\zeta = v)] \, dy = cv, \tag{44}$$

and, for the first scenario of system behavior, we get:

$$A(z) = c \int_0^z vr(V - z + v) dL(v), \tag{45}$$

and the function $S(z)$ has the form (30). For example, if demand volume has an exponential distribution $L(x) = 1 - e^{-fx}$, $f > 0$, and $r(x) = 1 - x/V$ for $x \in [0; V]$, we obtain, taking in consideration (30), that

$$S(z) = \frac{fz + e^{-fz} - 1}{fV}, \quad A(z) = \frac{c}{f^2V} \left[fz - 2 + (2 + fz)e^{-fz} \right]. \tag{46}$$

Thus, all components of formulas (35)–(38) are determined.

For the system under the second scenario, we get:

$$A(z) = cr(V - z) \int_0^z v \, dL(v), \tag{47}$$

and $S(z)$ is determined by (39). For the exponentially distributed demand volume and the same function $r(x)$ as above, we obtain:

$$S(z) = \frac{z}{V} \left(1 - e^{-fz} \right), \quad A(z) = \frac{cz \left[1 - (1 + fz)e^{-fz} \right]}{fV}. \tag{48}$$

7 Numerical Examples

In this section we present numerical examples illustrating theoretical results. Let us consider, firstly, the case of the demand service which does not depend on the volume. Assume that volumes of entering demands are exponentially distributed with parameter $f = 1$ and that the accepting function is a linear one, i.e. $r(x) = = 1 - \frac{x}{V}$, $x \in [0, V]$. For $V = 8$, $n = 5$ and two different values of the traffic load: $\rho = 0.70$ and 1.00, we obtain the stationary queue-size distributions as shown in Figs. 1 and 2 (for the first and second scenarios of system's operation). The values of the loss probability equal 0.350704 ($\rho = 0.70$) and 0.409832 ($\rho = 1.00$) for the first scenario, and 0.252966 and 0.318422 for the second scenario, respectively.

In Figs. 3 and 4 we present similar results for the case $V = 15$ and $n = \infty$ (we give successive probabilities from 0 to 10). For the case of infinite number of independent servers we get the loss probability equal to 0.302237 (for $\rho = 0.70$) and 0.363743 (for $\rho = 1.00$) in the case of the first scenario of the system operation. Similarly, for the second scenario, we have 0.243615 and 0.307163, respectively.

Moreover, in Figs. 5 and 6 the case of the service time being proportional to demand volume is presented, where the coefficient of proportionality $c = 1$, $a = 0.70$ and 1.00, and the remaining system parameters are the same as in Figs. 3 and 4. For the first scenario we get loss probabilities equal to 0.339437 ($a = 0.7$) and 0.396688 ($a = 1$). Similarly, for the second scenario of the system's operation we obtain, 0.269370 and 0.338323, respectively.

Lastly, on Figs. 7 and 8 the stationary queue-size distribution is shown for the case of $c = 1.5$, two different values of arrival intensity, namely $a = 2$ and $a = 1$ and, as in Figs. 5 and 6, taking $V = 8$, $f = 1$ and $n = 5$. The obtained values of the loss probability are 0.593461 (first scenario) and 0.228903 (second scenario) for $a = 2$ and, respectively, 0.468284 and 0.0074844 for $a = 1$.

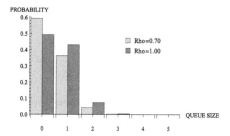

Fig. 1. 5 servers, first scenario (independent)

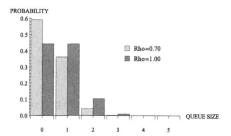

Fig. 2. 5 servers, second scenario (independent)

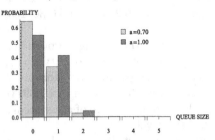

Fig. 3. Infinite number of servers, first scenario (independent)

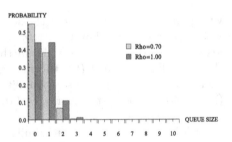

Fig. 4. Infinite number of servers, second scenario (independent)

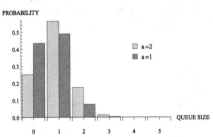

Fig. 5. 5 servers, first scenario (proportional), $c = 1$

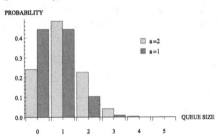

Fig. 6. 5 servers, second scenario (proportional), $c = 1$

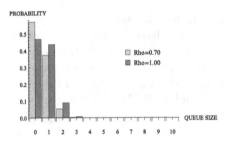

Fig. 7. 5 servers, first scenario (proportional), $c = 1.5$

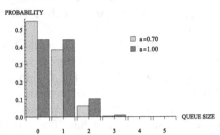

Fig. 8. 5 servers, second scenario (proportional), $c = 1.5$

8   Conclusions

In the paper a multi-server Erlang loss queueing system with Poisson arrival stream, finite memory capacity and $n \leq \infty$ independent service stations is considered, in which volumes of the arriving demands and their service times are dependent random variables. An AQM-type non-increasing accepting function is implemented for controlling the incoming flow of demands. Two possible acceptance rules are considered: in the first one the arriving demand is being accepted for service with probability depending on its volume and on the volume of all demands being processed at the arrival epoch; in the second one the accepting probability is independent on the volume of the arriving demand. Using supplementary variables' technique, the stationary queue-size distribution and the loss probability are obtained for both scenarios of the system behavior. Three different special cases are discussed and illustrated via numerical examples. The results obtained in the paper can be used for estimating of total demands capacity characteristics in the nodes of computer and telecommunication networks.

References

1. Floyd, S., Jacobson, V.: Random early detection gateways for congestion avoidance. IEEE/ACM Trans. Netw. **1**(4), 397–412 (1993)
2. Kempa, W.M.: On main characteristics of the $M/M/1/N$ queue with single and batch arrivals and the queue size controlled by AQM algorithms. Kybernetika **47**(6), 930–943 (2011)
3. Liu, S., Basar, T., Srikant, R.: Exponential RED: a stabilizing AQM scheme for low- and high-speed TCP protocols. IEEE/ACM Trans. Netw. **13**, 1068–1081 (2005)
4. Tikhonenko, O.M.: Generalized Erlang problem for service systems with finite total capacity. Probl. Inf. Transm. **41**(3), 243–253 (2005)
5. Tikhonenko, O.M.: Queueing systems of a random length demands with restrictions. Autom. Remote Control **52**(10, pt. 2), 1431–1437 (1991)
6. Tikhonenko, O.: Computer systems probability analysis. Akademicka Oficyna Wydawnicza EXIT, Warsaw (2006). (in Polish)
7. Tikhonenko, O., Kempa, W.M.: The generalization of AQM algorithms for queueing systems with bounded capacity. In: Wyrzykowski, R., Dongarra, J., Karczewski, K., Waśniewski, J. (eds.) PPAM 2011. LNCS, vol. 7204, pp. 242–251. Springer, Heidelberg (2012). doi:10.1007/978-3-642-31500-8_25
8. Tikhonenko, O., Kempa, W.M.: Queue-size distribution in M/G/1-type system with bounded capacity and packet dropping. In: Dudin, A., Klimenok, V., Tsarenkov, G., Dudin, S. (eds.) BWWQT 2013. CCIS, vol. 356, pp. 177–186. Springer, Heidelberg (2013). doi:10.1007/978-3-642-35980-4_20
9. Tikhonenko, O., Kempa, W.M.: On the queue-size distribution in the multiserver system with bounded capacity and packet dropping. Kybernetika **49**(6), 855–867 (2013)
10. Zhou, K., Yeung, K.L., Li, V.O.K.: Nonlinear RED: a simple yet efficient active queue management scheme. Comput. Netw. **50**(18), 3784–3794 (2006)

Queueing Systems with Demands of Random Space Requirement and Limited Queueing or Sojourn Time

Oleg Tikhonenko[1]([✉]) and Pawel Zajac[2]

[1] Faculty of Mathematics and Natural Sciences, College of Sciences,
Cardinal Stefan Wyszyński University in Warsaw,
Ul. Wóycickiego 1/3, 01-938 Warsaw, Poland
o.tikhonenko@uksw.edu.pl
[2] Institute of Mathematics, Czestochowa University of Technology,
Al. Armii Krajowej 21, 42-200 Czestochowa, Poland
pawel_zajac@vp.pl

Abstract. We investigate queueing systems with demands of random space requirements and limited buffer space, in which queueing or sojourn time are limited by some constant value. For such systems, in the case of exponentially distributed service time and Poisson entry, we obtain the steady-state demands number distribution and probability of demands losing.

Keywords: Queueing system · Buffer space capacity · Demand volume · Demands total volume · Queueing time · Sojourn time

1 Introduction

We consider queueing systems with demands of a random space requirement (volume) and a limited buffer space capacity. This means that each demand is characterized by some non-negative random indication named the demand space requirement or demand volume ζ. The total sum $\sigma(t)$ of volumes of all demands present in the system at arbitrary time instant t is limited by some constant value V, which is named a buffer space capacity of the system. Such systems have been used to model and solve the various practical problems occurring in design of computer and communication systems. They were widely studied in the literature (see, e.g., [3,5–12]).

In our work, we study queueing systems in which demands are also "impatient". In other words, they can leave the system during their waiting in the queue, or even during their servicing.

Such systems are the models of some real processes. E.g., systems of information transmission often deal with the process of messages information reduction. The outdated messages can be removed from the system before their transmission. In this case, we have the system with limited queueing time. The typical

P. Gaj et al. (Eds.): CN 2017, CCIS 718, pp. 380–391, 2017.
DOI: 10.1007/978-3-319-59767-6_30

example of the system with limited sojourn time is a radar for airplanes supervision. This installation is characterized by limited area of servicing. The problem is to keeping up with the service of all airplanes that are in this area during some constant or random amount of time.

In the classical queueing theory (for demands without volume), systems with limited queueing or sojourn time were investigated, e.g., in [2]. We consider systems with queueing or sojourn time limited by some constant value τ. In the systems under consideration, the buffer space and the number of waiting places in the queue can be also limited.

For the systems under consideration, we obtain the steady-state demands number distribution and the probability that a demand is lost or leaves the system because of queueing or sojourn time limitation.

This work is organized as follows. In Sect. 2, we give the mathematical description of the models and introduce some necessary notations. In Sect. 3, we investigate the system with limited queueing time. In Sect. 4, we investigate the system with limited sojourn time. Section 5 contains concluding remarks.

2   Models Description

Consider the $M/M/n/m$-type queueing system with identical servers and FIFO service discipline. Let a be the intensity of demands entrance flow, μ be the parameter of service time. Each demand has some random volume ζ which does not depend on the volumes of other demands nor on the demand arriving epoch. Let $L(x) = \mathsf{P}\{\zeta < x\}$ be the demand volume distribution function and $\sigma(t)$ be the sum of volumes of all demands present in the system at time instant t. The values of the process $\sigma(t)$ are limited by the constant value V (buffer space capacity). Let us denote by $\eta(t)$ the number of demands present in the system at time t. Let a demand having the volume x arrive to the system at epoch t. Then, it will be accepted to the system if $\eta(t^-) < n + m$ and $\sigma(t^-) + x \leq V$. In this case, we have $\eta(t) = \eta(t^-) + 1$, $\sigma(t) = \sigma(t^-) + x$. In opposite case, the demand will be lost and $\eta(t) = \eta(t^-)$, $\sigma(t) = \sigma(t^-)$. If t is the epoch when a demand of volume x leaves the system, we have $\eta(t) = \eta(t^-) - 1$, $\sigma(t) = \sigma(t^-) - x$.

Assume that demand service time ξ does not depend on its volume ζ. In the system with limited queueing time, a waiting demand leaves the system immediately if its queueing time achieves the value τ. In this system, a demand on service never be lost. In the system with limited sojourn time, a demand will be lost if its sojourn time achieves the value τ.

3   Queueing System with Limited Buffer Space and Limited Queueing Time

3.1   Process and Characteristics

Let $\xi_j(t)$ be the length of time interval from the moment t to the moment when the jth demand leaves the queue (starts its service or is lost),

$j = n + 1, \ldots, \eta(t)$. Let $\sigma_j(t)$ be the volume of the jth demand. It is clear that $\sigma(t) = \sum_{j=1}^{\eta(t)} \sigma_j(t)$.

The system behavior is described by the following Markov process:

$$\begin{cases} (\eta(t); \sigma_j(t), j = 1, \ldots, \eta(t)), & 1 \leq \eta(t) \leq n, \\ (\eta(t); \sigma_j(t), j = 1, \ldots, \eta(t); \xi_l(t), l = n+1, \ldots, \eta(t)), & n < \eta(t) \leq n+m. \end{cases} \quad (1)$$

Let us introduce the functions to characterize the process (1). First, we introduce the functions

$$P_k(t) = \mathsf{P}\{\eta(t) = k\}, \ k = 0, 1, \ldots, n+m. \quad (2)$$

For $k = 1, 2, \ldots, n+m$, we introduce the functions

$$G_k(t, x) = \mathsf{P}\{\eta(t) = k, \sigma(t) < x\}. \quad (3)$$

It is clear that, for such k, we have $P_k(t) = G_k(t, V)$.

For $n + 1 \leq k \leq n + m$, we introduce the functions $R_k(t, x, y_{n+1}, \ldots, y_k)$ and $H_k(t; y_{n+1}, \ldots, y_k) = R_k(t, V, y_{n+1}, \ldots, y_k)$ with the following probability sense:

$$R_k(t, x, y_{n+1}, \ldots, y_k) \mathrm{d}y_{n+1} \ldots \mathrm{d}y_k$$
$$= \mathsf{P}\{\eta(t) = k; \sigma(t) < x; \xi_j(t) \in [y_j; y_j + \mathrm{d}y_j), j = n+1, \ldots, k\}, \quad (4)$$

$$H_k(t, y_{n+1}, \ldots, y_k) \mathrm{d}y_{n+1} \ldots \mathrm{d}y_k$$
$$= \mathsf{P}\{\eta(t) = k; \xi_j(t) \in [y_j; y_j + \mathrm{d}y_j), j = n+1, \ldots, k\}. \quad (5)$$

3.2   Steady-State Demands Number Distribution

It is easily to show that, for the functions (2)–(5), the following differential equations hold:

$$P_0'(t) = -aP_0(t)L(V) + \mu P_1(t); \quad (6)$$

$$P_1'(t) = aP_0(t)L(V) - a \int_0^V G_1(t, V - x)\mathrm{d}L(x) - \mu P_1(t) + 2\mu P_2(t); \quad (7)$$

$$P_k'(t) = a \int_0^V G_{k-1}(t, V - x)\mathrm{d}L(x) - a \int_0^V G_k(t, V - x)\mathrm{d}L(x)$$
$$- k\mu P_k(t) + (k+1)\mu P_{k+1}(t), \ k = 2, \ldots, n-1; \quad (8)$$

$$P_n'(t) = a \int_0^V G_{n-1}(t, V - x)\mathrm{d}L(x) - a \int_0^V G_n(t, V - x)\mathrm{d}L(x)$$
$$- n\mu P_n(t) + H_{n+1}(t, 0); \quad (9)$$

$$P'_k(t) = a \int_0^V G_{k-1}(t, V - x) dL(x) - a \int_0^V G_k(t, V - x) dL(x)$$

$$- \int_{y_k=0}^\tau \int_{y_{k-1}=0}^{y_k} \cdots \int_{y_{n+2}=0}^{y_{n+3}} H_k(t, 0, y_{n+2}, \dots, y_k) dy_k \dots dy_{n+2}$$

$$+ \int_{y_{k+1}=0}^\tau \int_{y_k=0}^{y_{k+1}} \cdots \int_{y_{n+2}=0}^{y_{n+3}} H_{k+1}(t, 0, y_{n+2}, \dots, y_{k+1}) dy_{k+1} \dots dy_{n+2},$$

$$k = n + 1, \dots, n + m - 1; \tag{10}$$

$$P'_{n+m}(t) = a \int_0^V G_{n+m-1}(t, V - x) dL(x)$$

$$- \int_{y_{n+m}=0}^\tau \int_{y_{n+m-1}=0}^{y_{n+m}} \cdots \int_{y_{n+2}=0}^{y_{n+3}} H_{n+m}(t, 0, y_{n+2}, \dots, y_{n+m}) dy_{n+m} \dots dy_{n+2}. \tag{11}$$

Assume that at least one of the values V and m is finite. Therefore, for $\rho = a/(n\mu) < \infty$ the steady state exists for the system under consideration, i.e. $\eta(t) \Rightarrow \eta$ and $\sigma(t) \Rightarrow \sigma$ in the sense of a weak convergence, where η and σ are a steady-state number of demands present in the system and their steady-state total volume, respectively. Hence, the following finite limits exist:

$$p_k = \lim_{t \to \infty} P_k(t) = \mathsf{P}\{\eta = k\}, \ k = 0, 1, \dots, n + m; \tag{12}$$

$$g_k(x) = \lim_{t \to \infty} G_k(t, x) = \mathsf{P}\{\eta = k, \sigma < x\}, \ k = 1, 2, \dots, n + m; \tag{13}$$

$$r_k(x, y_{n+1}, \dots, y_k) = \lim_{t \to \infty} R_k(t, x, y_{n+1}, \dots, y_k), \ k = n + 1, \dots, n + m; \tag{14}$$

$$h_k(y_{n+1}, \dots, y_k) = \lim_{t \to \infty} H_k(t, y_{n+1}, \dots, y_k), \ k = n + 1, \dots, n + m. \tag{15}$$

Then, from (6)–(11), we obtain the following equations for the steady-state functions (12)–(15):

$$0 = -ap_0 L(V) + \mu p_1; \tag{16}$$

$$0 = ap_0 L(V) - a \int_0^V g_1(V - x) dL(x) - \mu p_1 + 2\mu p_2; \tag{17}$$

$$0 = a \int_0^V g_{k-1}(V - x) dL(x) - a \int_0^V g_k(V - x) dL(x)$$

$$- k\mu p_k + (k + 1)\mu p_{k+1}, \ k = 2, \dots, n - 1; \tag{18}$$

$$0 = a \int_0^V g_{n-1}(V - x) dL(x) - a \int_0^V g_n(V - x) dL(x) - n\mu p_n + h_{n+1}(0); \tag{19}$$

$$0 = a \int_0^V g_{k-1}(V - x)\mathrm{d}L(x) - a \int_0^V g_k(V - x)\mathrm{d}L(x)$$

$$- \int_0^\tau \int_0^{y_k} \cdots \int_0^{y_{n+3}} h_k(0, y_{n+2}, \ldots, y_k)\mathrm{d}y_k \ldots \mathrm{d}y_{n+2}$$

$$+ \int_0^\tau \int_0^{y_{k+1}} \cdots \int_0^{y_{n+3}} h_{k+1}(0, y_{n+2}, \ldots, y_{k+1})\mathrm{d}y_{k+1} \cdots \mathrm{d}y_{n+2},$$

$$k = n + 1, \ldots, n + m - 1; \tag{20}$$

$$0 = a \int_0^V g_{n+m-1}(V - x)\mathrm{d}L(x)$$

$$- \int_0^\tau \int_0^{y_{n+m}} \cdots \int_0^{y_{n+3}} h_{n+m}(0, y_{n+2}, \ldots, y_{n+m})\mathrm{d}y_{n+m} \cdots \mathrm{d}y_{n+2}. \tag{21}$$

In the steady state, the following boundary conditions hold:

$$a \int_0^V g_n(V - x)\mathrm{d}L(x) = h_{n+1}(0); \tag{22}$$

$$a \int_0^V g_k(V - x)\mathrm{d}L(x)$$

$$= \int_0^\tau \int_0^{y_{k+1}} \cdots \int_0^{y_{n+3}} h_{k+1}(0, y_{n+2}, \ldots, y_{k+1})\mathrm{d}y_{k+1} \cdots \mathrm{d}y_{n+2}, \tag{23}$$

$$k = n + 1, \ldots, n + m - 1.$$

It can be easily shown by direct substitution that the following functions are the solution of Eqs. (16)–(21) for which the boundary conditions (22) and (23) hold:

$$r_k(x, y_{n+1}, \ldots, y_k) = p_0 \frac{\mu^{k-n}(n\rho)^k}{n!} e^{-n\mu y_k} L_*^{(k)}(x), \; k = n + 1, \ldots, n + m, \tag{24}$$

whereas we get:

$$h_k(y_{n+1}, \ldots, y_k) = p_0 \frac{\mu^{k-n}(n\rho)^k}{n!} e^{-n\mu y_k} L_*^{(k)}(V), \; k = n + 1, \ldots, n + m.$$

For the functions $g_k(x)$, we have:

$$g_k(x) = p_0 \frac{(n\rho)^k}{k!} L_*^{(k)}(x), \; k = 1, 2, \ldots, n; \tag{25}$$

$$g_k(x) = \int_0^\tau \int_0^{y_k} \cdots \int_0^{y_{n+2}} r_k(x, y_{n+1}, \ldots, y_k)\mathrm{d}y_k...\mathrm{d}y_{n+1}$$

$$= p_0 \frac{\mu^{k-n}(n\rho)^k}{n!} L_*^{(k)}(x) \int_0^\tau e^{-n\mu y_k}\mathrm{d}y_k \int_0^{y_k} \mathrm{d}y_{k-1} \cdots \int_0^{y_{n+2}} \mathrm{d}y_{n+1} \tag{26}$$

$$= p_0 \frac{n^n \rho^k}{n!} \left[1 - e^{-n\mu\tau} \sum_{j=0}^{k-n-1} \frac{(n\mu\tau)^j}{j!} L_*^{(k)}(x) \right], \; k = n + 1, \ldots, n + m.$$

Finally, we obtain:

$$
p_k = \begin{cases}
p_0 \dfrac{(n\rho)^k}{k!} L_*^{(k)}(V), & k = 1, \ldots, n, \\[3mm]
p_0 \dfrac{n^n \rho^k}{n!} \left[1 - e^{-n\mu\tau} \displaystyle\sum_{j=0}^{k-n-1} \dfrac{(n\mu\tau)^j}{j!} \right] L_*^{(k)}(V), & k = n+1, \ldots, n+m.
\end{cases}
$$

From the normalization condition $\sum_{k=0}^{n+m} p_k = 1$, we have:

$$
p_0 = \left\{ \sum_{k=0}^{n} \frac{(n\rho)^k}{k!} L_*^{(k)}(V) + \frac{n^n}{n!} \sum_{k=n+1}^{n+m} \rho^k \left[1 - e^{-n\mu\tau} \sum_{j=0}^{k-n-1} \frac{(n\mu\tau)^j}{j!} \right] L_*^{(k)}(V) \right\}^{-1}.
$$

3.3   Loss Probability

Let A be the event that an arbitrary arriving demand is accepted to the system and served completely. The probability of this event can be calculated as follows:

$$
\mathsf{P}\{A\} = p_0 L(V) + \sum_{k=1}^{n-1} \int_0^V g_k(V-x) \mathrm{d}L(x) + \left(1 - e^{-n\mu\tau}\right) \int_0^V g_n(V-x)\mathrm{d}L(x)
$$

$$
+ \sum_{k=n+1}^{n+m-1} \int_0^V \int_0^\tau \cdots \int_0^{y_{n+2}} r_k(V-x, y_{n+1}, \ldots, y_k)
$$

$$
\times \left[1 - e^{-n\mu(\tau - y_k)} \right] \mathrm{d}L(x)\mathrm{d}y_k \ldots \mathrm{d}y_{n+1},
$$

whereas, taking into consideration formulae (24) and (25), we obtain:

$$
\mathsf{P}\{A\} = p_0 \left\{ \sum_{k=0}^{n-1} \frac{(n\rho)^k}{k!} L_*^{(k+1)}(V) \right.
$$

$$
\left. + \frac{n^n}{n!} \sum_{k=n}^{n+m-1} \rho^k \left[1 - e^{-n\mu\tau} \sum_{j=0}^{k-n} \frac{(n\mu\tau)^j}{j!} \right] L_*^{(k+1)}(V) \right\}.
$$

It is clear that the loss probability (or the probability that a demand is lost at its arriving epoch or is not served completely) can be determined as

$$
P_{loss} = 1 - \mathsf{P}\{A\}.
$$

4   Queueing System with Limited Buffer Space and Limited Sojourn Time

4.1   Process and Characteristics

Let us assume that demands are served according FIFO discipline. Then, accepting ones can leave the system only during their servicing (see, e.g., [2]). For this system, we denote by $\gamma_l(t)$ the length of time interval from the moment t to the moment when the lth server releases from service of demands accepting to the system before the moment t if this server is busy at this moment. We assume that $\gamma_l(t) = 0$ if the lth server is free at this moment, $l = 1, \ldots, n$. Later on, we shall numerate busy (at time instant t) servers according to appropriate values $\gamma_l(t)$.

Then, the system behavior is described by the following Markov process:

$$(\eta(t); \sigma_j(t), j = 1, \ldots, \eta(t); \gamma_l(t), l = 1, \ldots, \max(\eta(t), n)). \qquad (27)$$

We shall characterize the process (27) by the following functions:

$$P_k(t) = \mathsf{P}\{\eta(t) = k\}, \ k = 0, 1, \ldots, n + m, \qquad (28)$$

$$R_k(t, x, y_1, \ldots, y_k)dy_1 \ldots dy_k$$
$$= \mathsf{P}\{\eta(t) = k; \sigma(t) < x; \gamma_l(t) \in [y_l; y_l + dy_l), l = 1, \ldots, k\}, \ k = 1, \ldots, n, \qquad (29)$$

$$R_k(t, x, y_1, \ldots, y_n)dy_1 \ldots dy_n$$
$$= \mathsf{P}\{\eta(t) = k; \sigma(t) < x; \gamma_l(t) \in [y_l; y_l + dy_l), l = 1, \ldots, n\}, \ k = n + 1, \ldots, n + m, \qquad (30)$$

$$H_k(t, y_1, \ldots, y_k)dy_1 \ldots dy_k$$
$$= \mathsf{P}\{\eta(t) = k; \gamma_l(t) \in [y_l; y_l + dy_l), l = 1, \ldots, k\}, \ k = 1, \ldots, n, \qquad (31)$$

$$H_k(t, y_1, \ldots, y_n)dy_1 \ldots dy_n$$
$$= \mathsf{P}\{\eta(t) = k; \gamma_l(t) \in [y_l; y_l + dy_l), l = 1, \ldots, n\}, \ k = n + 1, \ldots, n + m. \qquad (32)$$

It is clear that $H_k(t, y_1, \ldots, y_k) = R_k(t, V, y_1, \ldots, y_k)$ for $1 \leq k \leq n$ and $H_k(t, y_1, \ldots, y_n) = R_k(t, V, y_1, \ldots, y_n)$ for $n + 1 \leq k \leq n + m$.

Taking into consideration the established way of servers numeration, let us introduce the functions

$$G_k(t, x) = \mathsf{P}\{\eta(t) = k, \sigma(t) < x\}$$
$$= \int_0^\tau \int_0^{y_k} \ldots \int_0^{y_2} R_k(t, x, y_1, \ldots, y_k)dy_k \ldots dy_1, \ k = 1, \ldots, n, \qquad (33)$$

$$G_k(t, x) = \mathsf{P}\{\eta(t) = k, \sigma(t) < x\}$$
$$= \int_0^\tau \int_0^{y_n} \ldots \int_0^{y_2} R_k(t, x, y_1, \ldots, y_n)dy_n \ldots dy_1, \ k = n + 1, \ldots, n + m. \qquad (34)$$

It is clear that, for all $k = 1, \ldots, n + m$, we have $P_k(t) = G_k(t, V)$.

4.2   Steady-State Demands Number Distribution

For introduced functions, we can write the following equations:

$$P_0'(t) = -aP_0(t)L(V) + H_1(t,0); \tag{35}$$

$$P_1'(t) = aP_0(t)L(V) - a\int_0^V G_1(t, V-x)dL(x) - H_1(t,0) + \int_0^\tau H_2(t,0,y_2)dy_2; \tag{36}$$

$$
\begin{aligned}
P_k'(t) = &\ a\int_0^V G_{k-1}(t, V-x)dL(x) - a\int_0^V G_k(t, V-x)dL(x) \\
&- \int_0^\tau \cdots \int_0^{y_3} H_k(t,0,y_2,\ldots,y_k)dy_k \ldots dy_2 \\
&+ \int_0^\tau \cdots \int_0^{y_3} H_{k+1}(t,0,y_2,\ldots,y_{k+1})dy_{k+1} \ldots dy_2, \quad k = 2,\ldots n-1; \tag{37}
\end{aligned}
$$

$$
\begin{aligned}
P_n'(t) = &\ a\int_0^V G_{n-1}(t, V-x)dL(x) - a\int_0^V G_n(t, V-x)dL(x) \\
&- \int_0^\tau \cdots \int_0^{y_3} H_n(t,0,y_2,\ldots,y_n)dy_n \ldots dy_2 \\
&+ \int_0^\tau \cdots \int_0^{y_3} H_{n+1}(t,0,y_2,\ldots,y_n)dy_n \ldots dy_2; \tag{38}
\end{aligned}
$$

$$
\begin{aligned}
P_k'(t) = &\ a\int_0^V G_{k-1}(t, V-x)dL(x) - a\int_0^V G_k(t, V-x)dL(x) \\
&- \int_0^\tau \cdots \int_0^{y_3} H_k(t,0,y_2,\ldots,y_n)dy_n \ldots dy_2 \\
&+ \int_0^\tau \cdots \int_0^{y_3} H_{k+1}(t,0,y_2,\ldots,y_n)dy_n \ldots dy_2, \quad k = n+1,\ldots n+m-1; \tag{39}
\end{aligned}
$$

$$
\begin{aligned}
P_{n+m}'(t) = &\ a\int_0^V G_{n+m-1}(t, V-x)dL(x) \\
&- \int_0^\tau \cdots \int_0^{y_3} H_{n+m}(t,0,y_2,\ldots,y_n)dy_n \ldots dy_2. \tag{40}
\end{aligned}
$$

By the same way as in Sect. 2, we can write the steady-state equations for the number $p_0 = \mathsf{P}\{\eta = 0\}$ and the functions r_k, h_k and g_k (these functions do not depend on t) that are steady-state analogies of the functions R_k, H_k and G_k, respectively. It follows from Eqs. (35)–(40) that equations for the steady-state functions have the following forms:

$$0 = -ap_0 L(V) + h_1(0); \tag{41}$$

$$0 = ap_0 L(V) - a \int_0^V g_1(V - x)dL(x) - h_1(0) + \int_0^\tau h_2(0, y_2)dy_2; \qquad (42)$$

$$0 = a \int_0^V g_{k-1}(V - x)dL(x) - a \int_0^V g_k(V - x)dL(x)$$
$$- \int_0^\tau \ldots \int_0^{y_3} h_k(0, y_2, \ldots, y_k)dy_k \ldots dy_2$$
$$+ \int_0^\tau \ldots \int_0^{y_3} h_{k+1}(0, y_2, \ldots, y_{k+1})dy_{k+1} \ldots dy_2, \ k = 2, \ldots n - 1; \qquad (43)$$

$$0 = a \int_0^V g_{n-1}(V - x)dL(x) - a \int_0^V g_n(V - x)dL(x)$$
$$- \int_0^\tau \ldots \int_0^{y_3} h_n(0, y_2, \ldots, y_n)dy_n \ldots dy_2$$
$$+ \int_0^\tau \ldots \int_0^{y_3} h_{n+1}(0, y_2, \ldots, y_n)dy_n \ldots dy_2; \qquad (44)$$

$$0 = a \int_0^V g_{k-1}(V - x)dL(x) - a \int_0^V g_k(V - x)dL(x)$$
$$- \int_0^\tau \ldots \int_0^{y_3} h_k(0, y_2, \ldots, y_n)dy_n \ldots dy_2$$
$$+ \int_0^\tau \ldots \int_0^{y_3} h_{k+1}(0, y_2, \ldots, y_n)dy_n \ldots dy_2, \ k = n + 1, \ldots n + m - 1; \quad (45)$$

$$0 = a \int_0^V g_{n+m-1}(V - x)dL(x) - \int_0^\tau \ldots \int_0^{y_3} h_{n+m}(0, y_2, \ldots, y_n)dy_n \ldots dy_2. \qquad (46)$$

In the steady state, the following boundary conditions hold:

$$a \int_0^V g_1(V - x)dL(x) = \int_0^\tau h_2(0, y_2)dy_2; \qquad (47)$$

$$a \int_0^V g_k(V - x)dL(x) = \int_0^\tau \ldots \int_0^{y_3} h_{k+1}(0, y_2, \ldots, y_{k+1})dy_{k+1} \ldots dy_2,$$
$$k = 2, \ldots, n - 1; \qquad (48)$$

$$a \int_0^V g_k(V - x)dL(x) = \int_0^\tau \ldots \int_0^{y_3} h_{k+1}(0, y_2, \ldots, y_n)dy_n \ldots dy_2,$$
$$k = n, \ldots, n + m - 1. \qquad (49)$$

It can be shown by direct substitution that the following functions satisfy Eqs. (41)–(43) and boundary conditions (47) and (48):

$$r_k(x, y_1, \ldots, y_k) = p_0(n\mu\rho)^k e^{-\mu(y_1 + \cdots + y_k)} L_*^{(k)}(x),$$

$$h_k(y_1, \ldots, y_k) = r_k(V, y_1, \ldots, y_k) = p_0(n\mu\rho)^k e^{-\mu(y_1 + \cdots + y_k)} L_*^{(k)}(V),$$

$$g_k(x) = \int_0^\tau \int_0^{y_k} \cdots \int_0^{y_2} r_k(x, y_1, \ldots, y_k) dy_k \ldots dy_1 = p_0 \frac{(n\rho)^k}{k!}(1 - e^{-\mu\tau})^k L_*^{(k)}(x),$$

where $\rho = a/(n\mu)$, $k = 1, \ldots, n-1$.

The following functions satisfy Eqs. (45) and (46) and the boundary conditions (49):

$$r_k(x, y_1, \ldots, y_n) = p_0(n\mu)^n \rho^k (1 - e^{-\mu\tau})^{k-n} e^{-\mu(y_1 + \cdots + y_n)} L_*^{(k)}(x),$$

$$h_k(y_1, \ldots, y_n) = r_k(V, y_1, \ldots, y_n)$$

$$= p_0(n\mu)^n \rho^k (1 - e^{-\mu\tau})^{k-n} e^{-\mu(y_1 + \cdots + y_n)} L_*^{(k)}(V),$$

$$g_k(x) = \int_0^\tau \int_0^{y_n} \cdots \int_0^{y_2} r_k(x, y_1, \ldots, y_n) dy_n \ldots dy_1$$

$$= p_0 \frac{n^n \rho^k}{n!}(1 - e^{-\mu\tau})^k L_*^{(k)}(x),$$

where $k = n, \ldots, n+m$.

Then, for demands number distribution, we obtain the following relation:

$$p_k = g_k(V) = \begin{cases} p_0 \dfrac{(n\rho)^k}{k!}(1 - e^{-\mu\tau})^k L_*^{(k)}(V), & k = 1, \ldots, n-1, \\ p_0 \dfrac{n^n \rho^k}{n!}(1 - e^{-\mu\tau})^k L_*^{(k)}(V), & k = n, \ldots, n+m, \end{cases}$$

and, from the normalization condition $\sum_{k=0}^{n+m} p_k = 1$, we have

$$p_0 = \left[\sum_{k=0}^n \frac{(n\rho)^k}{k!}(1 - e^{-\mu\tau})^k L_*^{(k)}(V) + \frac{n^n}{n!} \sum_{k=n+1}^{n+m} \rho^k (1 - e^{-\mu\tau})^k L_*^{(k)}(V) \right]^{-1}.$$

4.3 Loss Probability

Let A be the event that an arbitrary arriving demand is accepted to the system and served completely. The probability $\mathsf{P}\{A\}$ of this event is determined as follows:

$$\mathsf{P}\{A\} = p_0 L(V)(1 - e^{-\mu\tau})$$

$$+ (1 - e^{-\mu\tau}) \sum_{k=1}^{n-1} \int_0^V \int_0^\tau \cdots \int_0^{y_2} r_k(V - x, y_1, \ldots, y_k) dL(x) \, dy_k \ldots dy_1$$

$$+ \sum_{k=n}^{n+m-1} \int_0^V \int_0^\tau \cdots \int_0^{y_2} r_k(V - x, y_1, \ldots, y_n) \left[1 - e^{-\mu(\tau - y_1)} \right] dL(x) \, dy_n \ldots dy_1.$$

If we substitute in this relation the obtained formulae for functions r_k, we get:

$$P\{A\} = p_0 \left\{ \sum_{k=0}^{n-1} \frac{(n\rho)^k}{k!} (1 - e^{-\mu\tau})^{k+1} L_*^{k+1}(V) \right.$$

$$+ \frac{n^n}{n!} \sum_{k=n}^{n+m-1} \rho^k (1 - e^{-\mu\tau})^k L_*^{(k+1)}(V)$$

$$- (n\mu)^n e^{-\mu\tau} \int_0^\tau \cdots \int_0^{y_3} y_2 e^{-\mu(y_2 + \cdots + y_n)} dy_n \cdots dy_2$$

$$\left. \times \sum_{k=n}^{n+m-1} \rho^k (1 - e^{-\mu\tau})^{k-n} L_*^{(k+1)}(V) \right\}.$$

For example, for $n = 1$, we have:

$$P\{A\} = p_0 (1 - e^{-\mu\tau}) L(V) + \sum_{k=1}^m \int_0^V \int_0^\tau r_k(V - x, y) \left[1 - e^{-\mu(\tau - y)}\right] dL(x) \, dy$$

$$= p_0 \left\{ (1 - e^{-\mu\tau}) L(V) + \left[1 - (1 + \mu\tau)e^{-\mu\tau}\right] \sum_{k=1}^m \rho^k (1 - e^{-\mu\tau})^{k-1} L_*^{(k+1)}(V) \right\},$$

and, for $n = 2$, we obtain:

$$P\{A\} = p_0 \left[(1 - e^{-\mu\tau}) L(V) + 2\rho (1 - e^{-\mu\tau})^2 L_*^{(2)}(V)\right]$$

$$+ 2p_0 \sum_{k=2}^{m+1} \rho^k (1 - e^{-\mu\tau})^{k-2} L_*^{(k+1)}(V)$$

$$- 4p_0 e^{-\mu\tau} \left[1 - (1 + \mu\tau)e^{-\mu\tau}\right] \sum_{k=2}^{m+1} \rho^k (1 - e^{-\mu\tau})^{k-2} L_*^{(k+1)}(V).$$

Then, the loss probability is equal to $P_{loss} = 1 - P\{A\}$.

5   Conclusions

In this paper, we investigate queueing systems with constant limitations of the demand total volume and queueing or sojourn time. We obtain formulae for demands number distribution and loss probability for the systems under consideration. The obtained formulae are not generally convenient for precise calculation, but the calculation is possible in some special cases (e.g., when a demand volume has gamma or uniform distribution). In other cases, we can use the numeric inversion of the Laplace transform [1,4].

The results obtained in the paper can be used for estimating messages number characteristics and loss probability in the information systems with limited area of servicing and limited buffer space.

References

1. Gaver, D.P.: Observing stochastic processes, and approximate transform inversion. Oper. Res. **14**(3), 444–459 (1966)
2. Gnedenko, B.V., Kovalenko, I.N.: Introduction to Queueing Theory. Birkhäuser, Boston (1989)
3. Morozov, E., Nekrasova, R., Potakhina, L., Tikhonenko, O.: Asymptotic analysis of queueing systems with finite buffer space. In: Kwiecień, A., Gaj, P., Stera, P. (eds.) CN 2014. CCIS, vol. 431, pp. 223–232. Springer, Cham (2014). doi:10.1007/978-3-319-07941-7_23
4. Stehfest, H.: Algorithm 368: numeric inversion of Laplace transform. Commun. ACM **13**(1), 47–49 (1970)
5. Tikhonenko, O.: Computer Systems Probability Analysis. Akademicka Oficyna Wydawnicza EXIT, Warsaw (2006) (In Polish)
6. Tikhonenko, O.: Determination of loss characteristics in queueing systems with demands of random space requirement. In: Dudin, A., Nazarov, A., Yakupov, R. (eds.) ITMM 2015. CCIS, vol. 564, pp. 209–215. Springer, Cham (2015). doi:10.1007/978-3-319-25861-4_18
7. Tikhonenko, O.: Districted capacity queueing systems: Determination of their characteristics. Autom. Remote Control. **58**(6), Pt.1, 969–972 (1997)
8. Tikhonenko, O.M.: Generalized Erlang problem for service systems with finite total capacity. Probl. Inf. Transm. **41**(3), 243–253 (2005)
9. Tikhonenko, O.M.: Queueing Models in Information Systems. Universitetskoe, Minsk (1990). (In Russian)
10. Tikhonenko, O.M.: Queueing systems of a random length demands with restrictions. Autom. Remote Control. **52**(10), pt. 2, 1431–1437 (1991)
11. Tikhonenko, O.: Queueing systems with common buffer: a theoretical treatment. In: Kwiecień, A., Gaj, P., Stera, P. (eds.) CN 2011. CCIS, vol. 160, pp. 61–69. Springer, Heidelberg (2011). doi:10.1007/978-3-642-21771-5_8
12. Tikhonenko, O.M.: Queuing systems with processor sharing and limited resources. Autom. Remote Control. **71**(5), 803–815 (2010)

Innovative Applications

Approaches for In-vehicle Communication – An Analysis and Outlook

Arne Neumann[1]([✉]), Martin Jan Mytych[1], Derk Wesemann[2],
Lukasz Wisniewski[1], and Jürgen Jasperneite[3]

[1] inIT - Institute Industrial IT, OWL University of Applied Sciences,
32657 Lemgo, Germany
{arne.neumann,martin.mytych,lukasz.wisniewski}@hs-owl.de
[2] OWITA GmbH, 32657 Lemgo, Germany
derk.wesemann@owita.de
[3] Fraunhofer Application Center Industrial Automation (IOSB-INA),
32657 Lemgo, Germany
juergen.jasperneite@iosb-ina.fraunhofer.de

Abstract. Electrical and electronic systems have been getting raising importance for innovations in the automotive industry. Networking issues are a key factor in this process since they enable distributed control functions and user interaction bringing together nodes from different vendors. This paper analyses available and emerging network technologies for in-vehicle communication from a requirements driven perspective. It reviews successful network technologies from other application areas regarding a possible deployment in vehicular communication and distinguishes passenger car and commercial vehicle sectors as far as possible. This contribution is oriented to the OSI reference model showing the state of the art and future opportunities at the level of the several communication layers with a focus on physical layer issues and medium access protocols and including information modeling aspects.

Keywords: In-vehicle networks · Controller Area Network · SAE J1939 · Isobus · BroadR Reach · Reduced Twisted Pair Gigabit Ethernet · Time Sensitive Networking · OPC UA

1 Introduction

Automotive systems became complex systems of a reasonable number of distributed electronic control units (ECUs) with even more sensors and actuators attached. In passenger cars the number of ECUs reached 70, processing about 2500 signal points already ten years ago [1] and their numbers are still growing. From the late 1980s on standardized serial communication protocols have been used to interconnect the ECUs and signals. This approach provides several advantages, including the following. Subsystems of different vendors become able to interact with each other, sensor data can be shared by different functions and the number of wires in a vehicle can be reduced in comparison to parallel

P. Gaj et al. (Eds.): CN 2017, CCIS 718, pp. 395–411, 2017.
DOI: 10.1007/978-3-319-59767-6_31

wiring of sensors, actuators and ECUs, which results in less costs for material and assembling, less weight and hence less fuel consumption of the vehicle.

There are many and various application functions utilizing the communication infrastructure of vehicles and new functions are evolving. For example, driver assistance systems improve towards autonomous driving, with truck platooning as a use case being deployed soon [2]. These functions impose requirements to both in-vehicle and car-to-X communication, where this paper focuses on in-vehicle networks. The application functions require different characteristics of the communication systems. For example, functions for driver assistance have a priority on determinism and functional safety, whereas other functionality, such as infotainment, has a priority on data throughput. As a consequence the communication structure consists of interconnected subsystems of heterogeneous technologies, which will be analyzed in this paper and opportunities for improvements will be discussed.

In other industries Ethernet-based technologies have been introduced successfully. For example in the industrial automation domain, Ethernet-based real-time protocols have been standardized and deployed in applications, where fieldbus protocols were used before. This development was primarily motivated by better capabilities for network management, maintainability and communication performance. In IEEE there are currently activities to specify extensions to the lower layers of the Ethernet protocol which may support its utilization at in-vehicle networks. The paper also aims to analyze where and under which conditions Ethernet-based technologies including their currently developed extensions can support in-vehicle communication and how a migration could be done.

This paper reviews the requirements for communication networks in the domains of passenger cars and of commercial vehicles. It gives an overview of available technologies and discusses their applicability in commercial vehicles. The focus of the paper will be on network technologies, which enable a broad range of vehicular applications while technologies for specialized applications will be only briefly dealt.

2   In-vehicle Networks

2.1   Automotive Networks and Topologies

Vehicular networks started with the controller area network (CAN, ISO 11989-2), developed in 1983 and presented in 1987, defining layers 1 and 2 of the OSI reference model [3]. It basically offers a linear bus topology, which greatly reduces the wiring efforts in cars. In addition to CAN as a universal solution, other vehicular communication systems have been developed for more specialized applications. The local interconnect network (LIN, ISO 17987-1 to -7) focuses on small networks mainly for discrete I/O signals with low bandwidth requirements. LIN implements a master-slave-topology offering a low-cost, single wire solution compared to CAN-enabled devices. In the other direction, FlexRay was introduced 2000, offering benefits over CAN in means of bandwidth, real-time capability, redundancy and functional safety. The driving aspect was the advent of X-by-wire technologies, which needed a higher reliability and safety rating. FlexRay offers a redundant connection between nodes and supports both star and bus topology. The ability to support time-critical closed loop control application in conjunction the resulting higher cost and complexity of the components has preferred FlexRay's usage to engine, steering and advanced driver assistance systems (ADAS). Media Oriented Systems Transport (MOST) was developed exclusively for telematics and multimedia applications and is utilized only in the infotainment system. Comprehensive surveys about the outlined network technologies can be found in the literature [3–6].

These core standards are still a subject for improvements. For example for CAN there are SAE J2284/3 (High-Speed CAN for Vehicle Applications at 500 kbit/s) aiming at high transmission rate and higher allowable node count, and SAE J2411 (Single Wire CAN Network for Vehicle Applications) providing a simplified variant for low requirements regarding bit rate, bus length and robustness. As a disadvantage, sometimes compatibility issues arise, such as for CAN FD (flexible data-rate) [7].

Upon these communication layers, a number of protocols and standards have been developed for network control and data exchange. For CAN, this includes general purpose protocols like ISO 11898-4 (TTCAN, Time-Triggered Communication on CAN), industry-specific protocols like CANopen, SAE J1939 and ISOBUS [3,4] and protocols for special purpose vehicles, mainly derived from CANopen like EnergyBus (pedelecs, E-bikes), CleANopen (municipal vehicles) and FireCAN (DIN 14700, for external firefighter equipment). LIN does not comprise diverse higher layer protocols, but is most often terminated with a gateway to connect to an overlying CAN network. FlexRay, as a safety-critical subsystem, allows a diagnostic function via gateway, but also includes no diverse higher layer protocols. An outstanding application layer protocol is On-board Diagnostic (OBD) specifying self-diagnostic and reporting capability to assist the vehicle owner and repair technician. The development of OBD began in the 1980s driven by legal requirements for continuous emission surveillance during the entire lifetime of a vehicle. There are several standards for OBD, some of them contain both protocol and data object definitions. At the beginning ISO 14230 (Road vehicles - Diagnostic communication over K-Line, DoK-Line) gain importance, also known as KWP2000 and referring to ISO 15031-5 (Road vehicles - Communication between vehicle and external equipment for emissions-related diagnostics). Its CAN-based version ISO 15765-3 (Road vehicles - Diagnostic communication over Controller Area Network, DoCAN) has been widely implemented but never released by ISO. The most recent standard for OBD is ISO 14229 (Road vehicles - Unified diagnostic services, UDS). It focuses on application data and services, decoupling them from the lower layers. UDS provides data and services with the same semantics as the ODB standards based on KWP2000 and extend them but the representation is not compatible. This collection is not complete, there are additional standards about OBD, such as definitions by SAE or about communication to external equipment.

The different areas of preferred application for each bus system has led to a heterogeneous network structure, so far with only few needs for interconnection; each segment is mainly designed to work standalone, exchanging mainly status information with other networks. Nowadays the bus segments are usually connected by a centralized gateway. A typical network architecture is shown in Fig. 1, while additional topology examples can be found in [3]. Other approaches focus on the introduction of backbone networks for different application areas and different positions in the vehicle as described in [5,8].

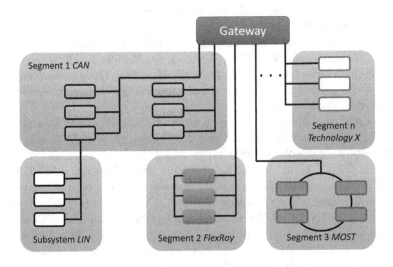

Fig. 1. Typical vehicular network architecture

Almost all available literature about vehicular network architectures aims at passenger cars, while the commercial vehicles sector is inadequately represented in the related work.

2.2 Communication Requirements and Applications

With more driver assistance, autonomous driving and other integral functionality, there are variable applications demanding information exchange between the ECUs, sensors and actuators. These applications also require diverse qualities of service, mainly determined by the update rate of information. In [9] update interval requirements at commercial vehicles are given, such as tire pressure: 10 s, battery current: 1 s, cruise control information: 100 ms and electronic transmission controller information: 10 ms. This fact of variable requirements in conjunction with the large number of communication nodes and the limited bus capacities led to a functional separation of bus segments into subsystems. The most typical subsystems comprise the power train bus (engine and gear control), the chassis system (e.g. anti-lock brake system) and driver assistance (e.g. electronic stability program), the body and comfort electronics (e.g. air condition), and the infotainment (audio and navigation). The power train bus needs to be generally accessible for diagnostics including emission monitoring to fulfill legal regulations, while all other bus system diagnostics is depending on manufacturer-specific tools.

The requirements imposed by the applications on the communication network comprise determinism, fault tolerance, data throughput and functional safety. Security requirements have to be managed for components, which grant access from outside components. This becomes a raising issue since car-to-X communication and infotainment connectivity pose growing challenges. In [10] an overview about the subsystems and their priorities of communication requirements is given. Table 1 extracted from this source summarizes the assignment of the different requirements to the subsystems.

Table 1. Automotive subsystems and their major requirements acc. to [10]

	Fault tolerance	Determinism	Bandwidth	Flexibility	Security
Chassis	Yes	Yes	Some	No	No
Airbag	Yes	Yes	Some	No	No
Powertrain	Some	Yes	Yes	Some	No
Body and comfort	No	Some	Some	Yes	No
X-by-wire	Yes	Yes	Some	No	No
Multimedia / Infotainment	No	Some	Yes	Yes	No
Wireless / Telematics	No	Some	Some	Yes	Yes
Diagnostics	No	Some	No	Yes	Yes

An advantage of the segmented topology is, that dependability of the communication for critical applications can be achieved by taking into account only a small number of components interconnected by a single segment. Additionally, every single bus segment can be configured in a way that is exactly matching the specific application requirements. On the other hand, the application functions of the vehicles become more complex and require information exchange across several bus segments. The therefore necessary communication paths involve segment transitions, which lead to an additional, resource demanding load for those ECUs acting as gateways between the bus segments. The number of cross-segment functions and gateways influences the efficiency of the overall network topology.

2.3 Influences by Upcoming Power Concepts on In-vehicle Networks

Vehicles like heavy duty road trains, buses and equipment for forestry and agriculture are characterized by a big number of auxiliary aggregates. These auxiliaries comprise compressors, fans, hydraulic pumps for servo-assisted steering, lifts etc. Nowadays they are usually driven directly by the combustion engine. The available power budget is coupled to the speed of the engine and cannot be steered on demand. Therefore the aggregates have to be scaled in a way that they can be operated at low engine speed. As a consequence, weight and size of the aggregates raise, decreasing the efficiency of the vehicle and resulting in a higher fuel consumption. In contrast to this, electrically powered drives allow a flexible power supply management which can adjust the power to operate an aggregate depending on the individual demand. Hence, the introduction of electric drives for the auxiliaries has a high potential to increase their efficiency. An accompanying effect of this concept is a significant increase of the number of communication nodes and signal points of the in-vehicle network, since the power management will require information exchange among the electrically powered auxiliaries and between the auxiliaries and other vehicle equipment. In contrast to the passenger cars with a mostly static configuration, in the context of commercial vehicles the communication topology is more dynamic due to the often changing of truck/trailer or tractor/implement combinations. Especially upon the initial composition of such a combination the exchange of device descriptions of the auxiliaries can become necessary which will cause a high amount of data to be transferred. Even this scenario does not happen very often, it is a some minutes lasting procedure when realized by conventional in-vehicle networks. Additionally there is the challenge of introducing many instances of the same or at least of a similar device type to the vehicle network. This opportunity will become important especially for modular devices and it shows a lack of scalability of the current communication standards. Requirements coming from this use case may exceed not only the number of physical nodes but also the number of logical addresses of a network segment when the modules shall be addressed individually. Another issue is about the information model. The standardized information models for commercial vehicles, for example SAE J1939 [9] describe only the common available data objects and do not allow a dynamic management of the object pool. Currently, additional objects can only be described in a proprietary way, which increases the engineering effort for the information exchange.

3   Physical Layer Aspects

In this chapter, the state of the art technologies which are most widespread in the automotive domain will be described. These are CAN (ISO 11989-2) for the passenger car sector and SAE J1939 as a CAN-based adaptation for the commercial vehicle sector. In contrast to this, a state of the art technology which is widespread in other domains will be introduced, Ethernet 100BASE-TX. Relevant criteria are robustness, bit rate, number of nodes, network extension and topology in order to fulfill application requirements on fault tolerance, bandwidth and scalability. With Ethernet 100BASE-T1 and Ethernet 1000BASE-T1 two emerging technologies will be described, which are promising candidates to enable Ethernet based protocols on a physical layer being as simple and reliable as nowadays solutions.

Currently used in-vehicle networks have different physical layer characteristics because of their design and application area. Upcoming technologies and concepts like ADAS or in-vehicle power concepts require network systems with a higher bandwidth to handle the amount of data. Ethernet is generally regarded as a next in-vehicle network for future development. A comparison of physical layer characteristic between CAN and Ethernet is shown in Table 2.

Table 2. Physical layer characteristic of CAN and Ethernet based communication

	CAN 2.0	SAE J1939	Ethernet 100BASE-TX	BroadR-Reach® (100BASE-T1)	Ethernet 1000BASE-T1
Standardization	ISO 11898-2	SAE J1939	IEEE 802.3 Clause 25	IEEE 802.3bw	IEEE 802.3bp
Possible topologies	Bus	Bus	Star	Star	Star
Max. transfer speed	1 Mbit/s	250 kbit/s	100 Mbit/s	100 Mbit/s	1 Gbit/s
Max. cable length	40 m for 1 Mbit/s	40 m	100 m	15 m	15 m
Transmission media	Copper, twisted pair	Copper, shielded twisted pair	Copper, 2 unshielded twisted pair	Copper, single unshielded twisted pair	Copper, single unshielded twisted pair

Typical CAN applications range from engine control and diagnostics to comfort electronics, with different bandwidths being employed. Typically, the range below 125 kbit/s is regarded as low-speed CAN oder CAN B, and the range from 125 kbit/s up to 1 Mbit/s is regarded as high-speed CAN, or CAN C. The CAN A class defines a bandwidth of 10 kbit/s or lower, historically used for diagnostic purposes. The maximum line length is depending on the chosen bandwidth. This limitation arises from the propagation time of the signal on the medium combined with the need for CAN to sample the received data exactly bit-synchronously. All bus nodes need to see the same bit value at the same point in time. Fault-tolerance is achieved by the use of differential signaling and the insertion of stuff-bits after 5 consecutive identical bit values, guaranteeing a state transition occurrence for synchronization. For commercial vehicle, the Society of Automotive Engineers (SAE) defines a communication protocol standard named J1939. It uses CAN as physical layer and is widely in use. Compared to CAN 2.0, SAE J1939 sets some limitations for the physical layer. The standard defines a maximum transfer speed of 250 kbit/s with a maximum cable length of 40 m, being below the allowed rating of up to 1 Mbit/s on 40 m distance for CAN, and a bus topology with a maximum number of 30 physical nodes.

However, new technologies and concepts need an enhancement of the physical layer. Nowadays, Ethernet is a widely used point to point communication technology. With 100BASE-TX it is possible to transfer data with 100 Mbit/s over a maximum cable length of 100 m. Due to the requirements on electromagnetic interference (EMI) and radio frequency interference (RFI) in the automotive market, Ethernet 100BASE-TX could not be used as an in-vehicle communication network. In addition to that limitation, 100BASE-TX uses 2 unshielded twisted pair cable, which would increase the overall cable weight and cost. To compensate the disadvantages of Ethernet in physical layer for automotive, new PHY's are ready for operation or in development. BroadR-Reach® supports 100 Mbit/s transfer speed over a single unshielded twisted pair cable which meets automotive EMI requirements [11] and is standardized as 100BASE-T1 in IEEE 802.3bw-2015 [12]. BroadR-Reach® has been used in a real-time Ethernet in-car backbone project, where BroadR-Reach® became a part of the Ethernet backbone system [13]. Also the applicability of BroadR-Reach for use with an industrial Ethernet protocol has been approved in [14]. But future challenges like uncompressed video for ADAS would need more bandwidth [15]. Hence, the next generation for Ethernet in the automotive field is under development. The standardization of a 1000BASE-T1 PHY in IEEE 802.3bp is currently in progress [16]. The 1000BASE-T1 PHY supports a maximum transfer speed of 1 Gbit/s in full duplex mode over a single unshielded twisted pair cable with a maximum length of 15 m. First PHY's on the basis of IEEE802.3bp draft are introduced [17]. Another point for the trend of Ethernet as in-vehicle communication system is the possibility to support voltage and current levels over a single twisted pair Ethernet link. Currently the 1-Pair Power over Data Lines (PoDL) Task Force defines under IEEE802.3bu a standard for that feature [18]. The deployment of these Ethernet based physical layers in the commercial vehicle sector is more challenging in comparison to passenger cars. A main reason is the topology extent beyond 15 m which requires components for signal refreshing. Beside this, the harsher environment induces higher requirements for ingress protection and overall robustness of connectors and may cause signal refreshing too.

4   Medium Access Control Aspects

Medium access control is most relevant to fulfill application requirements on determinism, transmission latency and data throughput. Here, the data link layer methods for medium access control of CAN and Ethernet as state of the art technologies will be briefly discussed. Time sensitive networking (TSN) targets to the real-time capability of Ethernet. Relevant TSN specifications will be described, as they can contribute to cover a broad range of vehicular requirements.

4.1   State of the Art Protocols

CAN specifies an asynchronous, event based medium access protocol. The communication is message-oriented, with a given identifier being assigned to a certain information, but not to a specific device. The number of devices on a bus is theoretically not limited, while the number of possible message identifiers depends on the their length. Two types of identifiers, 11 bit and 29 bit, are available to choose. CAN follows the Carrier Sense Multiple Access Collision Resolution (CSMA/CR) scheme, where each network node is allowed to send data when it detects an idle state at the medium. The messages are prioritized by their identifier, i.e. in case of conflicts an arbitration occurs and the message coming up with the higher order identifier being sent successfully. The arbitration reduces the number of retries and avoids a stop of data transfer due to congestions [3]. Although CSMA/CR represents a non-deterministic method, determinism can be reached for messages holding the highest priority. To fulfill application requirements on latency of transmission, a serious engineering effort regarding the assignment of priorities and update intervals of data objects is necessary and simulation and test of the network configuration is recommended. A detailed analysis about schedulability in CAN Networks is provided in [19].

The IEEE 802.1 Ethernet standard utilizes Carrier Sense Multiple Access with Collision Detection (CSMA/CD) for medium access and was originally not designed to transport any time sensitive traffic and hence does not provide determinism. After introduction of the IEEE 802.1Q, providing the possibility to assign a defined priority level to a particular message by using the Virtual LAN (VLAN) field, many proprietary industrial protocols were developed, e.g. Ethernet/IP, PROFINET RT, SERCOS and many other, which were build upon this feature. Due to limits of the priority based communication, some additional functionalities to further improve the real-time efficiency were introduced. These are: TDMA based communication (e.g. PROFINET IRT), polling based communication (Powerlink) or summation frame communication (EtherCAT). All mentioned approaches, allow to achieve high real-time performance, however it require modification of the original IEEE Ethernet MAC [20]. Beside this development of industrials protocols for Ethernet to transfer sensitive traffic a pool of establishments from the automotive area, like BMW and Daimler AG, developed a in-vehicle communication protocol, known as FlexRay, to handle the requirements like real-time communication.

4.2 Time Sensitive Networking

Due to the rapid evolution of the IT technology, especially the entertainment sector, such as high quality audio and video streaming, demands for real-time communication followed to the establishment of a new IEEE working group, Audio Video Bridging (AVB). The aim was to further enhance the real-time capabilities of the Ethernet standard. The suitability of AVB for particular vehicular use cases already has been proven by simulation [21]. Due to the high interest of the industry in this activity, the focus of the group has been broadened by including industrial application requirements [22]. At the same time, the name of the working group has been changed to more generic Time Sensitive Networking (TSN). An important aspect for this activity was to offer low costs devices, which require a minimal configuration effort to achieve plug-and-play functionality [23]. In case of in-vehicle communication, the plug-and-play functionality is not of major importance. It is due to the fact that in opposite to the industrial automation, in cars, the installed in-vehicle network infrastructure remains unchanged. More important is the spectrum of traffic classes that can be supported by the TSN technology. It allows to satisfy demands in terms of high throughput required by multimedia or infotainment systems, but also provide high determinism and availability, thus enabling support of control loops and safety critical functions. Having one system supporting different traffic flows, would help to significantly reduce the complexity of the current in-car communication infrastructures, and open the possibility for the future functionalities, such as highly sophisticated ADAS. The focus of TSN is very broad, therefore multiple sub-groups has been established to deal with a particular aspect. The most relevant for in-car communication are listed below:

- Timing and synchronization aspects:
  - timing and synchronization IEEE 802.1.AS
- Quality of service aspects and resource reservation:
  - stream reservation protocol IEEE 802.1Qat and the further extension IEEE 802.1Qcc
  - path control and reservation mechanisms IEEE 802.1Qca
- Forwarding and queuing mechanisms
  - forwarding and queuing enhancements for Time-Sensitive streams IEEE 802.1Qav
  - deterministic communication through time aware shaper IEEE 802.1Qbv and cycling queuing and forwarding shaper IEEE 802.1Qch
  - frame pre-emption IEEE 802.1Qbu
- Reliability
  - seamless redundancy IEEE 802.1CB
  - redundancy mechanisms included in IEEE 802.1Qca

There are several papers currently available, which try to evaluate some of the TSN amendments in the in-car communication context. In [24], authors investigated the worst case behavior of three different shapers, namely Burst Limiting Shaper (BLS), Time Aware Shaper (TAS) and Peristaltic Shaper (PS) using analytical calculation and simulation. According to the authors in [24], the best performance in terms of latency and latency jitter had the TAS, however require a lot of configuration efforts. BLS offers a compromise between performance and configuration efforts. The PS offers the easiest configuration, however the worst performance as comparing to other shapers. An additional deep investigation of the worst case latency provided by BLS in a typical automotive setup was conducted in [25]. Authors shown that in some cases it is better to use the IEEE 802.1Q than TSN + BLS. In order to efficiently use BLS some additional filtering functionality is required. The same authors analysed the effect of TSN with frame preemption (IEEE 802.3br) to worst-case end-to-end latency in [26]. Their experiments in a typical automotive setup show, that latency guarantees for time-critical traffic can be significantly improved while preemptable traffic only slightly degrades. In [27], authors investigated bandwidth allocation ratio for the scheduled traffic (IEEE 802.1Qbv), while adjusting the Maximum Transmit Unit (MTU). They have shown that using two time sensitive flows it is possible to achieve cycle times of $250\mu s$ for the MTU size of 109 bytes. A survey in [28] provide a broad overview about the Ethernet-based communication with the focus on IEEE AVB. It discusses especially the scheduled traffic and presents simulation results, where offset scheduling TAS were combined to achieve an temporal isolation from other kinds of traffic. The fault-tolerance aspects of TSN were investigated in [29]. Authors compared two different approaches aiming to guarantee seamless redundancy. They pointed out that the current seamless redundancy mechanisms provided by TSN lacks of flexibility in terms of stream reconfiguration and mechanisms for automatic stream reservation. Despite of all advantages, TSN increases the configuration overhead of a network. In [30] an ontology-based approach to support automatic network configuration of TSN is presented. The authors demonstrated the approach by modeling the TAS and came to the conclusion that the expressiveness of the ontology has to be further investigated. The several papers demonstrate that TSN is actually in focus, but it also shows a gap in the field of implementation to simulate the behavior of TSN. An easily accessible implementation of single protocols would be a benefit to gain insight of TSN and whose performance. After all it can be concluded that TSN is a prominent candidate for in-vehicle communication to handle future requirements. It supports different real-time classes, offers determinism and high reliability via seamless redundancy.

As a wrap-up of this chapter, Table 3 gives a summary about the access methods of the discussed communication technologies.

Table 3. Summary on medium access methods

	CAN	Ethernet	TSN
Basic access protocol	CSMA/CR	CSMA/CD	CSMA/CD
Additional measures	Priority based arbitration	–	Scheduling
Determinism	Restricted	No	Yes

5   Transport Protocol and Efficiency Aspects

In this chapter, the considerations are mainly driven by the application requirement of bandwidth. The state-of the-art technologies are compared regarding their performance to transfer different qualities of user data. For this communication layer no emerging technology is discussed, but new mappings with established protocols at higher layers open future opportunities.

The standard ISO 11898 for CAN does not specify higher protocol layers of the OSI reference model. CAN is limited to a maximum data object length of 8 octets and provides message oriented broadcasting without address information about the sender and receiver. This simplicity enables a small protocol overhead and allows short transmission times. Consequently, user data rates of approximately 7,5 KB/s for 1 octet payload and approximately 28 KB/s for 8 octets payload are possible, supposing a bus workload of 100%, according to [3, 31]. Although this throughput statement is not very impressive, it is sufficient for many applications regarding the update intervals of the required number of communication objects. In the domain of commercial vehicles, the widespread standard SAE J1939 defines transport protocols for segmented transport for both message-oriented broadcasts and node-oriented unicasts on top of the CAN layers. The transport protocols define an initial frame to announce the transmission and the user data length of a data frame is reduced to 7 by taking the first octet of the CAN payload for protocol information. The protocols shall not strongly interfere in the plain message exchange, therefore the standard defines a low CAN priority and a minimum frame gap of 50 ms. All these measures reduce the user data rate to below 140 bit/s and limit the application range to very low demanding functions.

While CAN based protocols show constraints which the upper limit of the payload size, Ethernet based protocols show a lack of efficiency considering the lower limit of the payload size. The payload size of an Ethernet frame is defined from 42 to 1500 octets, if the VLAN tag is used. When transferring control data of sensors and actuators, the user data length often will be between 1 and 4 octets and the remaining payload size needs to be filled by padding octets. Considering the overall Ethernet protocol overhead and the inter frame gap the gross ratio of net data becomes 1:84 for a single octet. When using a Ethernet bit rate of 100 Mbit/s, the net data rate in this worst case is still above 1 Mbit/s, again supposing a network load of 100%, which is significantly higher than CAN communication.

Consequently, the substitution of CAN based protocols by Ethernet based protocols can overcome bandwidth limitations for in-vehicle communication when transferring big sized data objects. In [32] an approach is published, where the SAE J1939 application protocols are mapped on top of a TCP/UDP stack. The authors claim the applicability for the power train segment in heavy duty vehicle networks, which still needs further investigation. Nevertheless, this contribution shows, that a changeover to more powerful network technologies is possible without essential modifications at the application interface.

6   Information Model Aspects

In this section, a well established information model of the commercial vehicle domain is discussed. To enable the easy integration of future application function a possible extension of this model, which preserves the existing application interface, is described in the second subsection.

6.1   Consideration About Available Information Models

The application layer protocol standards for in-vehicle communication SAE J1939 and ISO 11783, which are nowadays the mostly utilized standards for commercial vehicles, contain detailed information models to address vehicle components and their parameters. For example, one part of SAE J1939 comprehensively specifies parameters concerning typical components (e.g. engine, steering, collision sensors) and functions (e.g. speed control, air suspension control, aftertreatment) of an vehicle. The parameter description includes the unambiguous parameter identifier, information about name and acronym, data type, data range, affiliation to records for transmission and update intervals. This document provides a valuable contribution for the interoperability of the typical, widespread components and functions. On the other hand, the approach of SAE J1939 is difficult to manage in case of extending the model for new information object types or even for adding new instances of already existing object types.

Currently such a extensibility is rarely required, but upcoming application concepts like introducing modular electrical drives for auxiliaries will tighten the problem.

6.2   Potential Future Information Models

An object oriented modeling of application specific information structures can be used to improve the rigid information modeling provided by the current technologies for in-vehicle communication. OPC UA, a technology widely used mainly in the domain of industry automation, provides such an object oriented modeling. Currently, the OPC UA specification is being enhanced by PubSub, a new communication pattern according to the publisher/subscriber model enabling so called server based subscriptions [33]. In IEC 62541-3 the Address Space Model of OPC UA is defined. It can be considered as a meta model providing objects as the basis for any information model. The object elements are represented by nodes. These nodes comprise variables, methods or references to other objects. Additionally, IEC 62541-5 specifies nodes to be used for diagnostics and as entry points to server-specific nodes. As a result, an information model of an "empty" server is defined and the vendor of the component which is represented by the OPC server can customize it. As optional specification elements, predefined models for data access, alarms, history and others are available. Moreover, the information model can be changed during runtime of the server by adding or removing nodes. By this means, the OPC UA information model is independent from transport protocols and enable domain specific extendibility. For the deployment in in-vehicle networks, OPC UA needs the ability to be implemented on physical nodes with low resources. For this reason, OPC UA components need to be scaled down. In order to support this, the OPC UA specifications provide profiles, for example the OPC UA Nano Embedded device profile. Based on this profile, it is possible to scale down an OPC UA server to 15 KB RAM and 10 KB ROM [34], thus allowing implementation at the chip level of a resource limited device such as a sensor or an actuator.

To provide interoperability beyond this general information model and to ease the use of OPC UA in several domains, Companion Standards have been developed. For example, the specifications for building automation (in co-operation with BACnet), energy systems and management (participating in IEC TC 57

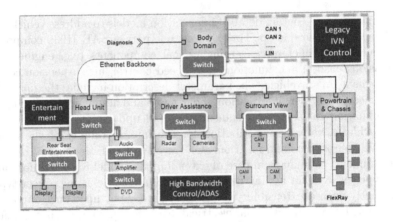

Fig. 2. Future architecture with Ethernet backbone acc. to [37]

"Power systems") or railways transportation shows that OPC UA already has been approached by applications outside the industrial automation. The utilization of the OPC UA Address Space Model as a wrapper of data models provided by the in-vehicle communication standards could be a step ahead to the required extendibility of the models. The existing information structures can be preserved and transformed into an object oriented approach as it is shown in [35] for building automation. At the same time the co-existence with information models of upcoming components and functions, which are not covered by the available standards in the vehicle domain, becomes possible. For example, SAE J1939 data in parallel to data according to the standard CAN in Automation DS402 for electric drives could be modeled and transferred on the same network.

7   Conclusion

This paper shows the current status of in-vehicle communication networks in the field of passenger cars as well as for commercial and heavy duty vehicle, and points at upcoming challenges. It depicts the future of Ethernet as in-vehicle communication system related to several parts of the OSI reference model. In summary, Ethernet will take place in the automotive market, see also [36]. However, ongoing developments and implementations show, that new network systems will not immediately replace, but rather supplement them. This strategy is beneficial especially for critical systems, where proven-in-use concepts contribute to the functional safety. The evolution of automotive Ethernet, according to [8], propose the implementation of Ethernet in three generations. The first generation already exist in high class vehicles. It uses 100BASE-TX Ethernet with Diagnostics over IP (DoIP) for on-board diagnostics and ECU's updates. Figure 2 given by the author of [37] illustrates the next generations. Second generation uses Ethernet as additional in-vehicle network to transfer the amount of data from camera systems for drive assistance and infotainment. Finally, the 3rd generation with the possibility to transfer 1 Gbit/s will implement Ethernet as a backbone system and change automotive wiring harness from heterogeneous to hierarchical homogeneous network by introducing a new network topology level.

In future, Ethernet in connection with TSN will be a possible approach for time relevant communication beside ADAS and infotainment. At the layer of information modeling, concepts incorporating dynamic and instantiable information object presentation like OPC UA can support the integration of new application functions.

References

1. Navet, N., Song, Y., Simonot-Lion, F., Wilwert, C.: Trends in automotive communication systems. Proc. IEEE **93**(6), 1204–1223 (2005)
2. Bishop, R., Bevly, D., Switkes, J., Park, L.: Results of initial test and evaluation of a driver-assistive truck platooning prototype. In: 2014 IEEE Intelligent Vehicles Symposium Proceedings, pp. 208–213, June 2014
3. Zimmermann, W., Schmidgall, R.: Bussysteme in der Fahrzeugtechnik - Protokolle, Standards und Softwarearchitektur, 5th edn. Springer-Verlag, Heidelberg (2014)
4. Navet, N., Simonot-Lion, F.: Automotive Embedded Systems Handbook. Industrial Information Technology Series. CRC Press, Boca Raton (2008). https://books.google.de/books?id=vB700Gb4RtkC
5. Zeng, W., Khalid, M., Chowdhury, S.: In-vehicle networks outlook: achievements and challenges. IEEE Commun. Surv. Tutorials **18**(3), 1–1 (2016)
6. Talbot, S.C., Ren, S.: Comparision of fieldbus systems can, TTCAN, FlexRay and LIN in passenger vehicles. In: 29th IEEE International Conference on Distributed Computing Systems Workshops, 2009, ICDCS Workshops 2009, pp. 26–31, June 2009
7. Cena, G., Bertolotti, I.C., Hu, T., Valenzano, A.: Improving compatibility between CAN FD and legacy CAN devices. In: 2015 IEEE 1st International Forum on Research and Technologies for Society and Industry Leveraging a better tomorrow (RTSI), pp. 419–426, September 2015
8. Hank, P., Müller, S., Vermesan, O., Keybus, J.V.D.: Automotive ethernet: in-vehicle networking and smart mobility. In: Design, Automation Test in Europe Conference Exhibition (DATE), 2013, pp. 1735–1739, March 2013
9. J1939 Surface Vehicle Recommended Practice; Part 71 Vehicle Application Layer, SAE International Std., June 2015
10. Nolte, T., Hansson, H., Bello, L.L.: Automotive communications-past, current and future. In: 2005 IEEE Conference on Emerging Technologies and Factory Automation, vol. 1, pp. 8–992, September 2005
11. Broadcom, "BroadR-reach® physical layer transceiver specification for automotive applications v3.0," Broadcom, Technical report (2014)
12. IEEE 802.3, working group for ethernet standards. http://www.ieee802.org/3/
13. Steinbach, T., Müller, K., Korf, F., Röllig, R.: Demo: real-time ethernet in-car backbones: first insights into an automotive prototype. In: 2014 IEEE Vehicular Networking Conference (VNC), pp. 133–134, December 2014
14. Banick, N.: Untersuchung des quelloffenen Ethernet Powerlink Stacks mit einer Zweidraht-Übertragungstechnologie für den Einsatz im Automobilbereich, Lemgo, January 2015

410   A. Neumann et al.

15. "Reduced Twisted Pair Gigabit Ethernet PHY - Call for Interest," IEEE 802.3 Ethernet Working Group, Technical report, March 2012. http://www.ieee802.org/3/RTPGE/public/mar12/CFI_01_0312.pdf
16. IEEE p802.3bp. 1000BASE-T1 PHY Task Force. http://www.ieee802.org/3/bp/
17. Marvell 1000BASE-T1 PHY. http://www.marvell.com/company/news/pressDetail.do?releaseID=7256
18. IEEE p802.3bu. 1-Pair Power over Data Lines (PoDL) Task Force. http://www.ieee802.org/3/bu/
19. Davis, R.I., Kollmann, S., Pollex, V., Slomka, F.: Controller area network (CAN) schedulability analysis with fifo queues. In: 2011 23rd Euromicro Conference on Real-Time Systems, pp. 45–56, July 2011
20. Wisniewski, L., Schumacher, M., Jasperneite, J., Schriegel, S.: Fast and simple scheduling algorithm for PROFINET IRT networks. In: 9th IEEE International Workshop on Factory Communication Systems (WFCS) 2012, pp. 141–144, May 2012
21. Alderisi, G., Caltabiano, A., Vasta, G., Iannizzotto, G., Steinbach, T., Bello, L.L.: Simulative assessments of IEEE 802.1 Ethernet AVB and time-triggered ethernet for advanced driver assistance systems and in-car infotainment. In: 2012 Vehicular Networking Conference (VNC), IEEE, pp. 187–194, November 2012
22. Imtiaz, J., Jasperneite, J., Schriegel, S.: A proposal to integrate process data communication to IEEE 802.1 Audio Video Bridging (AVB). In: 2011 IEEE 16th Conference on Emerging Technologies Factory Automation (ETFA), pp. 1–8, September 2011
23. Garner, G.M., Ryu, H.: Synchronization of audio/video bridging networks using IEEE 802.1AS. IEEE Commun. Mag. **49**(2), 140–147 (2011)
24. Thangamuthu, S., Concer, N., Cuijpers, P.J.L., Lukkien, J.J.: Analysis of ethernet-switch traffic shapers for in-vehicle networking applications. In: 2015 Design, Automation Test in Europe Conference Exhibition (DATE), pp. 55–60, March 2015
25. Thiele, D., Ernst, R.: Formal worst-case timing analysis of ethernet TSN's burst-limiting shaper. In: 2016 Design, Automation Test in Europe Conference Exhibition (DATE), pp. 187–192, March 2016
26. Thiele, D., Ernst, R.: Formal worst-case performance analysis of time-sensitive ethernet with frame preemption. In: 2016 IEEE 21st International Conference on Emerging Technologies and Factory Automation (ETFA), pp. 1–9, September 2016
27. Ko, J., Lee, J.H., Park, C., Park, S.K.: Research on optimal bandwidth allocation for the scheduled traffic in IEEE 802.1 AVB. In: 2015 IEEE International Conference on Vehicular Electronics and Safety (ICVES), pp. 31–35, November 2015
28. Bello, L.L.: Novel trends in automotive networks: a perspective on ethernet and the IEEE audio video bridging. In: Proceedings of the 2014 IEEE Emerging Technology and Factory Automation (ETFA), pp. 1–8, September 2014
29. Kehrer, S., Kleineberg, O., Heffernan, D.: A comparison of fault-tolerance concepts for IEEE 802.1 Time Sensitive Networks (TSN). In: Proceedings of the 2014 IEEE Emerging Technology and Factory Automation (ETFA), pp. 1–8, September 2014
30. Farzaneh, M.H., Knoll, A.: An ontology-based plug-and-play approach for in-vehicle Time-Sensitive Networking (TSN). In: 2016 IEEE 7th Annual Information Technology, Electronics and Mobile Communication Conference (IEMCON), pp. 1–8, October 2016
31. Traub, M.: Durchgängige Timing-Bewertung von Vernetzungsarchitekturen und Gateway-Systemen im Kraftfahrzeug -. KIT Scientific Publishing, Karlsruhe (2010)

32. Ruggeri, M., Malaguti, G., Dian, M.: SAE J 1939 over real time ethernet: the future of heavy duty vehicle networks. Society of Automotive Engineers (SAE), Technical report, September 2012

33. OPC Foundation. OPC UA is Enhanced for Publish-Subscribe (Pub/Sub). https://opcfoundation.org/opc-connect/2016/03/opc-ua-is-enhanced-for-publish-subscribe-pubsub/

34. Imtiaz, J., Jasperneite, J.: Scalability of OPC-UA down to the chip level enables "Internet of Things". In: 2013 11th IEEE International Conference on Industrial Informatics (INDIN), pp. 500–505, July 2013

35. Fernbach, A., Granzer, W., Kastner, W.: Interoperability at the management level of building automation systems: a case study for BACnet and OPC UA. In: 2011 IEEE 16th Conference on Emerging Technologies Factory Automation (ETFA), pp. 1–8, September 2011

36. Bello, L.L.: The case for ethernet in automotive communications. SIGBED Rev. 8(4), 7–15 (2011). http://doi.acm.org/10.1145/2095256.2095257

37. Hinrichsen, J.: The road to autonomous driving. In: Deterministic Ethernet Forum, Vienna, April 2015

An Approach for Evaluating Performance of Magnetic-Field Based Indoor Positioning Systems: Neural Network

Serpil Ustebay[1], Zuleyha Yiner[1], M. Ali Aydin[1], Ahmet Sertbas[1], and Tulin Atmaca[2(✉)]

[1] Department of Computer Engineering, Istanbul University, Istanbul, Turkey
{serpil.ustebay,zuleyha.yiner,aydinali,asertbas}@istanbul.edu.tr
[2] Laboratoire Samovar, Telecom SudParis, CNRS, Université Paris-Saclay, Evry, France
tulin.atmaca@telecom-sudparis.eu

Abstract. Indoor Positioning Systems are more and more attractive research area and popular studies. They provide direct access of instant location information of people in large, complex locations such as airports, museums, hospitals, etc. Especially for elders and children, location information can be lifesaving in such complex places. Thanks to the smart technology that can be worn, daily accessories such as wristbands, smart clocks are suitable for this job. In this study, the earth's magnetic field data is used to find location of devices. Having less noise rather than other type of data, magnetic field data provides high success. In this study, with this data, a positioning model is constructed by using Artificial Neural Network (ANN). Support Vector Machines(SVM) was used to compare the results of the model with the ANN. Also the accuracy of this model is calculated and how the number of hidden layer of neural network affects the accuracy is analyzed. Results show that magnetic field indoor positioning system accuracy can reach 95% with ANN.

Keywords: Magnetic-field indoor positioning systems · Neural network · Pattern recognition network · Cross entropy function · Performance · Accuracy · Support Vector Machines (SVM)

1 Introduction

Positioning systems, known as outdoor and indoor, become more prevalent with technological developments. These systems provide information for applications use location information like navigation, monitoring, tracking etc. Outdoor positioning systems operate with the GPS (Global Positioning System) signals coming from at least three satellites. GPS works with triangulation method which is based time of signal's time of arrival, angle, etc. In addition, GPS may be used for indoor positioning if necessary equipments are integrated inside buildings. However, line-of-sight transmission between receivers and satellites may

© Springer International Publishing AG 2017
P. Gaj et al. (Eds.): CN 2017, CCIS 718, pp. 412–421, 2017.
DOI: 10.1007/978-3-319-59767-6_32

not be efficient for an indoor environment due to lack of line-of-sight transmission in indoor. Indoor positioning system (IPS) is a system which uses radio waves, magnetic fields, or other sensory information collected by mobile devices in order to locate objects or people in a building. IPS is generally grouped as infrastructure-based and infrastructure-free. Since requiring some hardware pre-installed, infrastructure-based solutions have high cost comparing with infrastructure-free solutions. Positioning systems can be divided into sub-categories as shown in Fig. 1. During location determination, participants get into active group when information is directly transferred from the device from which they carry. The opposite of this situation is get into passive one [1]. To exemplify, the usage of Bluetooth signal of person's mobile phone in positioning is active classification. With acquisition of images through cameras in interior space and face recognition operations, learning of knowledge of where someone is located is a passive positioning. Furthermore, positioning is realized by using mathematical methods with respect to used technology [2].

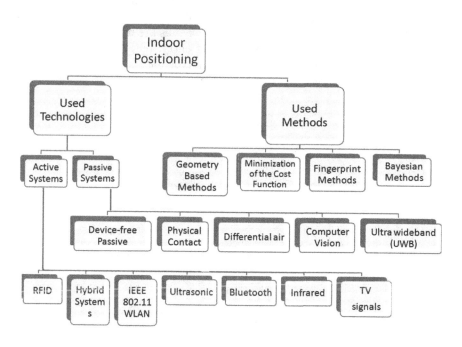

Fig. 1. Positioning technique taxonomy

Although used technologies are different, methods can be grouped under 4 basic headings. In geometry based methods, in order to calculate of distance between sender and receiver, the Angle of signals (AOA), arrival time (TOA), the time difference of reception and arrival (TDOA) are used. Fingerprint is a method based on power of signals and consists of two processes. In this method, Received Signal Strength (RSS) measurements are made from specific points within the building. A position estimation model is constructed based on the obtained Fingerprint signal map. Bayesian approach uses Bayesian inference techniques to calculate the position of a person or object according to probability distribution at time t.

Galván Tejada et al. [3] used 4 categories based on the technology that conduct between user and the environment. The first one is a location system that explicit technologies are used like Bluetooth and RFID. The second one is systems reuse sensory devices in the smart phones. Wi-Fi access points are important for series of sensors of specific locations that comprise specialized infrastructure. This is the third kind of indoor positioning categories. Lastly, the systems that use magnetic field or environmental audio as positioning.

Higtower and Borriello [4] grouped location systems with respect to ToA/TDoA (time of arrival and time difference of arrival use signal runtime between sender and receivers), AoA (the angle of incidence at receivers), the Received Signal Strength Indicator (RSSI) and Fingerprint. Fingerprint methods consist of offline and online phases. In the offline phase, Wi-Fi signal are measured and stored in a database with the location of their appearance. In this way fingerprint database is established. These signals are measured again in the online phase. After finding the best matching entry, positioning is performed. Earth magnetic field value is described by a vector has X, Y, Z features where X is for north, Y is for east and Z is for height [5] (Fig. 2).

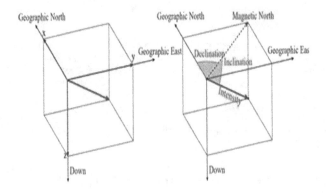

Fig. 2. Magnetic field vector [5]

Sensors are used to get magnetic field information. This information may be used in automotive, military, aviation and industrial areas. Advances in Micro Electro Mechanical Systems (MEMS) technology have allowed these sensors to be added to electronic devices such as smartphones, smart clocks and computer tablets. Applications called e-compass used in mobile phones can measure this new sensor data as short as 1nT. The data used by each device may be different because mobile phone manufacturers produce these sensors with 4 different approaches such as Hall Effect, Giant Magneto Resistance (GMR), Magnetic

Tunneling Junction (MTJ) sensing and Anisotropic Magneto Résistance (AMR) [6]. By the way, longitude, latitude and altitude information are entered by logging into the system online with the world's magnetic field calculator tool, which is obtained through joint participation of the United States National Geospatial-Intelligence Agency (NGA) and the United Kingdom's Defense Geographic Center (DGC). Tthe world's magnetic field data can be obtained without the need for any sensors [7] via these institutions.

Since the magnetic field data contains less noise than the WLAN signal data, it has been used in indoor positioning studies. Especially a strong magnetic field data prepared by the FingerPrint method allows the location of the person to be detected with an error rate of the centimeter level [8]. Also, it is possible to increase the accuracy by adding different technologies. In this study, a hybrid system was designed while the position of the person was found and magnetic field data was used to increase the accuracy of the results obtained [9]. Thus, false or missing measurement of the near-distance technology such as RFID, and false positioning of Wi-Fi signals is affected by the noise are prevented.

The location of person based on magnetic field data is determined by using Gaussian process with radial basis function (GausssprRadial), Single C5.0 Tree, Soft Independent Modeling of Classical Analogy (SIMCA), Multi-layer perceptron with Resilient Back propagation (Rprop), Bagged Classification and Regression Tree (CART) algorithms [10]. Results of this study show that magnetic field data provides good robustness and accuracy for buildings with low magnetic field variability.

This paper is organized as follows; Sect. 2 gives detailed information about magnetic field data is used. Section 3 describes artificial neural network. In Sect. 4, the implementation is introduced. Finally, the paper is concluded in Sect. 5.

2   Database for Implementation

In this paper, RFKONDB was used as a indoor signal strength map. RFKON is created by Sinem Bozkurt et al. [11]. which measurements are taken from the first floor of Eskisehir Osmangazi University Teknopark. This database is magnetic field based and constructed using 4 different mobile devices. Descriptions of devices are given in the Table 1.

Table 1. Devices used for measurements

Device ID	DEvice type	OS version
1	Samsung S4 Mini	Android .2.2
2	LG G3	Android 5.0
3	Sony Xperia Z2	Android .4.4
4	Samsung Galaxy Note 10.1	Android .4.2

Table 2. Magnetic field data set

Ref. point	Date	Device ID	x	y	Floor	Battery	X	Y	Z
1	01.07.2015 09:10	1	1.2	1.2	1	100	−13.02	5.87	−19.79
1	02.07.2015 09:10	2	1.2	1.2	1	100	−48.11	19.78	14.87
1	03.07.2015 09:10	3	1.2	1.2	1	100	−17.10	8.5	−22.79
1	04.07.2015 09:10	4	1.2	1.2	1	100	0.232	−17.10	−11.69

During measuring, 54 reference points is used and for each mobile device, 20 measurements were recorded at every reference point. Totally 4320 sample measurement is obtained for magnetic field database.

Magnetic field based dataset includes information about each data such as DeviceID, real world X,Y coordinates, Floor, Battery Level, and Magnetic Field X,Y, Z coordinates values. The database sample is given Table 2. While trying to evaluate the accuracy of the magnetic field based indoor location system, we just use reference points and Magnetic Field coordinates of each measure.

3   Models

In this work Support Vector Machines and Neural Networks are used to create an indoor positioning model.

3.1   Support Vector Machines (SVM)

Support Vector Machines (SVM) is a popular margin classifier which is widely used for linear classification problems. In linear classification problems, it is assumed that there exists an optimal separating line (1) discriminating samples of positive and negative groups in data space S.

$$f(x) = w_1 x_1 + w_2 x_2 + k = w^T x + k \tag{1}$$

$f(x)$ is defined by the weight vector w, and the shift amount k [12]. x is a sample in data space S, and is considered to be from positive class if $f(x) \geq 1$, or negative class if $f(x) \leq -1$. This can be shown by the general expression in (2). Here, c is the class label of x (positive if $c = +1$, or negative if $c = -1$)

$$c(w^T x + k) \geq 1 \tag{2}$$

The aim of SVM is to find optimal w vector of S' new data space that satisfies (2) for all samples in S. The method calculates the principal components of new data space S' by solving a quadratic optimization problem. By this way, samples that cannot be linearly discriminated in original data space S, can be linearly discriminated in new data space S' to which they are transferred.

Classical SVM works only on bi-class datasets which is a major disadvantage. Therefore, in this study, LIBSVM library [13] which is capable of multi-class classification was utilized.

3.2   Artificial Neural Network (ANN)

ANN is a computational model based on biological neural network. A neuron is a biological cell and process information in the brain. Axon and dendrite are connection branches between any two cells. A neuron receives signals from other neurons through its dendrites and transmits the signal to other neurons along the axon. Hereby the basic function of the nerve cell occurs.

In this study, we use Pattern Recognition Networks that are feed forward networks to evaluate accuracy of location system. This type of network can be trained to classify inputs according to target classes. The target data for pattern recognition networks consist of vectors of all zero values except for a 1 in element i, where i is the class they are to represent. In this kind of feed forward network, training function is Scaled Conjugate Gradient that updates weight and bias values according to the scaled conjugate gradient method. This method has advantage by reducing time-consuming line-search. Performance function of the network is Cross Entropy by default. Here is a basic Pattern Recognition Network in Fig. 3.

Pattern Recognition Network consists of an input layer, one or more hidden layer and an output layer. Neuron numbers in the hidden layer affect the learning relationship between input and output. By defining appropriate neuron in hidden layer, network will get better accuracy.

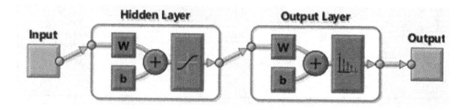

Fig. 3. Pattern recognition network

4   Implementation

In this study, we implemented a Neural Network localization model which contains 3 layers i.e. input, one hidden layer and output layer used. Magnetic field database is parted 80% training and 20% for test. Every input pattern has 3 features. Input pattern is applied to the input layer and the effect propagates through the network layer until an output is obtained. This process is repeated layer-by-layer until an error signal is generated which describes the contribution to each node in the network, relative to the common fault. After that, the actual output of the network is compared to the pending output and an error signal is calculated for each of the output nodes. Then, weights, defined by default initially, are adjusted with respect to calculated error. The process of finding

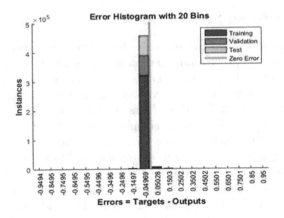

Fig. 4. Error histogram

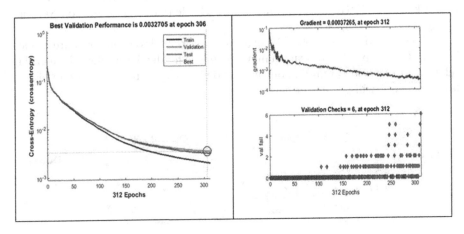

Fig. 5. Gradient changes with respect to epoch number (left). The error distribution of network (right)

proper weights such that for a given input pattern the network produces the desired output is defined as training.

Desired output is defined as training. Test data set is used for evaluating generalization error which indicates performance. The performance criterion is how well the artificial neural network can distinguish classes from each other through the given training set. For this; Test data are given to the generated neural network model and expected from it to find data classes. The classes which are estimated from the system are compared with test data set's real classes and accuracy is calculated as a percentage. We use Cross Entropy as an error measure to calculate the performance of the network.

It can be seen error value versus epoch plot in Fig. 4. As a result, the best accuracy is obtained at epoch 306 with using Cross Entropy performance

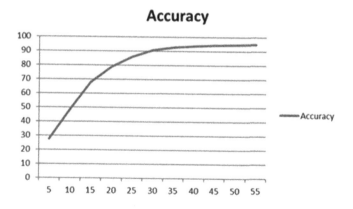

Fig. 6. Accuracy versus number of hidden layer neuron

function. Figure 5 (left) shows the gradient changes with respect to epoch number and (right). The error distribution of network.

In neural network, number of neuron in the hidden layer has important role on performance. The main problem becomes what the number of neurons must be. Defining large number of neurons increases the storage capacity of a network. Low number of neurons make network to have low performance. The plot for hidden layer neurons versus accuracy is given in Fig. 6.

As is seen after some changes in number of neurons, the performance curves a bit changes. It must be careful about the network memorizes far from learning. As seen in the Fig. 7, the accuracy is stabilized after some number i.e. 35.

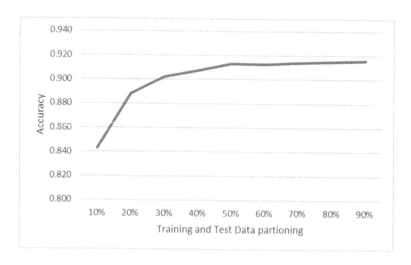

Fig. 7. SVM based localization model test results according to training and test data portioning rate

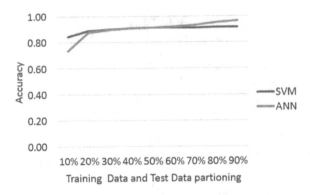

Fig. 8. SVM and ANN based localization model test results according to training and test data portioning rate

SVM based indoor localization method is used the linear kernel and the penalty value (C) was chosen as 1. Portioning defines division rate of training data and test data. Model was performed at 100 and mean accuracy was calculated and accuracy results are shown on Fig. 7.

Figure 8 shows comparative results of ANN and SVM based localization methods. Although the accuracy values of the algorithms are close, the highest accuracy value is obtained in the NN-based positioning model.

5   Conclusion

The main purpose of all positioning systems is to provide high accuracy, reliable results, and low cost construction. In many technologies, sometimes the more accuracy results require the more cost devices inside buildings. Nowadays, mobile devices have usage in order to estimate locations of individuals thanks to containing many sensors within itself. Using the correct sensor data and the correct positioning pattern will reduce system cost. Having less noise than other data types, magnetic field based data is more favorable.

The aim of this study was finding location of any object or device by using its magnetic field information. Two different localization models were created. First model is used ANN and second is used SVM. The results which are obtained by using neural networks are highly accurate rather than SVM based localization model. Magnetic field sensors which are integrated to mobile devices do not produces any hardware cost. For rising accuracy different kind of sensor data may be included to the localization system and tested afterwards. In future studies, we envisage to use a hybrid localization model with magnetic field data and RSSI data.

Acknowledgments. This work is also a part of the Ph.D. thesis titled "Design of an Efficient User Localization System for Next Generation Wireless Networks" at Istanbul University, Institute of Physical Sciences.

References

1. Pirzada, N., et al.: Comparative analysis of active and passive indoor localization systems. AASRI Procedia **5**, 92–97 (2013)
2. Seco, F., et al.: A survey of mathematical methods for indoor localization. In: IEEE International Symposium on Intelligent Signal Processing, WISP 2009. IEEE (2009)
3. Galván-Tejada, C.E., et al.: Evaluation of four classifiers as cost function for indoor location systems. Procedia Comput. Sci. **32**, 453–460 (2014)
4. Hightower, J., Borriello, G.: Location sensing techniques. IEEE Comput. **34**(8), 57–66 (2001)
5. Wikimedia. https://commons.wikimedia.org/w/index.php?curid=19810392
6. National Centers for Environmental Information. https://www.ngdc.noaa.gov/geomag/models.shtml
7. Online Calculators for the World Magnetic Model. https://www.ngdc.noaa.gov/geomag/WMM/calculators.shtml
8. Angermann, M., et al.: Characterization of the indoor magnetic field for applications in localization and mapping. In: 2012 International Conference on Indoor Positioning and Indoor Navigation (IPIN). IEEE (2012)
9. Ettlinger, A., Retscher, G.: Positioning using ambient magnetic fields in combination with Wi-Fi and RFID. In: 2016 International Conference on Indoor Positioning and Indoor Navigation (IPIN). IEEE (2016)
10. http://www.mdpi.com/1424-8220/15/7/17168/htm
11. Bozkurt, S., et al.: A novel multi-sensor and multi-topological database for indoor positioning on fingerprint techniques. In: 2015 International Symposium on Innovations in Intelligent SysTems and Applications (INISTA). IEEE (2015)
12. Alpaydin, E. (2013). Yapay Öğrenme. Boğaziçi Üniversitesi Yayınevi. ISBN-13: 978-6-054-23849-1.18. Lin., C.-C.C.-J. (2001)
13. LIBSVM: a library for support vector machine. http://www.csie.ntu.edu.tw/~cjlin/libsvm

Improvements of the Reactive Auto Scaling Method for Cloud Platform

Dariusz Rafal Augustyn[(✉)]

Institute of Informatics, Silesian University of Technology,
16 Akademicka St., 44-100 Gliwice, Poland
draugustyn@polsl.pl

Abstract. Elements of cloud infrastructure like load balancers, instances of virtual server (service nodes), storage services are used in an architecture of modern cloud-enabled systems. Auto scaling is a mechanism which allows to on-line adapt efficiency of a system to current load. It is done by increasing or decreasing number of running instances. Auto scaling model uses a statistics based on a standard metrics like CPU Utilization or a custom metrics like execution time of selected business service. By horizontal scaling, the model should satisfy Quality of Service requirements (QoS). QoS requirements are determined by criteria based on statistics defined on metrics. The auto scaling model should minimize the cost (mainly measured by the number of used instances) subject to an assumed QoS requirements. There are many reactive (on current load) and predictive (future load) approaches to the model of auto scaling. In this paper we propose some extensions to the concrete reactive auto scaling model to improve sensitivity to load changes. We introduce the extension which varying threshold of CPU Utilization in scaling-out policy. We extend the model by introducing randomized method in scaling-in policy.

Keywords: Cloud computing · Auto scaling · Custom metrics · Load balancing · Overload and underload detection

1 Introduction

Most of modern system architectures allow to use scaling capability provided by cloud platform. The cooperating components of the information system may be run in cloud environment on separated virtual machines called instances or service nodes. Inside the cloud, a load balancer can distribute a stream of requests among many operating service nodes. Cloud platform provides mechanisms (like software tools, APIs etc.) for managing such service nodes. Especially, these mechanisms allow to horizontal scaling-out and scaling-in by programmatic create/destroy a virtual server. This gives a possibility to apply some model of auto scaling [1], where the number of service nodes is adapted to a system load. Such approaches may be reactive [2,3] (they use information about current load and system state) or predictive [4–6] (they additionally use an extrapolation of load

© Springer International Publishing AG 2017
P. Gaj et al. (Eds.): CN 2017, CCIS 718, pp. 422–431, 2017.
DOI: 10.1007/978-3-319-59767-6_33

and system state in near future). The reactive auto scaling models are rather simple, but they may be applied to a poorly predictable load.

Obviously a scaling-in increases a cost of system. To measure the cost we may define a simple objective function:

$$MeanCost = \frac{1}{Time} \int_0^{Time} Number\_of\_service\_nodes(t)dt \qquad (1)$$

which evaluates a system respect to usage of service nodes during $Time$.

A decision of scaling-in or scaling-out may be taken according to assumed Quality of Service (QoS) requirements or system resource-based ones. A user may assume some high-level criterion of quality based on statistics (e.g. mean, high order quantile) of some application-level metrics like execution time of selected business service. The approach to auto scaling model which uses the application-level metrics will be denoted as CMAS (Custom Metrics Auto Scaling). A user may also define less intuitive low-level criterion based on statistics (e.g. mean) of a resource-level metrics like CPU Utilization of a service node. Such approach will by denoted as SMAS (Standard Metrics Auto Scaling). The approach proposed in [2] combines these two approaches.

The optimization problem to solve in auto scaling domain can be formulated as choosing such methods and values of their parameters to minimize the objective function subject to QoS requirements.

This paper focuses on extending the model and method of the reactive auto scaling module presented in [2]. In this paper we propose the following improvements of that method:

- the additional error-based criterion in determining overloaded state of system (Sect. 3),
- the method of obtaining limits for group CPU Utilization (that may cause better choosing the moment of launching scaling-out) adapted to number of currently launched virtual machine instances (Sect. 4),
- the more aggressive strategy of scaling-in based on a function probability of turning off a redundant service node (Sect. 5).

2   The Auto-scaled Distributed System Designed for AWS Cloud Infrastructure

In the considered model [2] a quality of service requirement is a constraint based on statistics for execution times of a selected business critical service. A user may explicitly set $T_{q\ acc}$ – a value of a threshold (a max value) for T_q - a high order quantile of execution times (the qth quantile). If $T_q > T_{q\ acc}$ than we assume that a system is overloaded. A user may also set MV_{acc} – a value of threshold (a min value) for MV – a mean value of execution times. If $MV < MV_{acc}$ than we assume that a system is not enough loaded.

In our work we consider a simple cloud-aware information system (Fig. 1) that consists of:

- load balancer which exposes service outside a cloud and internally distributes requests among service nodes,
- multiple (n) service nodes, that internally expose SOAP/WebServices to the load balancer; the so-called auto scaling group consists of such service nodes and load balancer,
- DaaS node (PostgreSQL Relational Database Service) which persists data.

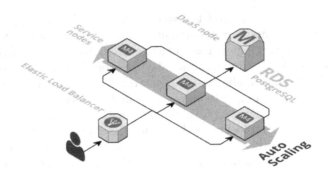

Fig. 1. The cloud-based architecture of the system: Elastic Load Balancer, and ($n = 3$) Elastic Compute Instances (service nodes), and Amazon RDS service (DaaS node).

The proposed in [2] software module that controls virtual machines (i.e. creates/destroys an instance of service node) is responsible for scaling-out/-in.

A scaling-out procedure uses an application-level custom metric – T_q and a AWS built-in resource-level metric – CPU Utilization. The module tries to use the estimator of T_q. If estimator of T_q exceeds $T_{q\ acc}$ the scaling-out should be performed. When estimators of T_q is not available (too less observations so we cannot positively verify at the assumed level that an estimator T_q belongs to the assumed confidence interval) the module uses $GroupCPUutil$ – a group CPU utilization (a mean of CPU utilizations of service nodes). If $GroupCPUUtil$ exceeds a $MaxCPUUtil$ that means that system is overloaded (but not by the selected business critical service) and the scaling-out should be performed, too.

A scaling-in procedure uses a application-level custom metric – MV and again a built-in resource-level built-in metric – CPU utilization. When estimator of MV is not available (too few observations what causes that it is not statistical confident) the module uses $GroupCPUUtil$. If $MinCPUUtil$ exceeds $GroupCPUUtil$ it means that the system is not loaded enough and the scaling-in should be performed.

The algorithm based on a custom metrics (T_q or MV) was called CMAS (Custom Model of Auto Scaling). The supplementary algorithm based on a built-in metrics ($GroupCPUUtil$) was called SMAS (Standard Model of Auto Scaling). Both CMAS and SMAS checks the conditions for T_q, MV, $GroupCPUUlil$ in some regular moments of time (determined by interval T_i). They launch scaling only if the condition is satisfied at least m times during last M tries (commonly $m > M/2$).

3 Analysis of System Efficiency

To describe an efficiency characteristic of a system, we consider to load it by a sequence of requests of selected business service. We assume the exponential distribution of intervals between subsequent requests with a mean value of intervals equals $1/\lambda$. The results of loading a system with $n = 1, 2, 3$ service nodes may look like those shown in Fig. 2.

Figure 2 presents how the mean value of execution time – $MV^{(n)}$, the qth quantile of execution times – $T_q^{(n)}$, the % of error requests per unit of time – $Err^{(n)}$ for $n = 1, 2, 3$ depend on increasing system load – λ.

In most cases, the error requests appear because nodes may be overloaded. This may happen either for service node or DaaS node. In fact, we may directly scale out in the system by multiply service nodes but we have no direct influence on scaling DaaS node. Thus we may expect that for overloaded system with many service nodes most of errors requests results from overloading of single DaaS node.

Quality of Service requirements define a not overloaded system where both criteria $T_q^{(n)} \leq T_{q\ acc}$ and $Err^{(n)} \leq Err_{acc}^{(n)}$ are satisfied. By increasing system load we may obtain the highest values of $\lambda - \lambda_{max}^{(n)}$ (blue color in Fig. 2) where $T_q^{(n)} \approx T_{q\ acc}$ and $Err^{(n)} \leq Err_{acc}^{(n)}$ for $n = 1$ (Fig. 2a, b), $n = 2$ (Fig. 2c), $n = 3$ (Fig. 2d).

In Figs. 2 and 3, the green color is used for marking acceptable operating points, the blue for boundary ones, and the red or brown for unacceptable ones.

We want to notice that the single criterion $T_q^{(n)} \leq T_{q\ acc}$ is not enough to determine a not overloaded system. When DaaS node becomes overloaded some time-out barrier may be crossed in communication between a service node and DaaS node. The architecture of the system should be adapted to such situations. Modern systems (see e.g. Repository of Electronic Medical Documentation – RepoEDM [2]) are based on a micro services architecture and supported by functionality which minimizes propagating of failure cascade and accelerates the backward information about time-outs (Hystrix[1]). So-called self-healing[2] mechanism (based on Hystrix/Eureka) reports that the service as unavailable so that subsequent requests do not run into the same timeouts. This results in very fast responses from error requests targeted to the overloaded DaaS node. This is illustrated in Fig. 3b where an empirical probability density function is bimodal. The execution times near the first local maximum (values close to $error\ MV^{(1)}$) correspond to error requests. The execution times near the second local maximum (values close to $corr.\ MV^{(1)}$) correspond to correctly processed requests. Although system is overloaded (Fig. 3b) and most of requests are processed incorrectly with time-outs (the mass near $error\ MV^{(1)}$ is greater than the mass near $corr.\ MV^{(1)}$), the mean value and the qth quantile are less relative to the ones from the not overloaded system (Fig. 3a).

---

[1] GitHub - Netflix/Hystrix (2016) https://github.com/Netflix/Hystrix.

[2] Hystrix and Eureka: the essentials of self-healing microservices (2016) https://www.dynatrace.com/blog/top-2-features-self-healing-microservices.

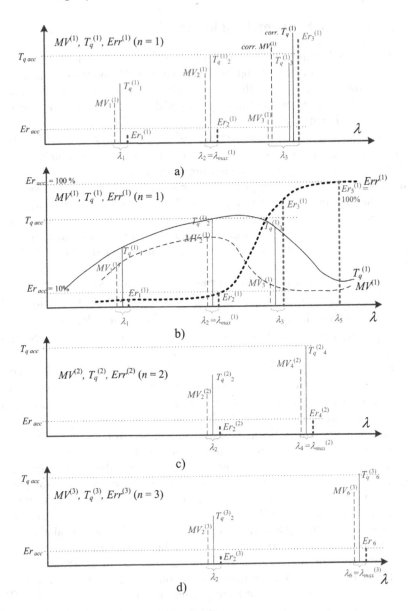

Fig. 2. Dependency between a load intensity λ and:
- the mean value of execution times – $MV^{(n)}$ (dashed line),
- the qth quantile of execution times – $T_q^{(n)}$ (solid line),
- the % of error requests per unit of time – $Err^{(n)}$ (fat dashed line);
(a) some operating points for a one-service-node system ($n = 1$),
(b) outlines of hypothetical courses for $MV^{(1)}, T_q^{(1)}, Err^{(1)}$ ($n = 1$),
(c) some operating points for a two-service-nodes system ($n = 2$),
(d) some operating points for a three-service-nodes system ($n = 3$). (Color figure online)

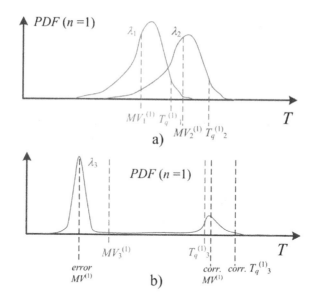

Fig. 3. Probability density function (PDF) of execution times T for a one-service-node system: (a) for a not overloaded system (green line for λ_1 intensity) and a boundary overloaded one (blue line for λ_2), (b) for an overloaded system (red line for λ_3). (Color figure online)

The satisfied condition $T_{q\,3}^{(1)} < T_{q\,2}^{(1)} = T_{q\,acc}$ in the $(3)^{th}$ operating point (red in Fig. 2a) may lead to incorrect conclusion that the overloaded system from Fig. 3b might be accepted as not overloaded. But it does not satisfy the Err-based criterion so finally it will be rejected according to QoS requirements.

4   Improvement of Scaling-Out in SMAS

SMAS model presented in [2] was based on an assumption that $MaxCPUUtil$ obtained for one-service-node system is enough accurate for a multi-service-nodes system. This assumption is only approximately valid because we can observe that CMAS and SMAS create instances in time differently even for the same load profile. Adapting $MaxCPUUtil$ values to n – the number of running service nodes – allows SMAS to behave almost the same like CMAS, i.e. we may observe situations when either SMAS or CMAS increases number of service nodes almost at the same moments of time.

We already noticed that a load of a single DaaS node may not be distributed like a load directed to many service nodes. During load increasing, CPU utilization of DaaS node will increase too and DaaS node becomes slower and the portion time of processing of a single request in DaaS node will increase too. Because we want to hold the same $T_{q\,acc}$ with increasing load the portion of time of processing in a service node should be decreased thus a service node should be faster and its CPU Utilization has to be lower.

We may experimentally find values of $MaxCPUUtil$ dependent on n. Values of $MaxCPUUtil(n)$ determine when the auto scaling module should switch the system from having n service nodes to $n+1$ ones. Those values may be obtained as means of CPU Utilizations of n service nodes in boundary operational points, i.e. for a load specified by $\lambda_{max}^{(1)}$ (when $n = 1$), ..., $\lambda_{max}^{(3)}$ (when $n = 3$), ...(Fig. 4). Such hypothetical shape of a decreasing dependency suggests that switching from n to $n+1 (n > 1)$ will be earlier (i.e. for smaller values of $GroupCPUUtil$) than it happens in the method from [2] where we had only a one and high value – $MaxCPUUtil(1)$ for all n.

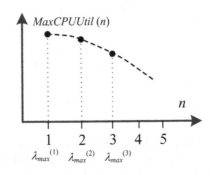

Fig. 4. $MaxCPUUtil$ threshold for SMAS adapted to n – the number of operating service nodes.

5   Improvement of Scaling-in in CMAS

According to a goal of minimizing the objective function $MeanCost$ (Eq. 1) some improvement of the scaling-in procedure was proposed. Let us remind rather conservative behavior in [2] – a scaling-in is needed when m times the criterion $MV < MV_{acc}$ is satisfied during M tries. In more detail, we obtain $\widehat{MV}$ (the estimator of MV) and verify that $\widehat{MV}$ is less than MV_{acc} at an assumed confidence level p. Such $\widehat{MV}$ we call confident. In [2], we only detect a fact of criterion satisfying but we do not use a difference value – $MV_{acc} - \widehat{MV}$.

To make the above-mentioned strategy of scaling-in more effective we propose to scaling-in when we satisfy the criterion J times where $1 < J \leq m$ but we will take into account only confident values of $\widehat{MV}$, too.

Although we will introduce some nondeterministic factor we want to hold a compatibility with the current strategy that satisfying the criterion m times launches scaling-in always i.e. with probability equals 1.

We do not want to fire scaling-in upon only a one try.

Let us denote as follows:

- $\widehat{MV}$ – confident estimator for $j \leq J$,
- $s = \sum_{j=1}^{J} \widehat{MV}_j$,
- $s_0 = J MV_{acc}$.

We introduce function of probability (p-function) of launching scaling-in as follows:

$$p(s, J) = \begin{cases} 0 \text{ for } s > s_0 \\ 0 \text{ for } J = 1 \\ 1 \text{ for } 0 \leq s \leq s_0 \wedge J = m \\ \dfrac{1 - \frac{1}{m-1}(J-1)}{0 - \frac{s_0}{m-1}(J-1)}(s - 0) + 1 \text{ for } 0 \leq s \leq \frac{s_0}{m-1}(J-1) \wedge J \in \{2, \ldots, m-1\} \\ \dfrac{\frac{1}{m-1}(J-1) - 0}{\frac{s_0}{m-1}(J-1) - s_0}(s - s_0) + 0 \text{ for } \frac{s_0}{m-1}(J-1) < s \leq s_0 \wedge J \in \{2, \ldots, m-1\} \end{cases} \qquad (2)$$

which is easier understandable using Fig. 5.

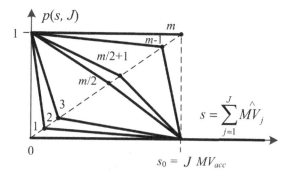

Fig. 5. Function of probability of launching scaling-in ($J = 1, \ldots, m$).

In new scaling-in method denoted by CMAS* we decided to scale-in with probability obtained from p-function (Eq. 2).

6   Some Experimental Results

In experiments we used real RepoEDM system described in [2] and run it in Amazon Web Services Cloud. The results illustrate SMAS/CMAS behavior after the improvements introduced in Sects. 4 and 5.

According to the idea from Sect. 4 for some assumed $T_{q\ acc} \approx 4000\,\text{ms}$ we obtain $MaxCPUUtil(1) \approx 82\%$, $MaxCPUUtil(2) \approx 71\%$, $MaxCPUUtil(3) \approx 63\%$. The differences among those values proved to be statistical significant at assumed confidence level ($p = 0.9$).

To evaluate the improvement of SMAS scaling-in we test both SMAS and CMAS with linearly increased λ in time during 20 min and with no load during 20 min. Test was repeated 10 times. The improved model is denoted by SMAS*.

Let us introduce the following coefficient:

$$Mean\left(\frac{\hat{T}_{q\ \text{CMAS}} - \hat{T}_{q\ \text{SMAS}}}{\hat{T}_{q\ \text{CMAS}}} 100\%\right). \qquad (3)$$

Experimentally, we obtained its value about 11% (for original SMAS with one $MaxCPUUtil$) and about 6% (after the improvement, for SMAS* with the series of $MaxCPUUtils(n)$). This result shows that the system under SMAS* becomes more similar to CMAS thanks to the scaling-out improvement.

To evaluate improvement of CMAS i.e. compare pure CMAS and CMAS with p-function of probability of lunching scaling-in (denoted by CMAS*) – we used a test profile defined by a sequence: constant λ during 20 min, and linear decreased to zero during 20 min, and 10 min no load. $m = 4$ and $M = 7$ were used in the experiment. Test was repeated 10 times.

For the following coefficient:

$$Mean\left(\frac{MeanCost_{\text{CMAS}} - MeanCost_{\text{CMAS*}}}{MeanCost_{\text{CMAS}}}100\%\right) \tag{4}$$

we obtained a value equals about 6% what shows a slight improvement in cost.

7   Conclusions

We rather expect poor effectiveness of load prediction during a process of mass migration of systems to cloud. Such process is complicated and it will depend on many technical factors, financial ones, or organizational ones. For such temporary situations we rather recommend a reactive model of auto scaling.

Although the idea of equivalency between custom-metrics-based QoS requirement (in CMAS) and resource-metrics-based QoS requirement (in SMAS) is not complicated but we did not meet such approach in known reactive models. We think that CMAS is well aligned to user expectations. But CMAS may not work sometimes (because of lack of metrics data) so it must be supported by adjusted SMAS.

In our work we provided a control module which implements proposed cooperative models of auto scaling (CMAS/SMAS). We also give a user a method and a software tool for finding the parameters of SMAS that are equivalent to given parameters of CMAS. The method and tool allow tuning parameter values for SMAS (adjusted to CMAS). These values may be later used in the proposed auto scaling control module.

Advantage of CMAS/SMAS approach results in its intuitiveness and simplicity comparing it to other more complex reactive models like [3] for example.

The paper presents some improvements of the reactive auto scaling model proposed in [2].

In the paper we justify the need for error metric (% of incorrectly processed requests per unit of time). This allows to minimize an impact of error requests on main QoS statistics (the qth quantile) based on execution times of all requests.

The first contribution is an extension of the scaling-out model that allows to early react on an increased load by using thresholds for group CPU Utilization that depend on the number of currently operating service nodes. Early turning on an additional node may cause better QoS (early try of overloading avoidance).

The second contribution is an extension of early scaling-in model where a function of probability of launching scaling-in (decreasing number of service nodes) was introduced, giving some nondeterministic solution. Early turning off a service node may cause a lower cost (early turned-off nodes does not load a budget).

The future work will concentrate on detail experimental verification the proposed extensions according to different load profiles.

We plan to verify a usefulness of introduction non-linear elements like hysteresis and dead zones into scaling-in/-out algorithms that operate on metrics.

References

1. Qu, C., Calheiros, R.N., Buyya, R.: Auto-scaling web applications in clouds: a taxonomy and survey. CoRR abs/1609.09224 (2016)
2. Augustyn, D.R., Warchal, L.: Metrics-Based Auto Scaling Module for Amazon Web Services Cloud Platform. In: Kozielski, S., Mrozek, D., Kasprowski, P., Małysiak-Mrozek, B., Kostrzewa, D. (eds.) BDAS 2017. CCIS, vol. 716, pp. 42–52. Springer, Cham (2017). doi:10.1007/978-3-319-58274-0_4
3. De Assuncao, D., Cardonha, M., Netto, M., Cunha, R.: Impact of user patience on auto-scaling resource capacity for cloud services. Future Gener. Comput. Syst. **55**, 1–10 (2015)
4. Jiang, J., Lu, J., Zhang, G., Long, G.: Optimal cloud resource auto-scaling for web applications. In: 13th IEEE/ACM International Symposium on Cluster, Cloud, and Grid Computing, CCGrid 2013, Delft, Netherlands, 13–16 May 2013, pp. 58–65 (2013)
5. Roy, N., Dubey, A., Gokhale, A.: Efficient autoscaling in the cloud using predictive models for workload forecasting. In: Proceedings of the 2011 IEEE 4th International Conference on Cloud Computing, CLOUD 2011, pp. 500–507. IEEE Computer Society, Washington, DC (2011)
6. Calheiros, R.N., Masoumi, E., Ranjan, R., Buyya, R.: Workload prediction using ARIMA model and its impact on cloud applications' QoS. IEEE Trans. Cloud Comput. **3**(4), 449–458 (2015)

Method of the Management of Garbage Collection in the "Smart Clean City" Project

Alexander Brovko$^{(\boxtimes)}$, Olga Dolinina, and Vitaly Pechenkin

Department of Information Systems and Technology,
Yuri Gagarin State Technical University of Saratov,
Saratov 410054, Russia
brovkoav@gmail.com

Abstract. This paper presents a solution of the problem of the route calculation for garbage removal trucks. The entire system architecture including algorithms of the route calculation, server and client part software, electronic devices on the garbage containers, mobile solution and collaboration with other routing services is presented. The dynamic network model and the optimization criterion on containers status and road traffic as well as the knowledge base which form the hybrid control system responsible for time of the garbage collection are introduced.

Keywords: Dynamic network · Hybrid control system · Smart City · Traffic flow · Knowledge base · Expert rules · Intellectual solution

1 Introduction

The management of garbage removal is a particular problem for all major cities, especially those overloaded by transport, with increasing traffic density and the amounts of garbage. In every city there are defined organizations and companies engaged in the collection and removal of garbage to refuse centers. All companies manage waste disposal according to a schedule or according to customer demands. However, there are situations when a garbage collection truck (GCT) arrives but garbage containers are half full; at the same time the GCT does not arrive to the real full garbage container. This is due to managers not accounting for the actual container content. This problem forms a part of a general theme of creating a healthy environment in modern cities, usually associated with "Smart Environment", described by attractive natural conditions, pollution, resource management and also by efforts towards environmental protection [1]. "Smart Environment" can be considered as a part of the "Smart City" technology. There are many definitions of the "Smart City" term [2]. An important component of the smart city concept is the use of the new information mobile technologies. This approach emphasizes the following definition as a city "combining ICT and Web 2.0 technology with other organizational, design and planning efforts to dematerialize and speed up bureaucratic processes and help to identify new, innovative solutions to city management complexity, in order to improve sustainability and livability [3]".

© Springer International Publishing AG 2017
P. Gaj et al. (Eds.): CN 2017, CCIS 718, pp. 432–443, 2017.
DOI: 10.1007/978-3-319-59767-6_34

There already exist specialised software and hardware that allows one to solve the problem of calculating a schedule for waste disposal. There are various approaches which use different types of detectors and allow for on-line control of the level of filling of garbage containers [4–8] but the problem in case of dynamic changes in the level of fullness of the containers and current changes of the real traffic situation in the city has not yet been solved. Currently, there are no solutions which take into account multiple optimization criteria simultaneously. The authors propose a solution using three optimization criteria for the task of garbage collection: minimum length of the route, processing for filled containers only, taking into account dynamic traffic situation.

2   Description of the Proposed Method

Special system termed "Smart Clean City" was developed to optimize garbage collection and manage this process. The system allows to carry out the following tasks:

1. to generate a message by the garbage containers to indicate when they are full;
2. to manage the process of sending a GCT for the garbage collection only if the containers are full;
3. to develop the optimal route for the garbage collection;
4. to distribute rationally the containers in the areas.

"Smart Clean City" allows one to solve the following urban, social and economic problem:

– increasing the economic efficiency of the company responsible for the garbage collection in terms of fuel, funding for equipment maintenance, optimization of staff, resources together with the amount of time dedicated to garbage collection;
– maintaining proper urban sanitation.

The system consists of two parts: software and signaling equipment. The technical part is represented by:

– equipment installed on each garbage site;
– equipment installed on every GCT.

Each area for garbage containers (AGC) is equipped with two types of electronic devices: one unit of bidirectional receiver/transmitter devices and sensors to determine the fullness of each container via a transmitter.

Each container is equipped with a vandal detecting device, located on the side wall of the container. It includes level sensors (both infrared and ultrasound) to monitor levels. A sensor is connected to a radio transmitter that transmits a signal indicating the level of each container to host receiver-transmitter unit; all elements being powered by an internal power supply. To save energy, the sensors are not active all of the time but with a frequency controlled by the microcontroller. This reduces the probability of a false signal transmission.

Data from the container transceiver unit is transmitted to the control room using a built-in GSM module. This makes it possible to transmit the received data via cellular communication. Information goes to the processing server and will be processed by the server software. The software system consists of client and server.

The tasks of the client part of the application:

– displaying routes graphically;
– client registration;
– warning about the need to empty the containers;
– notification of inability to continue work due to inevitable accidents;
– automatic authorization to get information depending on the area of truck's driver responsibility;
– automatic authorization to obtain client zone responsibility;
– periodically sending the information to the server about truck's position.

The server part of application:

– storage of information about the AGC: address, location, number and capacity of containers, date of the last maintenance;
– storage of information about garbage trucks: type, number of mobile devices, device ID, capacity;
– receive and process messages from the garbage sites;
– information about the fullness of the AGC in general and deciding on the need for removal of garbage from it;
– specification of the GCT for AGC needs to be cleaned, taking into account its current location;
– generating the route in accordance with the road traffic data;
– transmission the calculated route to a AGC for GCT;
– receiving information from the client application;
– dynamic updates on the status of AGCs for garbage in accordance with the actual situation;
– reporting and statistical analysis.

The software element provides the interface between the AGC, dispatch center and the garbage GCTs. Operation of the system takes place as follows:

1. Information about the status of the garbage containers levels in the area is transmitted to the central server where the software calculates the routes implemented. If the whole AGC is filled with more than 70%, the system decides to remove the garbage.
2. AGC is added to the list to be visited.
3. The garbage sites are represented as nodes of the city network (see details below). The system changes the weights of the edges on the basis of data traffic and road conditions. Traffic data are taken from the online road map service.

4. Server application calculates optimal routes on the basis of the time necessary for each GCT. Thus, the driver sees only the route to an area for the garbage containers, which needs to be cleaned.

System structure overview is shown in Fig. 1.

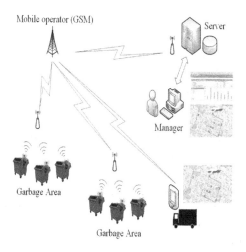

Fig. 1. Clean city system overview

The suggested formalization and method for solving optimization problem are original for the following reasons. Firstly, the network model of the transport system has dynamic nature [9]. Secondly, information about filling containers is handled automatically by the system during the development of the plan of the garbage collection.

2.1   Formal Network Problem Statement for Dynamically Optimal Route

To solve the problem let us define the weighted mixed network (there are directed and undirected edges)

$$G = (V, E, f, g, w) \tag{1}$$

where:

V – set of network vertices;

E – set of directed arcs (edges) corresponded to the city road network, arcs connects location of garbage containers, places of their discharge, home bases of GCT (garages) and linking them road network;

$f : V \times T \to R$ – vertex weight function at time t, which determines the time for passing the truck through the vertex;

$g : E \times T \rightarrow R$ – arc weight function at time t, which determines the time for passing the truck through the arc;

$w : V \times T \rightarrow R$ – vertex weight function at time t, which determines the amount of filled containers at the place of their location.

Vertices are superimposed on the map of the city road network. Arcs and edges correspond to the roads (with one-way and two-way traffic, respectively). V is defined as

$$V = V_1 \cup V_2 \cup V_3 \cup V_4, \tag{2}$$

where:

V_1 – network vertices correspond to the garbage site with containers;

V_2 – network vertices correspond to the solid domestic garbage dumps (SDGD);

V_3 – network vertices correspond to the garages location;

V_4 – vertices correspond to the connection points of the road segments (crossroads).

Let us define mapping f (temporal characteristics of road network vertices). Vertices' weight defines temporal characteristics of the GCT to pass through this vertex, weight is defined by the membership of the subsets V_1, V_2, V_3, V_4 and current time as follows:

$f(v,t)$ – time required for loading contents of containers on the specific garbage site for $v \in V_1$;

$f(v,t)$ – time to unload the GCT for $v \in V_2$;

$f(v,t) = 0$ for $v \in V_3$;

$f(v,t)$ – value that characterizes the delay of GCT at a crossroads (traffic light, unregulated crossroad), this value is determined on the basis of experimental data for $v \in V_4$.

Let us define the mapping g (time characteristics of a road network arcs). For all network arcs (edges) $e \in E$ $g(e,t)$ – value represents time to pass on route segment (depends on speed limit, quality of the roadway, segment road distance, traffic on this segment) defined for time moment t, determined by the current traffic situation.

Let us define the mapping w (number of filled containers). For all vertices from V_2, V_3, V_4 and for any time moment t the value $w(v,t) = 0$. For vertices that corresponds to the location of garbage containers the value of this function returns the number of full containers at time t.

$$w(v,t) = \begin{cases} 0, \text{ if } v \notin V_1 \\ \text{number of filled containers} \\ \text{at the site } v \text{ at time moment } t, \text{ if } v \in V_1 \end{cases}$$

Values for vertices and arcs of network at any time t are called markup. At the starting point of the network ($t = 0$) we name as initial markup. Dynamics of network changes is depended of actual traffic road situation, fullness levels of containers on garbage sites (varies in time), which results in a change of function values, change of markup.

Let the GCT has a capacity of L containers. At the initial moment ($t = 0$) on AGC v_j we have K_j filled containers that we need to take to a domestic garbage dump. It follows that $K_j = w(v_j, 0)$. Consequently, the GCT must visit each point of discharge (dump) at least S_j times, where

$$S_j = \left\lceil \frac{K_j}{L} \right\rceil \tag{3}$$

The total number of downloads–discharge cycles in this case is equal to

$$S = \sum_{1 \le j \le |V_1|} S_j.$$

The total number of filled containers is equal to

$$K = \sum_{1 \le j \le |V_1|} K_j.$$

Suppose that there is a single truck that collects garbage from all AGC and transfers it to the dump. In this case there are several possible optimization criteria. In this paper we consider only one for the time optimization, but the task is in reality is multi-objective one. We can consider other areas of analysis, such as maximizing the volume of handled garbage by trucks.

Designation: Let P – the route in network $G, U \subseteq V$. Designate $|P|$ as length of route P, and $|P|_U$ – the number of occurrences of vertices from set U in P. It is clear that for any route in the network $|P| = |P|_V$.

2.2   Problem Statement for One Truck

For given network find a route

$$P = v_0, v_1, v_2, \ldots, v_m$$

where:

1. $v_0 = v_m$; $v_0 \in V_3$ (GCT departs from garage at the beginning of the work and returns to the same garage after work completion);
2. $\forall_{v_i \in V_1} |P|_{\{v_i\}} = S_i$ (every AGC visited as many times as necessary to empty filled containers);
3. $|P|_{V_2} = S$ (dumps are visited S times – required amount of times for unloading filled containers);
4. $\sum_{i=1,m} (f(v_i, t_i) + g((v_i, v_{i+1}), t_i)) \rightarrow \min$, where the minimum is taken over all routes that satisfy conditions 1, 2, 3. t_i corresponds to time of events related to network vertices. It's clear that

$$t_1 < t_2 < \ldots t_m.$$

2.3 The Problem Generalization for n Trucks

Let the number of used garbage trucks is equal to n. For the given network there must be found n routes

$$P_i = v_0^i, v_1^i, v_2^i, \ldots, v_{mi}^i (i = 1, n)$$

and following conditions are satisfied

1. $\forall_{i=1,n} v_0^i = v_{mi}^i, v_o^i \in V_3$ (garbage trucks depart from garage at the beginning of the work and return to the same garage after work completion);
2. $\sum\limits_{i=1,n} |P_i|_{V_1} = S$ (every AGC is visited as many times as necessary to empty filled containers);
3. $\sum\limits_{i=1,n} |P_i|_{V_2} = S$ (dumps are visited S times – required amount of time for empty all filled containers);
4. $\sum\limits_{i=1,n} \sum\limits_{j=0,m_i} \left(f(v_j^i, t_j^i) + g((v_j^i, v_{j+1}^i), t_j^i) \right) \to \min$, where the minimum is taken over all routes that satisfy conditions 1, 2, 3. t_j^i corresponds to time of events related to network vertices for i-th truck. It's clear that for any $i = 1, n$

$$t_1^i < t_2^i < \ldots t_{m_i}^i.$$

If it is necessary to provide uniform load distribution for garbage trucks then one more condition should be added

5. Let $W_i = \sum\limits_{j=0,m_i} \left(f(v_j^i, t_j) + g((v_j^i, v_{j+1}^i), t_j) \right)$

then $\forall_{i \neq k, i, k=1, n} |W_i - W_k| \to \min$

Described below algorithm of building the path should be implemented with taking into account current traffic situation in the city. Information about time amount required for different segments of the route with current traffic conditions can be extracted from various online map services, for example using Google Maps Directions API [10]. This online service is available through an HTTP interface, with requests constructed as an URL string, using text strings or latitude/longitude coordinates to identify the location, along with API key. HTTP request to use Google Maps Directions API can contain some useful parameters, such as "waypoints" (intermediate points of the route which should be visited; up to 23 intermediate points for the business applications), "avoid" (objects which should be avoided in the route), "mode" (type of transport in use), and others. The response from the service is obtained as JSON array termed "routes", consisting of one or more segments "legs", depending on the presence of intermediate points in the request. Each segment of the route is described by using parameters "distance" (distance of the segment in meters), "duration" (time of driving in this segment in seconds), "duration_in_traffic" (time of driving calculated using statistical information and current traffic situation). These response parameters are taken into account in the optimal path calculation when the function $g(v, t)$ values are updating with the algorithm described below. The information obtained is then used to determine whether the calculated shortest route can lead the truck into a traffic jam. In this case the route have to be rebuilt, considering road situation.

Let us outline the algorithm of the dynamic route calculation for the truck which has the status "ready" and is making a request for the next AGC – "Get optimal route to AGC". The algorithm assumes the following GCT statuses and related information requests:

- (Status) Registration
- (Status) Ready
- (Status) Faulty truck
- (Status) On the route
- (Request) Get all day schedule
- (Request) Get optimal route to AGC
- (Request) Get optimal route to SDGD.

Algorithm of the dynamic optimal route calculation

```
ONE TRACK  SCHEDULING ALGORITHM
Input:  <Track Position>, <Service Type Request>
Output: Optimal route to AGC
If <Service Type Request> = <Get next container area> then
  /*Get statuses of AGC
  /*Statuses are <Filled>, <Maintenance>, <Cleaned>

 For all v ∈ V1 get Status(v) EndFor
   /*Get traffic situation
   UPDATE info on g(e,<Current time>)
   /*Get actual fullness levels info
   AGC_SET := ∅
   For all AGC v ∈ V1 do

   If Status(v) = <Filled> then
     Update info on w(v,t)
     AGC_SET := AGC_SET.Add(v)
   EndIf
   End For

   /*Select optimal route according to expert rules
   /*and the current values of fullness levels
   AGC_Next := GETOptimalAGC(AGC_SET)
   AGC_Next_Path := GETOptimalRouteToAGC(AGC_Next)
   Transfer data to the client application
   Status(AGC_Next) := <Maintenance>
End If
```

Method GETOptimalRouteToAGC uses the k-shortest simple paths search implementation of Yen's algorithm (loopless, one source) [11]. This algorithm has a computational complexity of $O(kn^3)$, where $O(n^2)$ is due to the shortest-path calculation. Here, n denotes the number of nodes in the road network model. The value of k is empirically set to 5. All built routes are ranked with using the expert rules which are presented in the knowledge base and describe the expert knowledge about the traffic situation in the considered period of time. The final list of routes ordered by their length and evaluation is given to the truck driver.

After taking the garbage from the container the installed sensor in the AGC updates their fullness status in the system, in case of a broken sensor the driver updates the status in the system manually and the AGC is marked for the sensor replacement (Fig. 2). Figure 3 shows making decision procedure.

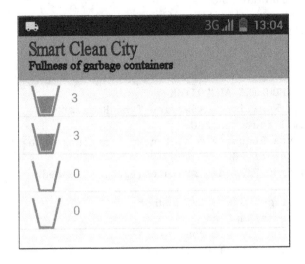

Fig. 2. The filling level of containers. Client application screen

Smart Clean City approach combines the described algorithm of building the optimal route with the intellectual approach represented by the knowledge base consisting of the rules:

$$pr_i : r_i : v_i : \text{If } a_j \text{ then } b_k \text{with the confidence} c_k, \qquad (4)$$

where:

$r_i \in \{R\}$ – the set of the rules,
$pr_i \in \{PR\}$ – the set of the priorities,
$v_i \in \{V\}$ – the set of V, see (2);
$a_j \in \{A\}$ – the set of the facts which represent the current situation,
$c_k \in \{C\}$ – the set of the linguistic variables,

where $C = \{\text{'possible', 'probable', 'most likely'}\}$, c_k represents the fuzzy variable described with the trapezoid function. Rules are formed by the experts (from

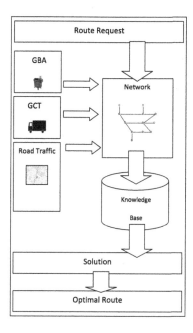

Fig. 3. The structure of the making decision

the traffic police or professional drivers) who are well acquainted with the traffic situation in the city. For example, in case of the traffic accident and corresponding traffic jam the experts could make the solution what step should be made – to change the other route or to wait. If the described algorithm of the building of the route tries to select the next node v_i but gets the message from the mobile maps service about the high load of the transport on the way to v_i and the knowledge base contains the rule r_i with the priority $pr_i \geq 80$, then the solution is made on the base of the selection of the r_i (to follow the algorithm or to select the other node or change the route to the new one).

Knowledge base consists of the rules the examples of which are presented below:

$80 : r_{32} : v_{4i} \in V_4$: if status (GCT) = "on the route" AND $f(v_i, t) > 20$ then recalculate AGC_Next
$100 : r_5 : v_{4j} \in V_4$: if status (v_{4j}) = "busy" then continue with calculated route with confidence "most likely"
$100 : r_{14} : v_{4j} \in V_4$: if status (V_4) = "busy" then use the calculated AGC_Next

3   Discussion

The described system "Smart Clean City" has been implemented in the October Region of the Saratov City (Russia) with the population about 1 million. A pilot exploitation of the system during the period from September 2015 till September 2016 demonstrated that the fuel saving achieved 21% by decreasing the time of the trucks being on the route in comparison with the standard manual route planning. The company responsible for the garbage collection has 24 trucks.

We evaluate the scheduling with synthetic and real time data by means of stochastic simulation in order to assess its performance. It is assumed that the region where the system was implemented has about 250 containers on 56 AGC. There are 2 dumps attached to the region. Each container has capacity 100 kg, while each GCT capacity is set to 5000 kg (actual capacity depends on the degree of compressibility of garbage). The results are shown in Fig. 4.

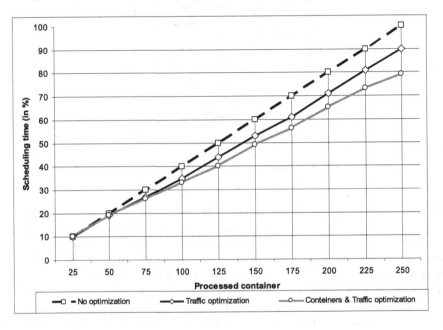

Fig. 4. Performance of the optimization algorithm

At the same time there was fixed a problem which has not been considered in the described system – lack of the proven information about the solidness of the garbage to be taken away from the container. It does not allow to take into consideration and include in the system the amount of the garbage to be loaded to the truck. The truck can take more compressed garbage. Information about the fullness of the containers without knowing the solidness of the garbage can not predict accurately the amount of the garbage which can be taken by the truck at the AGC.

4   Conclusion

System "Smart Clean City" allows one to manage garbage collection by using the hybrid control system based on the building of the route of taking away the garbage from the area of the garbage containers and using on-line information from the mobile application which collects information of the traffic jams and rules which can correct the calculated route. In this paper we present one optimization criterion for the time to empty all full garbage containers. Obviously the dynamic nature of the chosen mathematical model suggests using the other criteria as well for example, the "uniformity"of the garbage trucks loading that imposes additional restrictions on the algorithm for calculating routes. The advantage of the proposed system is the integration of information on the status of containers for garbage on special area with real-time traffic situation.

References

1. Global Innovators: International Case Studies on Smart Cities. Research paper number 135, October 2013. http://www.gov.uk/government/publications/smart-cities-international-case-studies-global-innovators
2. Anagnostopoulos, T., Zaslavsky, A., Medvedev, A., Khoruzhnikov, S.: Top-k query based dynamic scheduling for IoT-enabled smart city waste collection. In: Proceedings of the 16th IEEE International Conference on Mobile Data Management, Pittsburgh, US (2015)
3. Chourabi, H., Nam, T., Walker, S. Gil-Garcia, J.R., Mellouli, S., Nahon, K., Pardo, T.A., Scholl, H.J.: Understanding smart cities: an integrative framework. In: Proceedings of the 45th Hawaii International Conference on System Sciences, pp. 2289–2295 (2012)
4. Toppeta, D.: The Smart City Vision: How Innovation and ICT Can Build Smart, "Livable", Sustainable Cities. The Innovation Knowledge Foundation (2010). http://www.inta-aivn.org/images/cc/Urbanism/background%20documents/Toppeta_Report_005_2010.pdf
5. Optimising Waste Collection. http://www.enevo.com/
6. Kumar, N., Swamy, C., Nagadarshini, K.: Efficient garbage disposal management in metropolitan cities using VANETs. J. Clean Energy Technol. **2**(3), 258–262 (2014)
7. Kargin, R., Domnicky, A.: Routing the movement of road vehicles for the collection and disposal of waste. Roads Bridges "ROSDORNII" **28**(2), 92–102 (2012). (in Russian) Moscow
8. Doronkina, I.: Optimization of solid waste utilization. Serv. Russia Abroad **1**, 20 (2011). (in Russian)
9. Dolinina, O., Pechenkin, V., Tarasova, V.: Dynamic graph visualization approaches for social networks in educational organization. Vestnik SSTU **4**(62), 239–242 (2011). (in Russian) Saratov
10. Google Maps Directions API. http://developers.google.com/maps/documentation/directions/
11. Yen, J.: Finding the K shortest loopless paths in a network. Manage. Sci. **17**, 712–716 (1971)

Zone-Based VANET Transmission Model for Traffic Signal Control

Marcin Bernas[✉] and Bartłomiej Płaczek

Institute of Computer Science, University of Silesia,
Bedzinska 39, 41-200 Sosnowiec, Poland
marcin.bernas@gmail.com, placzek.bartlomiej@gmail.com

Abstract. The rising number of vehicles and slowly growing transport
infrastructure results in congestion issue. Congestion becomes an important research topic for transportation and control sciences. The recent
advances in vehicular ad-hoc networks (VANETs) allow the traffic control to be tackled as a real-time problem. Recent research works have
proven that the VANET technology can improve the traffic control at
the intersections by dynamically changing sequences of traffic signals.
Transmission of all vehicle positions data in real-time to a traffic lights
controller can generate a significant burden on the communication network, thus this paper is focused on the reduction of data transmitted to
a control unit by vehicles. The time interval between data transfers from
vehicles is defined by zones that are tuned for a given traffic control strategy using the proposed algorithm. The introduced zone-based approach
reduces the number of transmitted messages, while maintaining the quality of traffic signal control. The results of experiments firmly show that
the proposed method can be successfully used for various state-of-art
traffic control algorithms.

Keywords: Vehicular networks · Traffic signal control · Data
reduction · Congestion

1   Introduction

The last century was a place of very fast headway in motorization industry.
A vehicle, which was luxury good one hundred years ago, now becomes the
necessity to function in modern society. Rural areas, with constantly growing
population, are not prepared for this number of vehicles and in consequence traffic is disturbed by congestions. The traffic congestion is very costly phenomenon.
It causes substantial time losses for people and increases gasoline consumption
[1]. To tackle this issue, a reasonable solution is to increase the throughput of the
intersections, which are traffic bottlenecks. The throughput of intersection can
be increased by using traffic signal control [2]. Methods of traffic signal control
can be divided into two types: fixed-time control and traffic-responsive control
[3]. Traffic-responsive control proved to be more efficient than the fixed-time
approach; however, it requires reliable transfers of real-time traffic data [4]. This

© Springer International Publishing AG 2017
P. Gaj et al. (Eds.): CN 2017, CCIS 718, pp. 444–457, 2017.
DOI: 10.1007/978-3-319-59767-6_35

study is focused on decentralized traffic-responsive control strategies that are designed for urban networks with multiple intersections. One of such approaches was based on the Backpressure routing in computer networks [5]. The control strategy proposed by Helbing et al. [6] was based on self-organizing pedestrian traffic. Another approach analyses queue size of arriving vehicles (SOTL) [7]. Houli et al. [8] assumed that each traffic light controller is an agent that is able to learn to control the traffic light via interacting with the environment and neighbors. Recent works in this area utilize predictions of future traffic state in order to find optimal control actions LH [9]. All these methods assume that current traffic state can be monitored by means of sensors [10] (e.g., inductive loop detectors, cameras or radars). In recent years, VANET emerges as a reliable data source for the traffic control strategies. There is also a number of works related to VANET applications in traffic signal control. A case study of adaptive traffic light control algorithm and VANET data was presented in [11]. Majority of solutions is focused on a given traffic control method as fixed time traffic light control strategies [1] or adaptive one [5]. Up to date, many VANET-based models were proposed, however they were not popular in practical applications yet [12] and new data collection methods and control strategies are researched. There is lack of universal VANET models that can be used for any traffic signal control strategy. Therefore, this paper proposes a universal VANET communication scheme that could be implemented for most state-of-art traffic control strategies. The method assumes that each traffic control strategy requires data of defined precision [13], that can be obtained from fixed traffic areas (zones) with defined frequency. The proposed zones description aim is to find the optimal communication patterns, which will reduce number of transmitted messages and maintain a high traffic control quality. Finally, the algorithm to find optimal zones description for selected control strategy was proposed. In the following section the zones-based communication method for V2I and V2V VANET communication is presented in details. Section 3 describes simulation result obtained for three state-of-art traffic control strategies. Finally, in Sect. 4 conclusions are given.

2   Proposed Model

The proposed VANET-based communication model assumes that traffic signals controller also serves as a road side unit (RSU). Each RSU is able to communicate with vehicles via VANET. The RSU broadcasts periodically its own communication scheme and communication schemes of RSUs at neighboring junctions. The communication scheme defines zones with different data transmission frequencies. The vehicles moving from a previous junction to a next closest junction sends messages with frequencies assigned to the zones in which they are currently located. The message includes position and velocity of the sending vehicle. In this paper we assume, based on previous research concerned target tracking in WSN [14], that precise vehicle location data is especially important to make correct traffic control decision when vehicle is close to the junction. In this paper

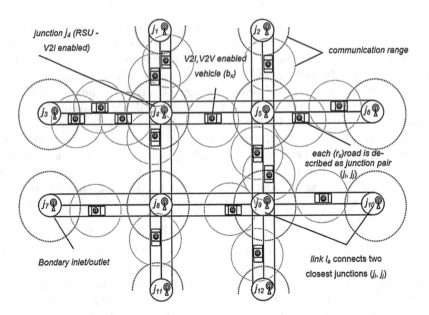

Fig. 1. Overview of VANET implementation.

we assume that precision of data obtained by traffic light controller is related to time interval between the transmissions of messages containing position updates. The overview of the proposed VANET-based communication model is illustrated in Fig. 1. The VANET communication model was based on WAVE implementation [15] and suppression strategy proposed in [16,17]. The communication frame is sent within one of the service channels. The vehicle, which passes a junction, obtains information about next junction on the way. The boundary inlets and outlets are treated as junctions and broadcast data as well. The RSU broadcasts the information directly to vehicles and this information is not forwarded further. In this model it was assumed that junctions, as in most real applications, are connected via wired infrastructure or cellular network, thus this communication burden is not considered. To simplify the model, each ith junction (j_i) is described by set of links L_i that connects inlet and outlet traffic with nearby junctions. Each link $l_a \in L_i$ of junction i is described as: $l_a = (j_i, j_k)$ or $l_a = (j_j, j_i)$ where: j_j, j_k – are nearby junctions or boundary RSUs. In this research it was assumed that the information about the traffic signal control strategy is not necessary. The control strategy is treated as a black box and it is described as a control function (C). Input of the function C is the traffic state dataset obtained from VANET (set D). D is a set of vehicles b_x registered on traffic lanes together with their positions and velocities. Elements of this set are tuples containing vehicle velocity, position and road id:

$$D_i(t, Z_i) = \{b_x\}, b_x = (p_x, v_x, r_x),$$ (1)

where: b_x - is vehicle registered at given road, p_x - is distance to the junction, v_x - velocity of vehicle, r_x - current road, x - is unique vehicle identifier (license plates or MAC of the communication device), Z_i - is the zone definition for a given junction described below. The quality of a selected control strategy is described by traffic delay (td), average speed (tv) and travel time (tt) of vehicles after t time steps. The traffic delay measure (td) is defined as a sum of delays of individual vehicles. The delay of a vehicle is calculated as a number of time steps (1 s) at which the velocity of the vehicle was equal 0 (vehicle was waiting in a queue). The average speed (tv) is measured as average velocities value for all vehicles. Finally, the travel time (tt) is a sum of time needed to pass through the monitored area. Thus, the performance of traffic control strategy (C) for junction i, based on data provided by VANET D function is defined as follows:

$$(td, tv, tt) = C_i(\bigcup_t D_i(t, Z_i)) \tag{2}$$

where: td, tv, tt - parameters of traffic control quality, C_i - control strategy for i^{th} junction, D_i - VANET based traffic state monitoring, t - considered time period, Z_i - zones definition. The function C for i^{th} junction can be any traffic signal control strategy that is dynamically adapted to the monitored traffic state. Thus, in this research three representative control strategies were used: LH [9] modification that selects optimal strategy based on predictions, a simple strategy that takes into account only current traffic data (SOTL) [7], and the Backpressure strategy [5] based on the routing algorithm for computer networks. The aim of this research is to find the optimal communication patterns (zones definition Z_i) that will reduce number of transmitted messages, while retaining a high traffic light control quality at the same time. The optimal communication scheme described by sequential zones (defined as Algorithms 1 and 2), is illustrated in Fig. 2. The zones definition was illustrated for a single road section. The vehicle has GPS device and based on localization determines the id of its current road (r_x) as well as the ids of junctions on the beginning and end of this road (3).

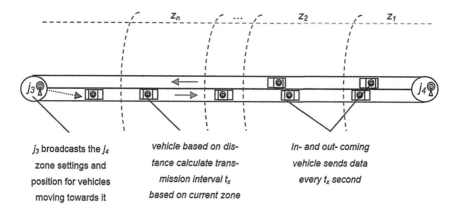

Fig. 2. Overview of the zones definition.

$$r_x = (j_s, j_e), \tag{3}$$

where: j_s, j_e–id of junctions at ends of current road. The vehicle sends data to both junctions using the zone settings. The zones are defined by an ordered set of distances from a junction (4).

$$Z_i = \{z_1 = (e_1, t_1), ..., z_j = (e_j, t_j), ..., z_n = (e_n, t_n)\}, \forall_{j \in 1..n-1} e_j < e_{j+1}, \tag{4}$$

1. Each ith junction is described by L_i and Z_i,
2. Obtain the nearest junction set $JS = \{j_j\}, j_j \in L_i$,
3. Send request, for zones definition, to junctions in JS set,
4. Store junctions zones, with links, as JZ set: $JZ = \{(j_j, Z_j)\}, j_j \in JS$.

#at each time step for junction i:
5. Set empty set $S'=\{\}$,
6. Broadcast set JZ to nearby vehicles,
7. Store all incoming messages from vehicles in set S (if data from vehicle already exist in S update its values),
8. For each message $b_x=(p_x, v_x, r_x=(j_s, j_e))$ from set S do:
a) Find $z_k=(e_k, t_k)$ in Z_i for which $p_x<e_k$ and ($p_x>e_{k-1}$ or $k=1$). If zone z_k exist then if $j_i=j_e$ then $g_x=[p_x, p_x-v_x \cdot t_{zi}]$ else $g_x=[p_x, p_x+v_x \cdot t_{zi}]$,
b) If g_x exist then $S' = S' \cup b'_x = \{rand(g_x), v_x, r_x\}$, where rand is a uniformly distributed random value from defined interval,
9. Remove all vehicles form S set for which $inf(g_x)<0$ or $sup(g_x)>e_n$, where e_n is in last defined zone: $z_n=(e_n, t_n), n=card(Z_i)$,
10. The algorithm returns the traffic state as S'.

Fig. 3. The RSU operations for junction i (Algorithm 1).

where: e_j - minimal Euclidean distance (ED) to junction, t_j - maximal time interval between position updates. Each zone has its time limits at which vehicle has to transmit data. The distances between junctions and vehicles are calculated using Euclidean distance (ED). The operations performed by the RSU at i^{th} junction are summarized in Algorithm 1 (Fig. 3). Algorithm 1 is able to process data obtained from VANET. Then it exchanges zones definition between nearby junctions (lines 1–4). The algorithm based on data obtained in lines (5–7) describes vehicle positions as intervals g_x (granules) in order to cope with uncertainty [13] (line 8a). Most control algorithms cannot tackle position uncertainty, thus degranulation procedure is performed (line 8b). The result of degranulation procedure returns one of the probable vehicle positions that are sent to traffic controller (line 9). Finally, the vehicles, which position intervals are moved to negatives or leave the zones are removed (line 10). The vehicle communication follows the simple procedure, defined by Algorithm 2 (Fig. 4). Algorithm 2 tracks vehicle position using GPS data, then based on obtained zone definition calculates the interval at which the communication should be performed for both junctions connected with a given road section (lines 6–7). The communication is only possible within the defined zone, thus if vehicle does not obtain the zone definition it does not broadcast its position. Both Algorithm 1 and Algorithm 2 are based on the zone definition. The zones are defined for specific junction, traffic

```
# initialize vehicle algorithm for bx: Set ts=0, tsmax=∞, te=0 temax=∞
1. ts=ts+1; te=te+1,
2. Read vehicle position cx and velocity vx based on GPS,
3. Receive all broadcast zones and store them in JZ set,
4. Determine current road rx=(js,je) based on localization [30],
5. Read zone definition Zs and Ze from JZ for js and je respectively,
6. If Zs is not empty then
    a) tsmax=∞,
    b) Calculate ED between vehicle and junction: ED(js,cx),
    c) For i=1 to card(Zs) do Zs(i)=zi=(ei,ti); If ED(js,cx) < ei then
       tsmax= ti, break,
7. If Ze is not empty then
    a) temax=∞
    b) Calculate ED between vehicle and junction: ED(je,cx)
    c) For i=1 to card(Ze) do Ze(i)=zi=(ei,ti); IF ED(je,cx) < ei then
       temax= ti, break,
8. If ts>=tsmax then ts=0, send set bx={px=ED(js,cx),vx,rx} to junction js.
9. If te>=temax then te=0, send set bx={px=ED(je,cx),vx,rx} to junction je.
```

Fig. 4. The zone-based vehicle communication (Algorithm 2)

conditions and traffic control strategy. The zones are selected by using Algorithm 3 and takes into consideration the relative traffic control effectiveness, for considered time period (t), measured by EF function (5).

$$EF(td, td', tv, tv', tt, tt') = \frac{1}{3} * (\frac{td - td'}{max(td, td')} + \frac{tv' - tv}{max(tv, tv')} + \frac{tt - tt'}{max(tt, tt')}), \quad (5)$$

where: (td, tv, tt) and (td', tv', tt') are quality measures returned by C function (Eq. 2). Algorithm 3 (Fig. 5) is divided into two phases. Firstly, the range at which data are vital for the selected traffic control strategy is found (lines 1–6). The initial simulation is performed for minimal distance (lines 1–3). Then the distance is extended by the value of $minDist$ parameter as long as the traffic control effectiveness (5) is increasing (lines 4–5). Line 6 was added to avoid local minimum, which can be registered in the first steps of algorithm. Then, in second phase, the obtained area is divided, using top-down strategy, into two zones with different message transmission intervals. The effectiveness of control strategy cannot decrease below a given threshold (α). If the division is not possible under given parameter assumption, the algorithm ends. The minimal length used to track vehicle is defined as $minDist$ and it is related with the used localization system and size of vehicles. The division algorithm (phase 2) was illustrated in Fig. 6. In first step (a) the performance of traffic control for single zone, with the maximum size and the most frequent transmissions (1 s), was calculated (C_i). Then, the zone is divided into two equal-length zones with various update time, i.e., 1 and 2 s (b). The traffic control performance for newly created zones is calculated. Then the performances for two zone settings are compared by the EF function. If the traffic control performance is not decreased below given threshold (α) the first zone (closer to junction) can be narrowed. In opposite situation (c) the first zone is enlarged. If the divided area is smaller than the defined threshold minDist (d) the division ends. The end of zone is determined

#phase 1
1. $F = max_{l_a \in L_i}(ED(l_a))$, step = minDist, X = step,
2. Define initial zone Z_i = {z_1 = $(X, 1)$},
3. Calculate the values of (td, tv, tt) = C_i (D_i (t, Z_i)) based on Eq. 2,
4. X = X + step;
5. If F > X then
a) Define initial zone Z_i' = {z_1' = $(X, 1)$},
b) Calculate values of (td', tv', tt')=C_i(D(t, Z_i')) based on Eq. 2,
c) If EF(td,td',tv,tv',tt,tt')>=0 then td=td', tv=tv', tt=tt', goto 4,
6. If F > X then
a) Define initial zone Z_i' = {z_1' = $(X+2*step, 1)$ },
b) Calculate values of (td', tv', tt')=C_i(D(t, Z_i')) based on Eq. 2,
c) If EF(td,td',tv,tv',tt,tt')>=0 then goto 4,
#phase 2
7. M_1=0, M_2=X, div=0, Z_i={},
8. If (M_2 - M_1)<=minDist goto 13,
9. M_3= (M_2 - M_1)/2+M_1;
10. Define initial zones Z_i'=$Z_i \cup$ { z_{div+1}=(M_3, 2^{div}), z_{div+2}=(M_2, 2^{div+1})},
11. Calculate the values of (td', tv', tt')=C_i(D(t, Z_i')) based on Eq. 2,
12. If EF(td,td',tv,tv',tt,tt') >= $-\alpha$ then M_2=M_3, goto 8,
    else M_1=M_3 , goto 8,
13. If EF(td,td',tv,tv',tt,tt') >= $-\alpha$ then $Z_i = Z_i \cup z_{div+1} = (M_1, 2^{div})$,
    else $Z_i = Z_i \cup z_{div+1} = (M_2, 2^{div})$,
14. div=div+1, IF (X$-$ M_3) > step then M_1= M_3, M_2=X, goto 8,

Fig. 5. The zones search Algorithm (3)

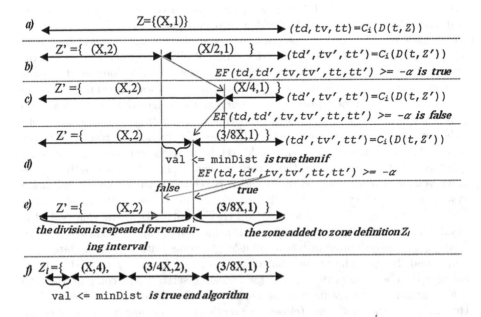

Fig. 6. The example of zone finding algorithm

by taking into account the value of EF function. After the first zone is found the rest of interval is further divided (e) to find the zones with higher transmission intervals (2, 4, 8). The algorithm ends, when the remaining interval is smaller than $minDist$ parameter.

3   Experiments

To illustrate the robustness of the proposed solution, three state-of-art traffic control strategies were used (LH [9], SOTL [7] and Backpressure [5]), with various datasets delivered from VANET. As input, the control strategies receive the data describing vehicle positions, velocity and road, in accordance with Algorithm 1. The control strategies were implemented in Matlab and integrated with SUMO simulation of road network containing four intersections (Fig. 7). Intensity of the traffic flow is determined for the network model by parameter q in vehicles per second. This parameter refers to all traffic streams entering the road network for $t = 1000$ s. At each time step vehicles are randomly generated with a probability equal to the intensity q in all traffic lanes of the network model. In this research the traffic intensity changes within a day to model the rural traffic characteristic $q = (0.05, 0.2)$. The initial experiment was conducted to find a borderline for analyzed method. Thus a constant transmit time for all area was researched - without zones. The average results of 20 simulations are presented in Fig. 8. The transmission range below 30 in case of SOTL and LH gives unpredictable results. For values between 70 and 110 both strategies are stabilized. In case of the Backpressure strategy, the best results are obtained for relatively small communication distance. Intuitively, the number of messages is growing with the communication distance and decreasing when the time interval between successive transmissions increases. Algorithm 3 was used to find the zones.

The zones was searched for $minStep = 10$ m and $\alpha = 0$ so no loss of control quality was allowed. The obtained results of 20 simulations are presented as box plot in Fig. 9 and compared with fixed interval of message transmission. The box plot shows mean (line inside a box), 1st/3rd quartiles (box) and

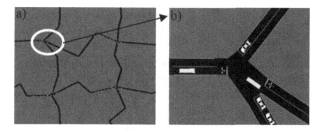

Fig. 7. The simulation test-bed in Sumo environment (a) road network (b) selected junction.

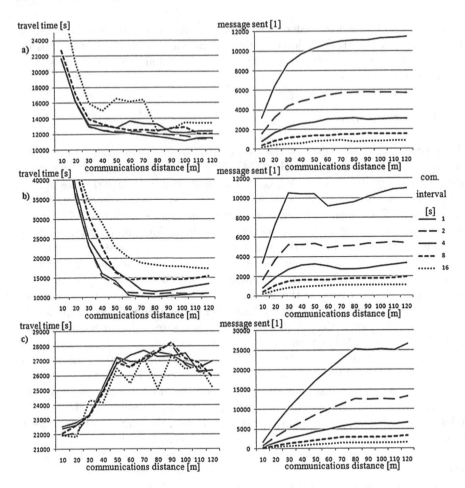

Fig. 8. Communication distance vs. travel time and number of messages: (a) SOTL, (b) LH, (c) Backpressure.

minimal/maximal values (error bars). The proposed zone model enables finding a robust solution despite the randomness of position of vehicles within each zone. It also allows maintaining the performance, while decreasing the number of sent messages. The Backpresure is least affected by random factors, while more precise algorithm are more vulnerable, however the results are tend to be more optimal. The top down strategy can give suboptimal solution, and thus the results were compared against the exhaustive search. The results of exhaustive zone search were presented in Table 1.

The last row of Table 1 shows the results obtained for the best zone selection in case when only one zone is used. It means that in the selected zone the transition interval is constant and equal to 1. The remaining rows in Table 1 show all results for which the optimal performance of the control strategies was

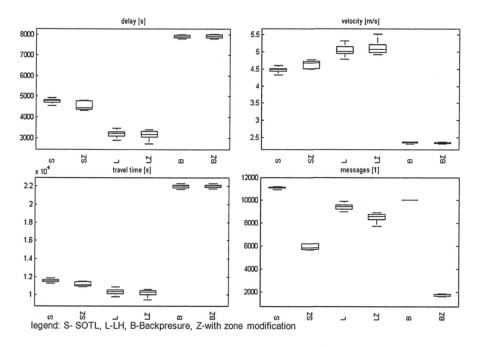

Fig. 9. The zone applied for state of art algorithms.

Table 1. The zones generated using exhaustive strategy.

	Zone(Z)	Delay[s]	Velocity[m/s]	Travel time[s]	Frames[1]
SOTL	(10,1),(30,2),(50,4),(110,8)	4631	4,57	11404	5698
	(10,1),(30,2),(70,4),(110,8)	4631	4,57	11404	5825
	(10,1),(30,2),(110,8)	4631	4,57	11404	5949
	(10,1),(50,2),(110,8)	4631	4,57	11404	6115
	(10,1),(50,2),(70,4),(110,8)	4631	4,57	11404	6319
	(10,1),(50,2),(110,4)	4631	4,57	11404	6376
	(10,1),(70,2),(90,4),(110,8)	4631	4,57	11404	6586
	(10,1),(90,2),(110,4)	4631	4,57	11404	6815
	(10,1),(110,2)	4631	4,57	11404	6916
	(110,1)	4708	4,56	11648	11602
LH	(30,1),(70,2)	2935	5,34	9891	7979
	(70,1)	3119	5,23	10123	9319
Backpress	(30,1)	7994	2,35	22206	10253
	(30,2)	7994	2,35	22206	5053
	(30,8)	7994	2,35	22206	1268
	(30,1)	7994	2,35	22206	10253

obtained. The solution with least messages sent is the one selected with proposed
Algorithm 3 for zone finding. In all three cases, the same distance gives the best
result in terms of several zones as for one zone. In case of LH strategy only one
interval was found.

The LH strategy performs the forecast of vehicles movement thus the precise
vehicle position are vital to its performance. Nevertheless, over 30 m the time
delay can be increased by 2 and does not influence the control strategy. In case
of SOTL algorithm, where the vehicles delays and queue lengths are the most
important features, the zone definition changes.

The least expensive communication assumes that the data collected at the
distance up to 10 m are vital and with the distance the precise position of vehicles is not that important. The last control strategy (Backpressure) balances the
amount of vehicles in separate traffic lanes, thus vehicle number is important
and positions of vehicles can be ignored, thus the tracking length is shortest.
The distance must be sufficient to register each vehicle. In this case the distance of 30 m is sufficient. The zones reflect the character of the traffic control
strategy. The more complex control strategy the more data it requires. The
proposed method of zones definition reduces the number of messages exchange
between vehicles and RSU by 50%, 14% and 10% respectively for Backpressure,
SOTL and LH strategy. Additional research was conducted to analyze the influence of α parameter on the algorithm 3 effectiveness (Fig. 10). The values on
X axis describes how many loops in first or second phase of the algorithm was

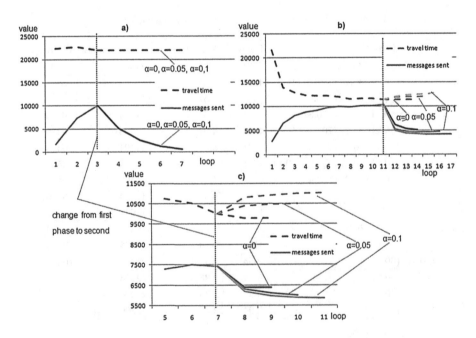

Fig. 10. Simulation results (a) Backpressure, (b) SOTL, (c) LHs.

executed. The results in Fig. 10 show that the number of messages and control quality decreases with increase of α parameter. For small values of $\alpha = (0, 0.1)$ the decrease of message number is especially visible for more complex algorithms (SOTL and LH). For $\alpha > 0.1$ the traffic control strategy is not optimal, thus the travel time of vehicles are longer and the number of messages sent does not decrease so rapidly. In case of the Backpressure method the data are already reduced significantly and further reduction has great impact on the control performance. Thus for small α values the result does not change.

In case of the more advanced control strategies, the zone analysis allows to find a balance between the transmission burden and traffic control quality.

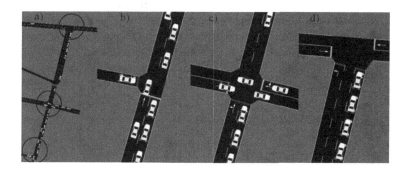

Fig. 11. The model of Francuska street: (a) overview, (b)–(d) junctions.

Table 2. The simulation results for Francuska street

	Control strategy	Distance [m]	Trans. interval [s]	Delay [s]	Velocity [m/s]	Travel time [s]	Messages [1]
Junction 1	SOTL	110	1	10680	3,02	29506	26151
	LH	30	1	11707	2,38	37198	18820
	Backpressure	120	1	22439	1,42	63220	60008
	SOTL (zone)	10, 30, 110	1,2,8	10019	2,89	30269	11826
	LH (zone)	20, 30	1,2	11742	2,38	37732	17837
	Backpressure (zone)	30, 120	1,8	21107	1,54	58368	29576
Junction 2	SOTL	50	1	8665	3,33	24874	16464
	LH	60	1	13726	2,38	35060	25591
	Backpressure	30	1	14964	1,52	53978	21081
	SOTL (zone)	30, 50	1,8	8626	3,29	24727	12998
	LH (zone)	50, 60	1,4	13762	2,38	34720	23539
	Backpressure (zone)	20, 30	1,8	14879	1,52	54462	19828
Junction 3	SOTL	90	1	3062	5,34	14080	12574
	LH	120	1	665	7,79	9409	7880
	Backpressure	80	1	1007	7,77	9500	8137
	SOTL (zone)	10, 90	1,8	2837	5,69	12915	2767
	LH (zone)	60,90,120	1,4,8	663	7,56	9736	7003
	Backpressure (zone)	30,80	1,4	995	7,79	9518	5021

Further experiments were conducted to verify the proposed approach for a realistic scenario of road network with various junctions. Three junctions were selected for these tests on the Francuska street in Katowice, Poland. The simulation model is presented in Fig. 11. The traffic volume was set based on the real traffic characteristics for this street during work days. As in the previous experiment, three traffic control strategies were considered with and without the proposed zone-based transmission method. The average results of 20 simulations for this scenario are presented in Table 2. The proposed zone-based transmission for all three traffic control strategies allowed decreasing the number of messages, while retaining the high quality of traffic control.

However, as in the first simulation scenario, the smallest decrease of message reduction was registered for LH strategy (8%), while the biggest reduction was observed in case of SOTL and Backpressure strategy (51% and 31% respectively).

The experimental results are promising and firmly show that the data transmission can be decreased, while not incorporating sophisticated suppression algorithms. However, the zones could be dynamically changed according to traffic intensity by using more sophisticated tracking and prediction mechanisms.

4   Conclusion

VANET is considered as useful source of input data for traffic signal control strategies. In this paper a zone-based transmission model is proposed for VANETs, which enables effective data collection for the traffic control applications. Three state-of-art traffic control strategies were investigated: SOTL, LH and Backpressure. The results show that it is possible to reduce the amount of messages sent by vehicles using various time intervals between data transmissions. The time intervals were selected based on distance to junction. The proposed method reduces the transmission burden by sending only data that are vital for given control strategy. The traffic control quality was measured by total delay, travel time and average vehicle speed. The proposed algorithm uses α parameter to balance the number of messages and the quality of traffic control. The proposed concept of zone-based transmission is promising. It reduces the number of messages sent between vehicles to minimum and allows the vehicle to be aware to traffic control strategy and data requirements. The future work will tackle with using multiple control strategy based on obtained data. Another research area will be focused on even further reducing the data transmission by implementing zone-dependent data suppression methods. Finally, the zones can be defined not only for all day traffic but for specific time periods, which could even further reduce data transmission.

References

1. Aslam, M.U., et al.: An experimental investigation of CNG as an alternative fuel for a retrofitted gasoline vehicle. J. Fuel Sci. Technol. Fuel Energy **85**(5–6), 717–724 (2006)

2. Wang, Q., Wang, L., Wei, G.: Research on traffic light adjustment based on compatibility graph of traffic flow. Intell. Hum. Mach. Syst. Cybern. (IHMSC) **1**, 88–91 (2011)
3. Płaczek, B.: A traffic model based on fuzzy cellular automata. J. Cell. Automata **8**(3–4), 261–282 (2013)
4. Qin, Z., Chao, P., Jingmin, S., Pengfei, D., Yu, B.: Cooperative traffic light control based on semi-real-time processing. J. Autom. Control Eng. **4**(1), 40–46 (2016)
5. Le, T., Kovács, P., Walton, N., Vu, H.L., Andrew, L.L., Hoogendoorn, S.S.: Decentralized signal control for urban road networks. Transp. Res. Part C Emerg. Technol. **58**, 431–450 (2015)
6. Helbing, D., Lämmer, S., Lebacque, J.-P.: Self-organized control of irregular or perturbed network traffic. In: Deissenberg, C., Hartl, R.F. (eds.) Optimal Control and Dynamic Games. Advances in Computational Management Science, vol. 7, pp. 239–274. Springer, USA (2005)
7. Cools, S.-B., Gershenson, C., D'Hooghe, B.: Self-organizing traffic lights: a realistic simulation. In: Prokopenko, M. (ed.) Advances in Applied Self-Organizing Systems. Advanced Information and Knowledge Processing, pp. 45–55. Springer, London (2013)
8. Houli, D., Zhiheng, L., Yi, Z.: Multiobjective reinforcement learning for traffic signal control using vehicular adhoc network. J. Adv. Sig. Process. **2010**, 7 (2010)
9. Płaczek, B.: A self-organizing system for urban traffic control based on predictive interval microscopic model. Eng. Appl. Artif. Intell. **34**, 75–84 (2014)
10. Choudekar P., Banerjee S., Muju, M.K.: Implementation of image processing in real time traffic light control. In: Proceedings of the 3rd International Conference on Electronics Computer Technology (ICECT), Kanyakumari, vol. 2, pp. 94–98 (2011)
11. Toor, Y., Muhlethaler, P., Laouiti, A., Fortelle, A.: Vehicle ad hoc networks: applications and related technical issues. IEEE Commun. Surv. Tutorials **10**(1–4), 74–88 (2008)
12. Kwatirayo, S., Almhana, J., Liu, Z.: Adaptive traffic light control using VANET: a case study. In: Proceedings of 9th International Conference on Wireless Communications and Mobile Computing Conference (IWCMC), Sardinia, pp. 752–757 (2013)
13. Abbas, M.K., Karsiti, M.N., Napiah, M., Samir, B.B.: Traffic light control using VANET system architecture. In: Proceedings of National Postgraduate Conference (NPC), Kuala Lumpur, Malaysia, pp. 1–6 (2011)
14. Song, M., Wang, Y.: Human centricity and information granularity in the agenda of theories and applications of soft computing. Appl. Soft Comput. **27**, 610–613 (2014). doi:10.1016/j.asoc.2014.04.040
15. Sun, M-T., Feng, W-C., Lai, T-H., Yamada, K., Okada, H., Fujimura, K.: GPS based message broadcasting for inter-vehicle communication. In: Proceedings of International Conference on Parallel Processing, pp. 279–286 (2000)
16. Płaczek, B., Bernas, M.: Uncertainty-based information extraction in wireless sensor networks for control applications. Ad Hoc Netw. **14**, 106–117 (2014)
17. Bernas, M.: WSN Power Conservation Using Mobile Sink for Road Traffic Monitoring. In: Kwiecień, A., Gaj, P., Stera, P. (eds.) Computer Networks. Communications in Computer and Information Science, vol. 370, pp. 476–484. Springer, Heidelberg (2013)

Author Index

Printed in the United States
By Bookmasters